22.56

D1223579

STUDIES IN THE PHILOSOPHY OF BIOLOGY

STUDIES IN THE PHILOSOPHY OF BIOLOGY

Reduction and Related Problems

Edited by
FRANCISCO JOSE AYALA
and
THEODOSIUS DOBZHANSKY

UNIVERSITY OF CALIFORNIA PRESS
Berkeley and Los Angeles

WITHDRAWN
ITHACA COLLEGE LIBRARY

University of California Press
Berkeley and Los Angeles, California

© The Macmillan Press Limited 1974
© Text selection and presentation
F. Ayala and T. Dobzhansky 1974
© Pages 259–284. K. R. Popper 1974

All rights reserved. No part of this publication
may be reproduced or transmitted, in any form
or by any means, without permission

ISBN 0-520-02649-7
LC 73-90656

Printed in Great Britain

Contents

Acknowledgments

The conference on 'Problems of Reduction in Biology' was made possible by the generosity of two private foundations. The Rockefeller Foundation hosted the conference at their Study and Conference Center in Villa Serbelloni, Bellagio, Italy. The Alfred P. Sloan Foundation provided a grant to cover travel and other expenses. We gratefully acknowledge their contributions. Dr Ralph W. Richardson, jr., Chairman of the Program Committee of Villa Serbelloni (Rockefeller Foundation), and Mrs Richardson were gracious hosts to the conference participants from 9 to 16 September 1972.

F.J.A.
Th.D.

Introduction

FRANCISCO J. AYALA

One of the outstanding characteristics of living matter is complexity of organisation. There is a hierarchy of complexity that runs from atoms and molecules, through cells, tissues, individual organisms, populations, communities and ecosystems, to the whole of life on earth. Different biological disciplines concentrate on the study of one or several levels of hierarchical organisation. The analytical method so successful in the scientific study of the physical world has also proved fruitful in biology. Phenomena at one level of complexity are illuminated by the study of the component elements and underlying processes. The most impressive achievements of biological research during the last few decades are those of molecular biology.

Because of the success of the analytical method, and in particular because of the spectacular accomplishments of molecular biology, some molecular biologists went so far as to contend that the only biological research which is ultimately significant, and indeed truly 'scientific', is that pursued at the molecular level. In contrast, some over-conservative biologists have claimed that molecular biology may be good physics, or good chemistry, but has contributed little to the understanding of the most important biological problems. Most biologists and philosophers of science reject these two contentions as extreme and, indeed, unreasonable. Yet there are genuine questions concerning what has been called the 'problem of reduction' in biology.

The problem—or 'problems', since many different questions can be raised—of reduction in biology have been the subject of much writing and discussion in recent years. Several conferences and symposia have dealt, at least in part, with problems of reductionism. Many of these writings and discussions have serious shortcomings. The questions discussed by philosophers are not always those of greatest concern to most scientists. Biologists defending the reductionist or antireductionist standpoint often misconstrue the position of their alleged opponents. Professor Theodosius Dobzhansky and I thought that there was need for consideration in depth of the problems of reductionism. We resolved to organise a conference dealing with these problems. We decided to invite a panel of distinguished philosophers and scientists representing a variety of viewpoints, and to give them maximum freedom to identify the relevant questions and to treat them as they saw fit. Position papers were requested for distribution among participants before the conference. Careful study of the position papers and abundant time for discussion would hopefully help to define the problems and to ascertain the grounds of agreement or disagreement. The conference was held at the Study and Conference Center of the Rockefeller Foundation in Bellagio, Italy, from 9 to 16 September 1972. The position papers served as focal points of the discussions. Although the focus of the conference was on the problems of reduction, the papers deal with many other questions in the philosophy of biology. The position papers make up the lion's share of the present volume. The discussions were generally not recorded, although comments written for publication by their authors have been incorporated. In the last working session of the conference, Professor Jacques Monod made an oral presentation on the subject of *Chance and Necessity*. The transcription of that address and ensuing discussion has been included in this volume.

Questions of reductionism arise in three different domains: ontological, methodological and epistemological. In discussions of reductionism, these three domains must be distinguished to avoid misunderstanding. The questions raised, and the answers given, may be different in different domains. A brief characterisation of the reductionist questions that arise in each domain is appropriate here, even at the risk of oversimplification.

From the *ontological* point of view the question of reduction is whether physicochemical entities and processes underlie all living phenomena. Substantive vitalists claimed that living processes are, at least in part, the effect of a nonmaterial principle or entity, which was variously called 'vital force', 'entelechy', '*élan vital*', 'soul', 'radial energy', or the like. As Dobzhansky says in his Introductory Remarks, 'most biologists are reductionists [ontologically, since they] see life as a highly complex, highly special and highly improbable pattern of physical and chemical processes'. Ontological reductionism implies that the laws of physics and chemistry fully apply to all biological processes at the level of atoms and molecules. Vitalism is now practically a dead issue in the philosophy of biology, particularly because of its sterility as a guide in biological research. A nonmaterial principle cannot be subject to scientific observation, nor lead to genuinely testable scientific hypotheses.

What I have called above the *methodological* domain encompasses questions concerning the strategy of research or the acquisition of knowledge. In the study of life phenomena, should we always seek explanations by investigating the underlying processes at lower levels of complexity, and ultimately at the level of atoms and molecules? or must we seek understanding from the study of higher as well as lower levels of organisation? Are there any general answers to these questions? or are different answers appropriate in different biological disciplines? Answers to these questions would be of considerable interest to biologists. Extreme reductionist and antireductionist positions have been characterised above. The extreme reductionist contends that the only biological explanations worth seeking are those obtained by investigating the underlying physicochemical processes. The extreme antireductionist might argue that such explanations do not truly belong to the realm of biology.

Questions concerning the strategy of research are explicitly considered in several contributions to this volume, and were the subject of much discussion during the conference. It seems that this is an area where much misunderstanding has occurred, and where agreement among biologists may be more nearly complete than appears on the surface. It became clear in the course of our discussions that most scientists who consider themselves methodological reductionists reject what I have characterised as extreme reductionism, while self-proclaimed antireductionists reject extreme antireductionism. I believe that all participants in the conference, and probably most scientists and philosophers, agree that the study of problems at a given level of complexity of the living world must proceed by exploring lower as well as higher levels of organisation. Consider the study of heredity. An extreme antireductionist might claim that knowledge of the molecular structure of DNA and of the molecular mechanisms of its replication, transcription and translation is of little biological interest. On the other hand, an extreme reductionist might argue that Mendelian genetics had no scientific import until the underlying molecular processes became known.

Epistemologically, the general question of reduction is whether the theories and experimental laws formulated in one field of science can be shown to be special cases of theories and laws formulated in some other branch of science. If such is the case, the former branch of science is said to have been reduced to the latter. This is the sense in which philosophers of science most often discuss questions of reduction.

Science is systematic organisation of knowledge about the universe on the basis of explanatory hypotheses which are genuinely testable. Science advances by developing gradually more comprehensive theories; that is, by formulating theories of greater generality which can account for observational statements and hypotheses which appear as *prima facie* unrelated. For example, the Mendelian principles of inheritance can explain observations that at first appear unrelated, such as the proportions in which characters are transmitted from parents to offspring, the preservation through the generations of discontinuity in traits with alternative states, and why sometimes features present in the parents do not show in their progenies. Knowledge about the process by which sex cells are produced, and about the behaviour of chromosomes during that process, was eventually shown to be connected with the Mendelian principles. The incorporation of such knowledge into the Mendelian theory made possible the explanation of additional observations, such as why certain traits are inherited independently of each other while other traits are transmitted together more often than not. Further discoveries have made possible the formulation of a unified theory of inheritance which explains many diverse observations including the overall constancy but occasional change of the hereditary information through the generations, the discreteness of biological species, and the adaptive nature of organisms and their features. The integration of scientific theories into a more comprehensive theory simplifies science and extends its explanatory power. The reduction of theories conforms to the goal of science.

Reduction of a whole branch of science to another has repeatedly been claimed in the history of science. During the last hundred years, several branches of physics and astronomy have been to a considerable extent unified by their reduction to a few theories of great generality, such as quantum mechanics and relativity. The reduction of thermodynamics to statistical mechanics was made possible by the discovery that the temperature of a gas reflects the mean kinetic energy of its molecules. A large sector of chemistry has been reduced to physics after it was discovered that the valence of an element bears a simple relation to the number of electrons in the outer orbit of the atom. None of these or other reductions has been completely successful; an unresolved residue has remained in every case (see K. R. Popper, Scientific Reduction and the Essential Incompleteness of All Science, in this volume). Yet these reductions are some of the most impressive accomplishments of science.

In biology, the binary character of reproduction at the cellular level, the relative constancy of the hereditary information through the generations, the mechanism of transmission of hereditary information from the nucleus to the cytoplasm, and other elements of the theory of inheritance, have been reduced to chemistry after the discovery of the structure and behaviour of certain molecules such as DNA, RNA and some enzymes. The spectacular success of this and other reductions has prompted in certain circles the conviction that the ultimate goal of all biological disciplines is to explain their theories and experimental laws as special cases of physical and chemical laws.

Ernest Nagel has formulated the two necessary and sufficient conditions to effect the reduction of one branch of science to another (Ernest Nagel, *The Structure of Science*, Harcourt, Brace and World, New York (1961); see also F. J. Ayala, Biology as an Autonomous Science, *American Scientist*, **56** (1968), 207–21). To accomplish the reduction of one branch of science to another, it must be shown that all the experimental laws and theories of the former are logical consequences of the theoretical constructs of the latter. This is the condition of *derivability*. To accomplish this deduction the laws of logic require that all technical terms used in the science to be reduced be redefined using terms of the science to which it is to be reduced. This is the condition of *connectability*.

It is clear that in the current state of scientific development a majority of biological concepts such as cell, organ, Mendelian population, species, genetic homeostasis, predator, trophic level, etc., cannot be adequately defined in physicochemical terms. Nor are there at present any class or classes of statements belonging to physics and chemistry from which every biological law could be derived. That is, neither the condition of connectability nor the condition of derivability is satisfied. These considerations make it clear that the reduction of all or even most of biology to the physicochemical sciences is premature at present.

Several contributions in this volume are concerned with questions of reduction in the epistemological sense. Some discuss the possibility or impossibility of reduction within certain sectors of biology; others claim that a reductionist programme can never be completely successful while pointing out that even incomplete reductions are valuable accomplishments; still others discuss notions like 'novelty', 'purpose', 'creativity', 'progress', which do not occur in, nor can be formulated using concepts of, the physicochemical sciences but are needed in biology; some papers develop patterns of explanation that may be distinctive of the biological disciplines.

There are many possible ways of sequentially ordering the papers in this volume. For the most part, they have been arranged in the same sequence in which they were discussed at the conference. Papers dealing with closely related topics or proposing similar points of view have been grouped together, followed whenever possible by papers advancing contrasting positions.

In the first contribution to this volume, G. Montalenti argues that two conflicting views, one reductionistic and the other antireductionistic, have existed since the most remote origins of biological studies. Among Greek philosophers, Aristotle had a holistic or vitalistic viewpoint, while Democritus advanced a mechanistic interpretation of life phenomena. The mechanistic approach has been most successful towards making new discoveries but is insufficient to account fully for biological phenomena. The laws of physics alone cannot explain biological processes above the molecular level of organisation. There is no need to introduce 'vital forces' or any other metaphysical entities, but principles of explanation not reducible to those of physics and chemistry are required. In the terminology which I have used above, Montalenti is an ontological reductionist, defends reductionism as the most successful research strategy, but argues against the possibility of reductionism in the epistemological sense.

Ernest Boesiger suggests that the greatest contributions of both Lamarck and Darwin were their attempts to provide a causal explanation of the origin and diversity of organisms, and of their adaptations. The adaptive or 'purposeful' character of organisms is explained by use and disuse (Lamarck) or by natural selection (Darwin). These patterns of explanation are distinctive of biology. They require an 'organismic' point of view and recognition of the relevance of the individual and not only of the 'class'. Boesiger shows that Lamarck's contributions to evolutionary theory are substantive in spite of his erroneous conception of the inheritance of acquired characteristics.

Gerald Edelman discusses the current selective theory of antibody formation. An extreme reductionistic approach led to the incorrect 'instructive' theory of antibody formation. The correct theory emerged when attention was shifted from the level of the antibody *per se* to the level of the cell. Methodologically, a two-tier approach involving reductionism (molecular level) and compositionism (levels of cell and organism) proved of greatest heuristic value.

Peter Medawar illustrates the notions of reduction and emergence by analogy with the hierarchy of geometries as conceived by Felix Klein. The empirical sciences may be ordered as follows: physics–chemistry–biology–ecology/sociology. Each science is in a sense a special case of the one which precedes it. Only a limited class of all possible interactions between molecules constitutes the subject matter of biology. Every statement that is true in one science (say, physics) is also true in the sciences following it (say, biology). Medawar argues against epistemological reductionism by pointing out that each science has 'concepts peculiar to and distinctive of [it], and not obviously reducible to the notions' of the preceding sciences. As we go along the sequence, the sciences become progressively richer in their empirical content and new concepts emerge. This enrichment occurs not *in spite of*, but precisely *because of* the progressive restriction and limitation of subject matter from one science to the next.

June Goodfield examines the philosophical outlook and methodological approach of several physiologists of the nineteenth and twentieth centuries. Some are epistemological reductionists; others, antireductionists. This seems to have made little difference in their scientific accomplishments. There are no methodological differences between scientists in the two groups. In their approach to experimental research all are compositionists as well as reductionists.

John Eccles states that reductionism is a necessary strategy of research. 'The programme of the neurobiologist is strictly mechanistic and reductionistic because we attempt to explain the neural events . . . solely in terms of physics and chemistry and their various developments in biophysics, biochemistry, neuropharmacology and neurocommunications.' Yet he argues that 'reductionism fails when confronted with the brain–mind problem'. Experiments with split-brain subjects have shown that conscious experiences arise only in the dominant (left) hemisphere of the brain. These experiments refute the psychoneural identity hypothesis. Eccles accepts free will and therefore is 'constrained, as a neuroscientist, to postulate that in some way, completely beyond my understanding, my thinking changes the operative patterns of neuronal activities in my brain'.

W. H. Thorpe argues against epistemological reductionism using Broad's notion of 'emergence'. To explain the behaviour of any whole 'we need to know the law or laws according to which the behaviour of the separate parts is compounded when they are acting together'. The laws that explain the behaviour of the whole cannot be derived from laws that explain the behaviour of the parts. The laws of physics and chemistry cannot account for the specific kinds of living things—that is, the particular configurations of matter in organisms. Among sectors of biology where reductionism is impossible, Thorpe discusses those dealing with the function of the nervous system in the brain of man and of higher animals.

Donald Campbell proposes that a conceptual model of 'unjustified variation–selective retention' applies not only to biological evolution but also to creative thought and scientific discovery. The process by which scientific concepts and theories first arise is not 'wise', 'foresighted' or 'preadapted', but rather is 'random', 'spontaneous' or 'blind'. He considers himself a 'physicalist, materialist and reductionist' because he believes that the purposiveness or teleology of systems can be explained by natural processes. Darwin's explanation of the adaptations of organisms by natural selection is 'compatible with the physical model of causation'. Campbell rejects, however, the reductionism of those 'who deny the existence of fit, design, purpose [and] emergent higher levels of organisation'.

Morton Beckner explicates the important distinction between hierarchically organised systems and hierarchically organised sciences. The existence of hierarchically organised systems is apparent in the multiple levels of increasing complexity found in the living world. The hierarchical organisation of sciences is, on the other hand, based on the logical relations that hold between theories, descriptions, conceptual schemes and other instances of language. A theory of a higher level is autonomous relative to a theory of a lower level if it is not reducible (in the epistemological sense) to the latter. 'The philosophical investigation of the concept of reduction . . . [has] been of great benefit in increasing our understanding of scientific methods.' Yet answers to questions about the reducibility of higher to lower-level theories have few or no methodological consequences with respect to scientific practice.

Donald Campbell's comment following Beckner's paper illustrates the ambiguity of the terms 'reductionist' and 'antireductionist' when they are used without qualification to describe the position of a scientist or philosopher. Campbell labels himself a reductionist, because he accepts ontological reductionism. But he denies the possibility of epistemological reductionism: 'Biological evolution . . . encounters laws . . . which are not described by the laws of physics and inorganic chemistry, and which will not be described by the future substitutes for the present approximations of physics and inorganic chemistry.' 'Description of an intermediate level phenomenon is not completed by describing its possibility and implementation in lower level terms.'

Dudley Shapere distinguishes two kinds of theories. *Compositional* theories provide explanations 'in terms of the constituent parts of the individuals'; *evolutionary* theories advance answers 'in terms of the time-development of items'. In the search for a compositional theory for a domain it often happens that the methods and concepts used in some other field are applied. Thus, the search for compositional theories often leads to the unification of fields. Such unification on occasion provides what has been called the reduction of one subject to another. If reduction is understood to require 'that the concepts of one area are definable in terms of those of the other [area], and the relationships of the former are deducible from those of the latter . . . [then] reduction would almost never have taken place'. The 'deductions' involved in the reduction of a branch of science to another are not strict, but involve all sorts of approximations, simplifications and idealisations. Reduction does not imply 'that the *field* reduced would be eliminated; for not only would its laws and individual events not be strictly deducible from those of the reducing theory; its methods, too, might still have much to offer which is inaccessible to those associated with the reducing theory'.

Henryk Skolimowski argues that 'Modern biology, and especially evolutionary biology, has proved time and again the insufficiency of the physical model of knowledge'. The hierarchy and complexity of living organisms are equivalent to nonreducibility; when we *comprehend* the function of organisms we go beyond physics and chemistry. To describe the process of evolution up to and including man, we need a new 'rationality' different from that developed by physical scientists. The new rationality requires the introduction into our language of 'open-ended concepts, growth concepts and normative concepts'.

Charles Birch notes that the physicist does not hesitate to attribute to atoms physical properties, like gravity, when he finds it necessary in order to explain the properties of matter in bulk, although 'gravity is not predictable from atomic and nuclear physics'. Consequently, he argues, the physicist should also ascribe to atoms 'proto-mental' properties, since these are needed to explain their function in elements composing the brain. Biology is not to be reduced to physics; rather 'we need a physics . . . which takes into account biological phenomena and is partly derived from the study of these phenomena'.

Bernhard Rensch advances the thesis that all events have a 'polynomistic' determination. They are determined not only by causal laws but also by laws of probability and logic, laws of conservation and others. The immediate correspondence between psychic phenomena and physiological brain processes can best be explained by what he calls 'panpsychistic identism', that is, by the 'identification of matter and mind. . . . This presupposes a protopsychical nature of matter.'

According to Karl Popper, scientific reductions are never completely successful; an unresolved residue is always left. Moreover, 'philosophical reductionism is . . . a mistake. It is due to the wish to reduce everything to an ultimate explanation in terms of essences and substances, that is, to an explanation which is neither capable of, nor in need of, any further explanation.' Nevertheless, we should continue to attempt reductions because we can learn an immense amount even from unsuccessful attempts at reduction. The number of interesting and unexpected results we may acquire on the way to our failure can be most fruitful for science. Successful reductions are among the greatest accomplishments of science. With respect to the mind–body problem, Popper proposes 'a form of *psychophysical interactionism*. This involves . . . the thesis that the physical world is not causally closed, but open to the world of mental states and events. . . . As a philosopher who looks at this world of ours, with us in it, I indeed despair of any ultimate reduction.' 'We live . . . in a world of emergent evolution; of problems whose solutions, if they are solved, beget new and deeper problems. Thus we live in a world of emergent novelty; of a novelty which, as a rule, is not completely reducible to any of the preceding stages.'

Ledyard Stebbins points out that the study of biological evolution is concentrated around two foci. One focus is the study of evolution at the level of the population, and uses methods which are quantitative, experimental and largely reductionistic. The other focus is concerned with evolution as a whole, 'a succession of events that took place over billions of years of time, and gave rise successively to living matter, organised cells, multicellular organisms . . . and finally man'. Scientists studying evolution as a whole must draw information from population genetics, systematics and palæontology; their explanations are for the most part extrapolations based on 'a broad knowledge of apparently disparate facts'. The student of evolution as a whole develops patterns of explanation which are compositionistic rather than reductionistic. Stebbins shows the use of the compositionistic approach in the formulation of explanatory theories concerning some fundamental evolutionary problems.

Theodosius Dobzhansky recognises that random processes are present in biological evolution, particularly at the level of mutation and sexual recombination. Yet natural 'selection puts a restraint on chance and makes evolution directional. . . . Selection increases the adaptedness of the population to its environments. It is responsible for the internal teleology so strikingly apparent in all living beings.' In evolution, chance and necessity are not alternatives. Consider the evolution of man. 'Was *Australopithecus* bound to evolve into *Homo erectus*, and this latter into *Homo sapiens*? The question must be answered in the negative. . . . We neither arose by accident, nor were we predestined to arise.' Mutation, sexual recombination and natural selection are linked together in a system which makes biological evolution 'the only process lacking intentionality and foresight which is nevertheless creative'. Dobzhansky shows that in evolutionary theory compositionist, as well as reductionist, patterns of explanation need to be used. Evolutionary theory requires explanatory notions that play no role outside the biological realm.

Francisco Ayala discusses the concept of biological progress, a notion which has explanatory value in biology but not in physics or chemistry. 'Progress has occurred in nontrivial senses in the living world because of the creative character of the process of natural selection.' The concept of progress contains two elements: one is descriptive—that directional change has occurred; the other is axiological—that the change represents an improvement or betterment. A value judgment needs to be made of what is better and what is worse, or what is higher and what is lower, according to some axiological standard. The choice of an axiological standard involves a subjective element, but should not be arbitrary. There is no criterion of progress which is 'best' in the abstract. Whether a criterion is valid depends on whether it leads to meaningful statements concerning the evolution of life. Different criteria of progress may be preferable in different contexts.

According to Jacques Monod, the 'Postulate of Objectivity' is the cornerstone of biology. The Postulate of Objectivity 'is the systematic or axiomatic denial that true knowledge can be obtained on the basis of assumptions or theories that pretend to explain things in the universe by final causes'. The Postulate is closely related to Popper's 'criterion of demarcation' to distinguish between a scientific statement and a statement outside the realm of science. The reductionistic approach of molecular biology has shown that the explanation of certain properties of organisms, such as invariance, homeostasis and self-organisation, 'is found at the level of molecules of sizes between 10 000 and a few million'.

The discussion following Monod's contribution deals with some of the issues raised by him, and with questions like the scientific character of evolutionary theory and the impact of 'reductionistic' science on ethics and sociology.

List of Participants

The conference on 'Problems of Reduction in Biology' was held in Villa Serbelloni, Bellagio, Italy, from 9 to 16 September 1972.

Francisco J. Ayala
Department of Genetics
University of California
Davis, California 95616 USA

Morton Beckner
Department of Philosophy
Pomona College
Claremont, California 91711 USA

Charles Birch
School of Biological Sciences
Zoology Building
Sydney, N.S.W., Australia

Ernest Boesiger
Laboratoire de Génétique
Expérimentale des Populations
Place Eugène Bataillon F-34
Montpellier, France

Donald T. Campbell
Department of Psychology
Northwestern University
Evanston, Illinois 60201 USA

Theodosius Dobzhansky
Department of Genetics
University of California
Davis, California 95616 USA

John Eccles
Department of Physiology
State University of New York at Buffalo
Amherst, New York 14226 USA

Gerald M. Edelman
The Rockefeller University
New York, New York 10021 USA

June Goodfield
Department of Philosophy
Michigan State University
East Lansing, Michigan 48823 USA

Peter Medawar
Clinical Research Centre
Watford Road
Harrow, Middlesex HA1 3UJ, England

Jacques Monod
Institut Pasteur
28 rue du Dr Roux
Paris 15, France

Giuseppe Montalenti
Istituto di Genetica
Facoltà di Scienze
Città Universitaria
00185 Roma, Italy

Karl R. Popper
Fallowfield
Manor Road
Penn, Buckinghamshire, England

Bernard Rensch
Zoologisches Institut der
Westfälischen Wilhelms-Universität
Badestrasse 9
44 Münster (Westf.), West Germany

Dudley Shapere
Department of Philosophy
University of Illinois
Urbana, Illinois 61801 USA

Henryk Skolimowski
Department of Humanities
University of Michigan
Ann Arbor, Michigan 48104 USA

Ledyard Stebbins
Department of Genetics
University of California
Davis, California 95616 USA

William H. Thorpe
Department of Zoology
University of Cambridge
Cambridge CB3 8AA, England

1. Introductory Remarks

THEODOSIUS DOBZHANSKY

The problems of reduction in biology are currently of considerable theoretical interest and practical significance. Explicitly or implicitly they underlie many discussions and disputes among biologists concerning research strategies that biologists should adopt. Most biologists, I believe all those gathered around this table, are reductionists to the extent that we see life as a highly complex, highly special and highly improbable pattern of physical and chemical processes. To me, this is 'reasonable' reductionism. But should we go farther, and insist that biology must be so reduced to chemistry that biological laws and regularities could be deduced from what we shall learn about the chemistry of life processes? This, I think, is 'unreasonable' reductionism. The most spectacular advances in biology in our time were unquestionably those in molecular biology. Yet it does not follow that organismic biology is from now on unproductive, or that all of us should work exclusively on molecular biology. Should not organismic and molecular biology both continue to develop, because one without the other can only give a distorted view of life? Should the philosophy of biology deal with organismic, or with molecular aspects, or with both?

We are, of course, not the first to appreciate the importance, and even the urgency, of these problems. In recent years, three symposia were held on theoretical biology, here at Villa Serbelloni, under the chairmanship of C. H. Waddington. Similar or related problems were discussed also at a symposium curiously entitled 'Beyond Reductionism', organised by A. Koestler at Alpbach. We are certainly not planning merely to recapitulate the discussions that took place at these symposia. My colleague F. J. Ayala and myself feel that it is advisable to have here, at the present symposium, a fresh panel of participants, and to approach the problems from different angles. We shall not be serving again the same dishes that were served at the other symposia. Having a fresh panel has a further advantage: if you invite Dr X who participated in the other panels, then Drs Y and Z are liable to feel hurt not being invited also. And to reiterate: we are not planning to tread again the same ground which the other symposia did.

Although copies of the papers submitted by the participants have been distributed to all of you in advance, one cannot assume that every one of us has a complete photographic memory. The authors should therefore present, in 15–20 minutes each, a summary of their papers, emphasising the points they consider most important or most likely to stimulate discussion. The presentation of each paper will be followed by a free discussion without time limit, except as necessitated by the meal schedule established by our hosts at the Villa. The discussions will not be recorded for publication, but the interlocutors who feel that this is desirable are invited to submit written statements, which will be published.

2. From Aristotle to Democritus via Darwin: A Short Survey of a Long Historical and Logical Journey

G. MONTALENTI

This is what happens in the generation of the bees according to tradition and to the facts known so far. But we have not sufficient observations on this matter. Should they become available, we should trust more the observations than the theory, and we should hold good the latter only if facts support it. (Aristotle, *De gen. Anim.*, III, 760, b)

The legacy of the Greeks

Democritus and Aristotle

G. G. Simpson (1963) maintains that instead of accepting the definition of science proposed by someone as 'thinking about the world in the Greek way' we should realise that science is *not* that way of thinking. On the contrary we should 'escape from the Greeks'. A few lines further, he completes his thought by referring to 'the Greek way, which became traditional in medieval Europe'.

I do not want to start a debate on this point, but I would like to emphasise that in the works of the Greek philosophers and naturalists we find already the main philosophical and epistemological leitmotifs which underlie our modern science. There is little doubt that the source of modern scientific thought—that is, the scientific attitude towards the world—is to be found in Greek philosophy.

Democritean and Aristotelian points of view are the two antithetic systems, the two opposite poles of the interpretation of natural phenomena. The first one represents, according to the philosophy adopted by the majority of modern scientists, the sound scientific attitude.

Dante (*Divina Commedia*) meets in the limbo the two great philosophers:

Aristotle, 'il maestro di color che sanno . . . tutti lo miran, tutti onor gli fanno', and 'Democrito, che il mondo a caso pone' (*Inf.*, IV, 131, 136).

It is true that the Aristotelian system, after many historical vicissitudes, which need not be recorded here, eventually became the more widely accepted *Weltanschauung* in the Middle Ages. It was successfully grafted upon the stem of the Christian philosophy by Thomas Aquinas, and thus it became the core of the official philosophical system of the Christian world.

The demolition of Aristotelism as a scientific system started in the Renaissance, and was successful in the physical disciplines. In biological sciences, however, Aristotelism lasted until the Darwinian revolution, which is still in progress in our days (Mayr, 1972). In fact, the Aristotelian viewpoint in biology still has a good number of supporters among modern biologists.

The ill fate of Democritean principles—atomism and causality (necessity) as contrasted to the holism and finalism of Aristotle—is undoubtedly due to the schematism of his interpretation of the world, which is expressed in lapidarian style by Dante ('il mondo a caso pone') as the result of a mere 'jeu du hasard'. But in spite of that, Democritean philosophy never died out in the course of history: it was transmitted to the modern world through the latin poet Lucretius Caro in his *De Rerum Natura.*

The whole course of events in the history of biology may be schematically represented as a continuous conflict between Aristotelian and Democritean—that is, vitalistic, or holistic, *versus* mechanistic—interpretations of the phenomena of life.

Thus it is not fair to say that 'the origin of science in the modern sense involved a revolt against thinking in the Greek way' (Simpson, 1963). It certainly was a revolt against thinking in the Aristotelian–scholastic way, which was the current system in the Middle Ages.

But was Aristotle himself really antiscientific in his way of thinking?

Aristotle and the principle of causality

Aristotle was essentially a biologist, with particular interest in animals. This is made clear by his enthusiastic approach to the study of animal life, by the extent of the treatises devoted to zoology, in relation to those concerned with inorganic nature, the ratio being approximately 3:2, and by the deep and precise knowledge he shows of the structure of animals and of many aspects of their life. Such knowledge has received the admiration of many modern naturalists.

The most striking character of biological phenomena is finalism. In biology Aristotle finds the most convincing evidence of the finality of natural phenomena he probably was looking for on theoretical grounds. The teleological interpretation, so easy and appropriate to biological facts, he transfers on to physical and cosmological phenomena, thus building up a teleological system of the whole world.

In the finalistic interpretation of biological phenomena, however, Aristotle meets with several difficulties, which compel him to overcome the rough and naive anthropocentric finalism of some of his predecessors (such as Xenophanes). He is obliged to analyse the concept of cause. As is well known, he admits four types of cause: the formal cause, the efficient cause, the final cause and the material cause. The first three categories are not very clearly differentiated, and they may well be considered as subspecies of the main one: the final cause. On the contrary the material cause is well defined as the opposite of the final cause, and is characterised by a mere passivity. So is matter out of which the sculptor makes the statue, whose form he has in mind. Matter is necessary in order to achieve the purposes that nature has to realise; but it is in some way reluctant to accept the form, and this explains the imperfections, and all the phenomena which modern biologists call dysteleological.[1]

Aristotle realised that it is not possible to explain everything as the result of the final cause. This could lead to erroneous conclusions, through different channels, namely (*a*) by inducing to neglect the principle of necessity, the only one admitted by Democritus, and contrasting with finalism; and (*b*) by the prevalence of the deductive argument, which may lead astray from factual evidence.

Perhaps the most explicit and clearcut example of what Aristotle means when he speaks of *necessity* is to be found in his treatise on *Physics*,[2] where he says 'Zeus does not send rain for the purpose of growing wheat, but by necessity, because the vapours must get cold, and hence become water and fall down'.

Necessity, which is closely connected with material cause, also explains individual variability. Again he gives an example: the eye has a function, hence it is produced by a final cause. But that the eye be grey or brown does not make any difference to the function; this character is not due to a final cause, but to mere necessity.[3] Thus individual variations are unimportant in respect to the Platonic idea or type.

In the introduction to his treatise on the *Parts of Animals*[4] and in several places in *On the Generation of Animals*[1] Aristotle tries to make clear the ideas about the relationship between necessity—that is, the principle of causality—and purpose or final cause. He does not succeed in getting at a clear definition of the problem: often he cannot avoid a state of rather confusing contradiction on this point. We can hardly blame him: the problem he is concerned with remained the central unsolved problem of biological

[1] *De gen. Anim.*, IV, 778a, and V, 789b. I have consulted the German translation of H. Aubert and Fr. Wimmer, *Aristoteles fünf Bücher von der Zeugung und Entwickelung der Thiere*, Engelmann, Leipzig (1860).

[2] *Phys.*, II, 198b, quoted from Th. Gomperz, *Griechische Denker*, Leipzig (1893–1902).

[3] *De gen. Anim.*, V, 1, 778a.

[4] *Aristotle on the Parts of Animals*, translated, with introduction and notes, by W. Ogle, Kegan Paul, Trench and Co., London (1882).

thought for many centuries to come. However, the main conclusion emerging from his analysis is that by far the most important cause in biological phenomena (and in physical phenomena as well) is the final cause, while material cause, or necessity, is entirely submitted to the former. Thus there is no doubt that Aristotelian philosophy is essentially teleological, as is traditionally admitted. The presence of necessity, in as much as it cannot be denied, constitutes a difficulty: the aporia is resolved by giving absolute predominance to finalism. This is represented, in biology, by the preexistence of a full and perfect idea (in the Platonic sense) of every organism, the *entelécheia*, which Hans Driesch reintroduced in biology at the end of the nineteenth century.

The second danger of a teleological interpretation could not escape the great naturalist's attention, namely that theory should lead to wrong interpretation of facts. On many occasions Aristotle warns against such an unscientific attitude, and asserts the need of trusting more the evidence of factual information than theoretical deductions. He is very explicit about this when summarising what is known about the reproduction of the bees.[5] In fact Aristotle himself was not always able to conform to this rule, which on the other hand was quite unfulfilled by his medieval followers, fanatics of the *ipse dixit*. But, going through his treatises on zoology, one gets the impression that he was really most of the time a good careful observer of facts, and that from these he tried to build up his theory.

Thus we may conclude that as a biologist Aristotle was a scientist much in the same sense as were practically all modern biologists until Darwin. His teleological interpretation, which was at that time—and for many centuries thereafter—the most plausible theory of the world, led him to build a grandiose and coherent system, which was going to last until modern times (Enriquez and Mazziotti, 1948).

Aristotle and the concept of chance

The concept of chance in Greek philosophy is rather confused, as much as the concept of 'probability'. The main confusion is due to the mingling up of the concept of chance with that of 'accident' as opposed to 'substance' (two terms which had a great importance in scholastic philosophy), and to the fact that often chance was connected with rare, exceptional occurrence.

Without going into the details of the discussion, I like to point out the definition which Aristotle gives of chance as 'the cause by accident of facts which are susceptible to become purposes'.[6] The famous examples are that of the creditor going to the public place for other purposes and meeting his debtor who happens to have the money to settle his debt, and that of the man ploughing his field and discovering a buried treasure.[7]

[5] *De gen. Anim.*, III, 760b; the sentence is reported as a motto to this article.
[6] *Phys.*, II, 5, 197a.
[7] *Phys.*, II, 5, 197a; *Metaph.*, 30.

Chance is devoid of any purpose—therefore we speak of good or bad fortune—but it may, by mere accident, give rise to teleological processes. This concept is very close to our concept of random mutations. But Aristotle does not see how it can be used in the explanation of biological phenomena. He is aware that Empedocles has tried to explain the finality of organisms by the survival of those which, by chance, were fit for life; but he ridicules such a rough biological theory. Had we seen the hybrid monsters such as 'bovine bodies with human heads', which Empedocles thinks are spontaneously generated, we should have considered them in the same way as we consider the monsters which are produced today: deviations from a previously established rule, and not phenomena preceding the establishment of the rule. Order and purpose cannot originate from disorder and purposelessness.

Democritus: atomism and causality

We encounter in Aristotle a statement of the main reason why the mechanistic interpretation of the world was unacceptable to him and to the majority of naturalists and philosophers of modern times. Causality or necessity, as he calls the principle upheld by Democritus, and chance alone cannot give a reasonable explanation of the harmony of the world, especially of the organic world, nor can they account for the undeniable finalism of living processes.[8]

In fact Democritus tried to explain the world by means of a purely mechanistic hypothesis based on the atomic structure and on movement of the atoms, which leads to a strictly materialistic and deterministic conception.

This would not mean—it seems to us—that the world is due to mere accidents: every effect has behind it its cause. The ultimate causes are the form and the movement of atoms, this being a sort of their natural status, without any other cause or principle external to them.

We might be surprised about Dante's reproachful statement that Democritus has attributed the world to the mere work of chance. But from the point of view of a medieval Aristotelian, as Dante was, the charge appears legitimate: how could one explain the marvellous harmony of the world without resort to the final causes, which are deliberately ignored in Democritean philosophy?

In conclusion, we may say that all the main themes of scientific interpretation were present in Greek philosophy. The mechanistic approach was developed by the atomists, and was upheld by their followers. But it was,

[8] There has been much discussion about the concept of final cause and teleology in Aristotle. Some authors have cast doubt upon Aristotle being a teleologist. For a recent survey *vide: Aristotele: Opere biologiche*, a cura di Diego Lanza e Mario Vegetti, Torino, UTET (1971), 504 ff, 526 ff. I have no doubt that the natural philosophy of the Greek thinker was essentially finalistic. This aspect was undoubtedly emphasised by Aristotle's scholastic interpreters.

since its beginning and during many centuries thereafter, the object of a constant stubborn fight sustained by the idealistic philosophy (from Plato, who never mentions Democritus, down to the Hegelians, to Croce and Gentile). The main reason why, on scientific grounds, mechanistic interpretation was not able to impose itself on biology is its inadequacy to account for finalism in the phenomena of life.

The solution of this long-lasting aporia has been possible only in recent times, in the light of the Darwinian revolution and the findings of molecular biology, as we shall see.

The attempts towards a physical interpretation of biological phenomena

It was comparatively easy in the scientific Renaissance to rid the physical disciplines of the principle of finality and to put them on the basis of a sound philosophy. The attempts towards extending the same principles to biology started early enough, practically with Descartes in the seventeenth century. In the subsequent course of the history of biology, we witness constant swings of the two opposite interpretations, namely the mechanistic and the vitalistic, with a great number of nuances in both categories.

Two facts emerge from a synthetic consideration of such historical events: (1) the mechanistic approach was the most fruitful of new discoveries and consistent progress; (2) the mechanistic approach has always failed to account fully for biological phenomena. The trend to reduce the phenomena of life to physical interpretation, as time passed and science progressed, was becoming more and more successful. Many phenomena formerly considered as irreducible were eventually explained by the principles of physics and chemistry. But the real end was never reached: an insoluble residue was always left. Hence the constant revival of vitalism. The last attempt to produce a vitalistic conception as a theory based on experimental evidence was made, as is well known, by Hans Driesch at the end of the last century (see Driesch, 1905).

Driesch's vitalism was a real revival of the old animistic theory of Aristotle, including the concept and the term *entelécheia*, meaning the perfect and complete idea of the organism, which exists before its actual material realisation.

Such coarse vitalism could not appeal very much to modern biologists, and in fact did not have a great success.

On the other hand the failure of an explanation in purely physicochemical terms of the complexity of vital phenomena and of their finalism was evident. The need to find an escape from this dilemma was indeed urgent. Several theories have been proposed in the first decades of the present century which refuse the strict reductionistic point of view and maintain that in order to explain life we should take into account first the complexity of the organism, as a whole. Such holistic or organismic theories (see Smuts, 1927; von Uexküll, 1928; J. S. Haldane, 1931; von Bertalanffy, 1932) enjoyed a certain popularity in the twenties and the thirties; undoubtedly they contain a grain of truth, as we shall see later on. But still the inadequacy of a physical explanation was indisputable. To say, as did for instance A. Meyer (1934, 1935), that 'the goal of biology should be to formulate its axioms, principles and laws in such a way that physical laws can be deduced from them by successive simplifications' seems a mere verbalism. Thus the old question remained open.

The modern solution

Perhaps modern biology has found the solution to it; it certainly has reached a stage in which a rational solution may be confidently foreseen. This is due to the following main achievements:

(1) the discovery of the dimension of time in biological phenomena;
(2) the 'Darwinian revolution' and the discovery of natural selection;
(3) the discovery of the individual as the most important element in evolution;
(4) the reaching of the roots of life at the molecular level;
(5) the identification of different levels of integration;
(6) the definition of chance, of indeterminacy and of creativity in evolution.

We shall now briefly discuss these points.

The discovery of the parameter of time

The world was supposed to be static in almost all the ancient cosmogonies. The exceptions of Heraclitus, and of the Epicurean philosophy as accepted by Lucretius, did not go very deeply into the time factor in historical biology and did not have great impact on the development of biological sciences. The discovery of the parameter of time in biology, which is the necessary prerequisite for a dynamic conception of the living world, was achieved, as is well known, in the eighteenth century, the most important protagonist of the discovery being Buffon. There is no need to go into details about the development of this idea. It is well known and has recently been illustrated by F. Jacob (1970).

The Darwinian revolution

The introduction of the parameter of time in biology was followed almost immediately by the idea of evolution. It is interesting to note, however, that the first complete and coherent theory of evolution formulated by Lamarck was not primarily concerned with the explanation of the succession of floras and faunas in geological time. On the contrary, the palæontological evidence was used by Cuvier *against* evolution. What Lamarck wanted to explain by his theory was adaptation, that is, purposiveness.

The Lamarckian theory of the mechanism of evolution was a combination of two principles, namely, internal drive towards progress, and inheritance of acquired characters. The first is to be rejected as a scientific explanation, because it has the same value as the *virtus dormitiva* of opium. The second is disproved by experimental evidence, and is not congruent with the considerable constancy of form and structure which dominates among living things.

The fact of evolution was firmly established by Darwin; but the major merit of the great biologist was the discovery of the main directional agent of evolutionary processes: natural selection. The concept of evolution by means of natural selection is the greatest conquest of modern science. It is thus quite legitimate to speak of 'Darwinian revolution' as E. Mayr (1972) did in his most penetrating article.

The most significant theoretical aspects of the Darwinian revolution are the new interpretation of biological finalism, and the recognition of the individual as the most important biological unit.

Teleology, conceived as the final cause of Aristotle, has been the ghost, the unexplained mystery which haunted biology through its whole history. The undeniable purposiveness of biological structures and functions frustrated many attempts towards a mechanistic interpretation of living phenomena. Natural selection is the key to the mystery, the salvation from teleology, this 'original sin' of all living beings. Natural selection is working as a *vis a tergo*, and leads to the making up and the differential survival of the most convenient combinations of genes, that is, the best adapted structures and functions. It is thus an interpretation of purposiveness in physical terms and in an evolutionary perspective. The adequacy of such a tool to account for the creative action of evolution, which was often denied by the opponents to the theory of natural selection, is most brilliantly and convincingly demonstrated by Dobzhansky (1954; Dobzhansky and Boesiger, 1968).

Finalism cannot be denied in biological affairs, and so its explanation as a result of selection gives a solution to a puzzle which was carried along from Aristotle up to the present time. In order to avoid confusion between the old and the new interpretation, biologists prefer to abandon the old term *teleology*, which still has a scholastic Aristotelian flavour, and substitute it with *teleonomy*.

The discovery of the 'individual'

The etymology of the word 'individual' is the same as that of 'atom': it means indivisible. The individual was thus conceived as the smallest unit which could not be split without losing its specific characters. In the typological concept of species (Mayr, 1969), individuals are the units representing the material realisation of the idea, of the type. They are in principle all alike, as the atoms of a given chemical element. The likeness of the general specific character is the only character that matters; individual differences (such as eye or hair colour, size of the body, form of the nose, etc.) are unessential. In the Aristotelian system they are due to material causes; hence they originate by necessity, and are not promoted by a final cause.

The same line of thought inspired Linnaeus when he said 'varietates laevissimas non curat botanicus'.

Evolution, on the contrary, points at individual differences as the basic element upon which selection can work. Thus, instead of considering as fundamental only the general characters, modern biology has recognised one of the most typical properties of life: individuality. E. Mayr (1961) has expressed it in the following words: 'the individuality . . . so characteristic of the organic world, where all the individuals are unique; all stages in the life cycle are unique; all populations are unique; all interindividual contacts are unique; all natural associations of species are unique; and all evolutionary events are unique'.

Individuality is often expressed by the word 'uniqueness'. I prefer individual and individuality. Some physical phenomena, such as the shape of a rock or the blowing of a wind, are unique, but not in the same sense as a living individual is unique. The uniqueness of populations and of biological associations (other entities which have been defined by modern biology, mainly as a result of an analysis which was stimulated by the theory of evolution) ultimately depends on the uniqueness of individuals.

The uniqueness of individuals, being the prerequisite of evolution by natural selection, is the most important characteristic of life, the one which differentiates more substantially living from nonliving things, physics from biology (Raffaele, 1905).

The roots of life: molecular biology

In the history of science one should be cautious in attributing modern ideas to old thinkers, and in considering as a bright prevision what may actually be mere fortuitous coincidence. But one cannot avoid being surprised by the fact that Democritus, who thought that atoms are endowed of only two qualities, size and form, also maintains that aggregates of atoms (molecules in modern scientific language) may differ not only by the form and size of their atoms, but also by their relative position, such as happens for the letters out of which words are composed.

Properties of atoms and their relative positions in the molecule are the basis of our understanding of the differences between chemical compounds, which are investigated by organic chemistry and molecular biology.

Molecular biology, the great achievement of our age, has reached to the very roots of life, the point of dichotomy of living from nonliving things.

Thus modern biology is now finally in position to discriminate between the two arrays of phenomena—physical and biological—at the molecular level and to test the validity of the old claim of the 'reductionists' to explain all biological phenomena by physicochemical principles.

It is often stated that reproduction is the most typical phenomenon characterising life. But at the molecular level we have many examples of molecules which, in proper conditions, can reproduce themselves and, in spite of that, cannot be considered alive (Giglio Tos, 1900).

What really matters to start life is—besides the faculty of reproduction—the appearance of individuality. As soon as a self-reproducing molecule becomes (by a process which has been called mutation) an individual different from the others of the same kind, natural selection starts operating and life begins.

Thus in individuality lies the real distinctive character of life, differentiating the biological from the physical world. Molecules of any nonliving compound are identical; their properties and behaviour can be fully described by physical laws, which usually are statistical. Even quantum indeterminacy admits a statistical description.

As soon as individuality appears, unique phenomena originate and the laws of physics become inadequate to explain all the phenomena. Certainly they are still valid for a certain number of biological facts, and they are indeed extremely useful in explaining a number of basic phenomena; but they cannot explain everything. Something escapes them, and new principles have to be established which are unknown in the inorganic world: first of all natural selection, which gives rise to organic evolution and hence to life.

Different levels of integration

As soon as life has started, a peculiar phenomenon intervenes: aggregation of molecules into more complex living units. Concomitantly two categories of molecules become differentiated in each individual: those which bear the genetic information, and the others which are functional in the individual but not in the hereditary transmission of the design. Weismann called the two entities, in multicellular organisms, *germen* and *soma*.

The reaching of this stage, the first level of integration, is perhaps at the present time the most difficult to explain. A multimolecular organism originates and it becomes partially isolated from the external world, being surrounded by a membrane which mediates all chemical exchanges with the environment. Several models of this fundamental step in biopoiesis have been proposed (J. B. S. Haldane, 1929; Bernal, 1968; Calvin, 1969; Oparin, 1969; Ponnamperuma, 1972). It is not essential to record them here.

The main point to be stressed is that the process of integration goes on, stepwise, throughout all the course of evolutionary history, leading to more and more complex units, like procaryotes, eucaryotes, unicellular and pluricellular, and to all their complications: corms, colonies, societies.

The process of successive integrations of biological units into units of a higher degree has been expounded by F. Jacob (1970) in the concluding chapter of his book. He has introduced the term *integron* for each unit of a given level. Integration means also organisation; thus, by successive integrations, a primitive simple organism gives rise to more complex ones, made of a number of differentiated parts, which are mutually interrelated in an organic ensemble.

The *primum movens* of all these processes, which proceeded along different pathways and eventually led to the astonishing differentiation we witness in the vegetable and animal kingdoms, is, according to the mechanistically minded biologists, natural selection.

What really needs to be pointed out in this connection is that each step, each integration level implies the appearance of new qualities, new relationships, which impose the need of new criteria of explanation. The principle of dialectical materialism is well known, that a change in quantity determines a qualitative change. This has been exemplified by saying that the sum of the properties of single grains of sand is not equal to the properties of a sand dune. It has also been often pointed out quite rightly that the properties of an organism are not simply equal to the sum of the properties of its parts.

In fact, if this principle is true for a mere physical accumulation of smaller entities, as in the case of the sand dune, *a fortiori* it is valid for an organism, in which the parts are mutually adjusted and functionally regulated in an organised structure.

Therefore at each level of integration new principles become active, and consequently new criteria have to be applied to find a scientific explanation of phenomena. At the molecular level the laws of physics and chemistry are the only ones that play a role. As the molecules become individually different and reproduce, the same laws are still valid, but a new law proper to living things, namely natural selection, appears and imparts a direction to physical phenomena. At successive levels new principles become operative. The fundamental physical laws such as the second law of thermodynamics and all the basic physicochemical laws are still valid, but cannot by themselves explain all biological phenomena.

This is the point where reductionism fails: elementary phenomena can be reduced to a physical explanation, and this is always fully valid in biology, but as soon as integrations above the molecular level occur, it is no longer sufficient.

This does not imply by any means either the introduction of vital forces or other metaphysical entities, nor does it mean that we should abandon the scientific method. The explanations we are looking for are always in the form of a cause–effect relationship, thus strictly adhering to scientific criteria; but the 'causes' and 'forces' implied are not only those known to physicists. Again, the example of natural selection, which is unknown in the physical world, is the most fitting. Others may be easily found in physiological, embryological and social phenomena. At this point one might recognise some validity in holistic or organismic arguments in so far as they do not have any metaphysical or finalistic (in the Aristotelian sense) implication.

Chance, the indeterminacy principle and creativity in evolution

Some events cannot be predicted as single occurrences, but only statistically: for example a single number, or even the colour red or black which will turn up in a roulette game, or the sex of a zygote after fertilisation. This does not mean that the single events are undetermined, but only that we are not in a position to know exactly all the sequence of causes which ultimately lead to the final result. Should we be able to see which kind of spermatozoon, whether bearing an X or Y chromosome, enters the egg, we would be able to predict exactly the sex of each egg immediately after fertilisation. Usually we cannot, for practical reasons, get such a level of knowledge; hence we must be content with the statistical laws, that is, with the evaluation of probabilities.

In the case of sex, the possibilities are only two; but as soon as we consider several traits, determined by a number of genes, the number of combinations becomes exceedingly large. If we take into account all the genes of an organism, even a simple one, the number of possible combinations becomes astronomically large. On the other hand the number of the individuals of a population is usually limited, and often it is rather small; hence the number of all possible combinations largely exceeds the number of those which are actually present as living individuals at any moment. For the more complex organisms, such as animals or plants, one may confidently say that the number of potential possibilities is much larger than the number of individuals which have existed on the earth in the past and at present.

This makes a substantial difference between physical and biological events. The former may be at least statistically predictable, while events of life have a great deal of unpredictability, even in statistical terms. Both are determined, according to the principles of classical physics (we shall mention later the principle of indeterminacy of quantum physics), but in biology the principle of Laplace cannot be applied.

Laplace, as is well known, declared, on the basis of a strict determinism, that an Intelligence who would know at a given instant all the forces at work and the exact position of all the particles of the world, would be able to reconstruct exactly the world at any moment of the past, and to predict exactly the future.

This may be true—according to classical physics—because physical events are individually or statistically predictable. It is not true for biological events: the enormous number of possibilities as compared to the limited number of actual realisations leaves a great range of unpredictability. Not even the hypothetical superior Intelligence could know which array of combinations would actually be available to the work of selection during the whole of the geological lifetime of a species.

In this context the question has been debated whether the principle of indeterminacy of quantum mechanics may be applied to biology. The indeterminacy principle implies a new conception of chance or random events, inasmuch as randomness is not an appearance due to our incomplete knowledge of causal relationships but an essential primary property which escapes determinism.

One possible answer to the question is the following. Indeterminacy exists at the atomic and molecular levels; above these levels the determinism of classical physics enters in action. Gene mutation is a molecular event and lies in the realm of the principle of indeterminacy. Thus the basic elementary phenomenon of evolution is a random undetermined process. As soon as it has occurred, and the mutated allele begins to work making itself explicit at phenotypic level, indeterminacy ceases and classical determinism alone dominates the subsequent series of events.

This point of view, which has been put forward by some physicists, may well be true at least for some categories of mutations (such as those arising from tautomeric changes). But the essential point is, in my opinion, that even for events occurring in the deterministic realm of physics, such as chromosome and gene recombination, chromosomic and genomic mutation, there is quite a lot of room for unpredictability, in the sense defined above. Thus the possibility of prediction in biological sciences, especially in population genetics and evolution (to say nothing of behavioural sciences) may be much less precise and strict than in physical sciences, even at structural levels which are subject to the deterministic principle of classical physics.

In connection with the discussion on chance and randomness, two items deserve a particular mention, namely (1) whether such complex structures as organic bodies can be produced by chance, and (2) how evolution by natural selection can have a really creative action and produce novelties.

First, the specious but fallacious argument of the 'typist monkeys' should be contradicted. This argument has been taken over again and again in different forms. It has been presented with the precision of probability calculation by E. Guye, and reproposed as such by Lecomte du Noüy, by Blandino, and by practically all those who want to reject, usually on philosophical and/or religious grounds, the mechanistic interpretation of life (Guye, 1922; Lecomte du Noüy, 1939; Blandino, 1964).

It is obvious, of course, that a team of monkeys working on typewriters will never produce the *Divina Commedia*. Nor will the random combination of atoms produce, by chance, an elephant or even a bacterium. Let us accept Guye's assumption that even the production of a protein molecule by random assembly of atoms has such a high degree of improbability that we can consider its formation practically impossible.

The fallacy of the argument may be easily demonstrated. The process postulated to explain biopoiesis is gradual. We know that relatively simple organic molecules are formed by means of merely physicochemical forces. These elementary molecules may be assembled to form more complex molecules such as amino acids. This step has recently been shown in the laboratory, as is well known (Miller, 1955). From this stage on, the formation of more complex chemical units under the action of mere physical forces becomes much less improbable. As soon as the property of reproduction is established and selection begins to play its role, more and more complex and coordinated structures arise, and evolution proceeds.

Thus the old tenet of Aristotle, that harmony and order cannot arise as a result of chance, which is also the tenet of the vitalistic schools of any time, is not demonstrated by probabilistic reasonings which wear apparently respectable clothes, but are actually founded on unsound basic statements.

The second question is whether natural selection has any power in building something new, that is, whether it has a creative efficiency.

One of the old objections to the Darwinian theory was in fact the alleged merely negative effect of natural selection. Selection, it was said, consists in an elimination of the less fit individuals; hence it cannot build anything new, it merely exerts a choice among what is offered by nature and discards some combinations, thus favouring others. In the epic time of post-Darwinian controversies, this line of thought was expressed as the *Ohnmacht der natürlichen Zuchtwahl* to which the selectionists opposed the *Allmacht der natürlichen Zuchtwahl.*

As I said before, Th. Dobzhansky (1954) has devoted to this subject much thought and factual exemplification, and has demonstrated how selection works not only as an eliminative force, but as a creative force. I will not go into details on this subject so brilliantly expounded by Dobzhansky. I merely want to say that anyone who wants to understand the mechanism of evolution as it is conceived nowadays, on the basis of our knowledge of genetic structure at its different integration levels, must abandon the rather crude picture current among those who know only a smattering of genetics. In order to get a deep insight into this problem one must penetrate to the structure and functions of genes not only as single, independent units, but as the interacting elements of a highly complex organic structure, upon which selective forces work. He should also realise that individuals are, from an evolutionary point of view, the elements of a larger, complex unit, the population, which is the field of action of the selective forces (Mayr, 1972).

Summary and concluding remarks

(1) From a bird's-eye view of the history of biology from the Greeks to the present time, one may see that the main fundamental problems about the interpretation of nature and life are to be found already in Greek philosophy, and recurrently in subsequent historical periods.

(2) Structural and functional complexity of organisms, and above all the finalism of biological phenomena, have been the insuperable difficulty, the insoluble aporia preventing the acceptance of a mechanistic interpretation of life. This is the main reason why in the competition of Aristotelian and Democritean interpretations the former has been the winner, from the beginning to our days.

(3) All the attempts to establish a mechanistic interpretation were frustrated by the following facts: (*a*) the inadequacy of physical laws to explain biological finalism; (*b*) the crudeness of the physical schemes for such fine and complex phenomena as the biological ones; (*c*) the failure of 'reductionism' to realise that at each level of integration occurring in biological systems new qualities arise which need new explanatory principles that are unknown (and unnecessary) in physics.

(4) Modern biology has reached the solution, or at least has entered the way towards a solution, of the old aporias by means of (*a*) the Darwinian revolution and the discovery of natural selection as the driving force of evolution and a scientific explanation of teleonomy; (*b*) the reaching of the molecular level, where the roots of life are to be found, and the dichotomy between nonliving and living systems begins; (*c*) the understanding that successive levels of integration, such as occur in the whole living system which has evolved on the earth, imply the appearance of new qualities which need different explanatory principles.

(5) Pure reductionism of biological phenomena to physical principles is inadmissible. Does that mean that we have to renounce a scientific explanation of biological phenomena? Certainly not: the important principle in order to maintain the interpretation of biological facts into a scientific frame is that scientific criteria—that is, the search for cause–effect relationships—must not be abandoned, and no appeal should be made to metaphysical entities or preordained designs which cannot be admitted in science. The fact that some events—especially evolutionary happenings—are not predictable does not mean that they cannot be scientifically interpreted. Unpredictability in a Laplacian sense is a characteristic of many life phenomena, due to the fact that the number of possible combinations by far exceeds the number of those which are really present at a given moment, in a given population of individuals.

(6) The inadequacy of pure reductionism does not mean either that living systems escape the obedience to fundamental physical laws, such as the second law of thermodynamics. Living systems are subject to physical laws, inasmuch as they are made of the same chemical elements to be found in the inorganic world. Life—that is, reproduction, mutation and function—implies some other principles and laws, which are not in opposition but complementary to the physical ones.

At the end of this survey, we must put forward some questions for further consideration. Is the mechanism of evolution which has been elaborated by the geneticists on the basis of the Darwinian theory—that is, the system random mutation–selection–genetic drift—efficient enough to account for the whole history of biological evolution on our planet? Is natural selection able to account for the appearance and establishment of the main characteristics of life, namely, as J. Monod (1970) defines them, teleonomy and invariance?

The first question is answered in the negative by some biologists, among whom are also some geneticists. As to the possible alternatives, excluding all vitalistic or telefinalistic theories (such as that of P. Lecomte du Noüy, 1939) for the reason given above, we might mention Goldschmidt's (1955) assumption of 'systemic mutations'. These are big mutations, the existence of which is so far purely hypothetical, which would change abruptly the fundamental structure of an organism, giving rise to a new *phylum*. As long as their existence is not demonstrated, the majority of geneticists do not want to have recourse to such hypothetical phenomena.

Some geneticists, excluding those who still adhere to vitalistic interpretations, have cast doubts on the omnipotence of natural selection as a force capable of promoting such an astonishing variety of forms and the rise towards complexity that is evident in biological evolution. Selection is at the moment the only instrument available for such a purpose. Several authors, including Mayr, Dobzhansky, Monod and Jacob, already cited, have produced factual as well as speculative evidence to support the claim that natural selection can account for all the known facts. This line of work, being the only one allowing a scientific interpretation of biological phenomena, should not be abandoned, and research should be carried on along that line. This will lead, it is hoped, to the clarification of doubtful points in the current theory, or perhaps to the discovery of new facts or principles which may better account for the main steps of evolution.

References

Bernal, J. D. (1968). *Origin of Life*. World Publishing, Cleveland.
Bertalanffy, L. von (1932). *Theoretische Biologie*. Berlin.
Blandino, G. (1964). *Problemas y teorias sobre la naturaleza de la vida*. Razón y Fe, Madrid.
Calvin, M. (1969). *Chemical Evolution*. Oxford University Press, New York and London.
Dobzhansky, Th. (1954). Evolution as a creative process. *Caryologia*, Suppl., 435–49.
Dobzhansky, Th. and Boesiger, E. (1968). *Essais sur l'Evolution*. Masson, Paris.
Driesch, H. (1905). *Der Vitalismus als Geschichte und als Lehre*. Leipzig.
Enriquez, F. and Mazziotti, M. (1948). *Le Dottrine di Democrito di Abdera*. Zanichelli, Bologna.
Giglio Tos, E. (1900). *Les Problèmes de la Vie: essai d'une interpretation scientifique des phénomènes vitaux*, I. *La substance vivante et la cytodiérèse*. Turin.
Goldschmidt, R. (1955). *Theoretical Genetics*. University of California, Berkeley.
Guye, C. E. (1922). *L'Evolution physico-chimique*. Paris.
Haldane, J. B. S. (1929). The origin of life. *The Rationalist Annual*, 148–53.
Haldane, J. S. (1931). *The Philosophical Basis of Biology*. London.
Jacob, F. (1970). *La Logique du Vivant: une histoire de l'hérédité*. Gallimard, Paris.
Lecomte du Noüy, P. (1939). *Il Tempo e la Vita*. Torino.
Mayr, E. (1961). Cause and effect in biology. *Science*, **134**, 1501–6.
Mayr, E. (1969). *Principles of Systematic Zoology*. McGraw-Hill, New York.
Mayr, E. (1972). The nature of the Darwinian revolution. *Science*, **176**, 981–9.
Meyer, A. (1934). *Ideen und Idealen der biologischen Erkenntnis*. Leipzig.
Meyer, A. (1935). Idee des Holismus. *Scientia*.
Miller, S. L. (1955). Production of some organic compounds under possible primitive earth conditions. *J. Am. Chem. Soc.*, **77**, 2351.
Monod, J. (1970). *Le Hasard et la Nécessité*. Ed. du Seuil, Paris.
Oparin, A. (1969). *Biogenesis and Early Development of Life*. Academic Press, New York.

Ponnamperuma, C., ed. (1972). *Exobiology*. North-Holland, Amsterdam.
Raffaele, F. (1905). *L'Individuo e la Specie*. Sandron, Palermo.
Simpson, G. G. (1963). Biology and the nature of science. *Science*, **139**, 81–8.
Smuts, J. C. (1927). *Holism and Evolution*. London.
Uexküll, J. von (1928). *Theoretische Biologie*. Berlin.

Discussion

Rensch

Although *natural selection* is normally defined in a special biological sense, it seems to be possible to use this term also for several inorganic processes. When for instance different chemical compounds become mixed, certain molecules will only select particular other molecules to which they have an affinity, so that a new type of compound or aggregation originates. Or when a competition between gravitational forces of different strengths arises, the stronger one will attract the weaker ones. Selection also takes place when a competition exists between different types of human tools, engines, working methods, social systems, scientific assumptions and so on. Normally the better ones will replace the less good ones (Rensch, 1959, 1970: *Homo sapiens*). I mention these relations because they show that selection can be regarded as a more general or even *universal law*. This may be important as only a limited number of general universal laws exist.

Montalenti

I believe that the term 'natural selection' should be limited to entities which are endowed with the properties of self-reproduction and mutation, that is, genetic variation. Perhaps it may also be extended merely as an analogy to tools, engines, working methods, etc., which are designed by man and made and reproduced by him. I do not think that the concept of natural selection can be applied to inorganic processes. It is a matter of definition, of course, but to attribute to the term a wider meaning would generate confusion.

3. Evolutionary Theories after Lamarck and Darwin

ERNEST BOESIGER

Introduction

Both Lamarck and Darwin made great contributions to the theory of evolution. Darwin's hypothesis is well known, but that is not so for the ideas of Lamarck. Ever since Darwin there existed and there still exists a strong antagonism between the followers of the two main theories, which often seem to them mutually exclusive.

In some countries, and especially in France, a majority of biologists were and still are opposed to the Darwinian hypothesis; persistent opposition to Darwinism in France has several specific reasons. One of them is the influence of the philosophy of Descartes and of its perverted nineteenth-century form, presented by A. Comte. Another is the not yet definitively abolished survival of the strictly anti-evolutionistic ideas of the powerful Cuvier through the smoother oppositions to evolution of Quatrefages or Lacaze-Duthiers. And when it was no more possible to reject the theory of evolution, many French biologists opposed Lamarck to Darwin, in spite of the fact that Lamarck was rejected during his lifetime, that his publications remained practically unknown after his death, and even now his real ideas are almost ignored in France and elsewhere. More generally there exists concealed opposition against the principle of the theory of evolution and especially against Darwin's theory. Under these circumstances it is not astonishing that the discussions between evolutionists and anti-evolutionists, and between neo-Lamarckians and neo-Darwinians, have often been scholastic and rather sterile. Observations and experimental data played a minor role in these discussions until recently.

Nearly all Darwinists or neo-Darwinists reject completely the ideas of Lamarck. Some exceptions exist nevertheless. Some Darwinists try to have a new look at Lamarck's work and to give some credit in the formation of modern evolutionary theories. Such are Simpson (1964), Ernst Mayr (1972) in his article 'Lamarck revisited', and Boesiger (1971).

Mayr (1972) states: 'We can now study him without bias and emotion and give him the attention which this major figure in the history of biology clearly deserves. . . . A truly penetrating study of Lamarck is still a *desideratum.*'

But the purpose of this paper is not a new study of the ideas of Lamarck and of Darwin in the light of the modern knowledge, and even less a historical study of all neo-Lamarckians and neo-Darwinians. My aim is to show that some fundamental ideas of both Lamarck and Darwin should be adapted to the actual state of our knowledge and integrated in a synthesis of the theory of evolution. After all, the founders of the biological theory of evolution formulated in the thirties, on a wealth of biological data, especially in the field of genetics, a basically Darwinian theory of evolution. By doing so they elevated Darwin's theory of natural selection to a higher level of integration.

In the last forty years followers of the creators of the biological theory of evolution of the thirties have studied the action of natural selection in experimental and natural populations, the genetical constitution of natural populations, and the relations between the composition of natural populations and the ecological conditions of their habitats.

The integration of all this new knowledge in a coherent, modernised biological theory of evolution is still an urgent need. The inhibitory effect of older 'classical' schools has to be broken, as Simpson (1964) stated so clearly:

> The Neo-Lamarckians knew and overemphasised the fact that adaptation is pervasive in nature and essentially purposeful in aspect, as if the environment had forced and the organism had sought adaptation. The neo-Darwinians knew and overemphasised the fact that the more or less adaptive status of variations is influential in determining the parentage of a following generation. The geneticists knew and overemphasised the fact that new hereditary variants arise abruptly and, as far as we know and as far as adaptive status is concerned, at random. . . . What was necessary was synthesis, bringing together the facts and theories of all the schools, accepting those mutually consistent and reciprocally reinforcing.

One of the conditions for the creation of a new synthesis is elimination of many false ideas about the founders of the schools, Lamarck and Darwin, which have been propagated by neo-Lamarckians on the one side and by neo-Darwinists on the other side.

Simpson (1964) has written about neo-Lamarckism: 'This is the ironic joke: that the theory to which Lamarck's name became and still remains attached and to which all his posthumous fame is due is fundamentally different from what he himself intended. . . . Most [neo-Lamarckians] did not even read his work.' For the sake of the progress of the theory of evolution we have to abandon the intellectual inbreeding of schools.

Lamarck

Let us try to restate briefly the main points of Lamarck's theory of evolution. Grassé (1971a) says of Lamarck: '. . . son oeuvre demeure peu et mal connue'. His ideas are often presented incorrectly, even by his followers. In many cases only the false hypothesis of the transmission of acquired characters by heredity is quoted.

First of all, we have to insist on the fact that, in contrast to many neo-Lamarckians, Lamarck *was not a vitalist*! Lamarck says in 1802: 'Je suis convaincu que la vie est un phénomène très naturel, un fait physique, à la vérité un peu compliqué, et n'est point un être particulier quelconque.'

Lamarck rejects the ideas of some philosophers about the existence of a vital principle, about a 'soul' of organisms. The only solid knowledge about organisms comes from the study of the laws of nature.

It is not relevant for our purpose to discuss the religious views of Lamarck, but in comparison with many neo-Lamarckians it is interesting to note that Lamarck does say clearly that the transformation of organisms and all phenomena of life are produced by nature, by natural laws, which we can study objectively by scientific means. Even if he speaks about a 'sublime Author', the 'first cause of everything', he adds immediately that it is possible to understand nature and its laws by itself.

In contrast to many neo-Lamarckians, *Lamarck was not a finalist* (see Vachon *et al.*, 1972):

> C'est une véritable erreur que d'attribuer un but, une intention à la nature dans ses opérations. Elle n'en saurait avoir puisqu'elle n'est pas une intelligence, un être particulier. Elle fait nécessairement tout ce qu'elle fait; et si les résultats de ses actes paraissent souvent des fins prévues et combinées, c'est parce que dirigée partout par des lois constantes.
>
> C'est surtout dans les corps vivants qu'on a cru apercevoir un but aux opérations de la nature. Ce but cependant n'y est, là comme ailleurs, qu'une simple apparence et non une réalité. En effet, dans chaque organisation particulière de ces corps, un ordre des choses préparées depuis longtemps par les causes qui l'ont graduellement établi ne fait qu'amener par des développements successifs ce qui nous paraît être un but, et ce qui n'est réellement qu'une nécessité.

In contrast to many neo-Lamarckians, *Lamarck was not a dualist*. There is no fundamental difference between the living and the non-living, living bodies and mineral substances. There is no duality between matter and spirit: 'Le sentiment lui-même n'est qu'un phénomène résultant des fonctions d'un système capable de le produire. . . . Le physique et le moral ne sont sans doute qu'une seule et même chose.'

Lamarck was a materialist. He clearly refuses to search for 'first causes' or for metaphysical forces. The causes acting now and observable by us now are quite sufficient for the maintenance of life and for the evolution of organisms. This materialistic standpoint of Lamarck leads him to the creation of a psychophysiology based on the physiology of the nervous system which served him as an explanation of the transformation of animals, through the feeling of a need (*besoin*) for the adaptation to new changed conditions.

The greatest merit of Lamarck, besides his outstanding contribution to the classification of invertebrates, was the unambiguous statement of the transformation of plants and animals and of the evolution of all organisms by natural laws, from the very first organisms to the most complex organisms, including man.

Hodge (1971) has noticed that Lamarck never speaks of the origin of species, or of evolution. But Lamarck speaks about the greatest secret of nature, which is now discovered, that of the 'origine de tous les corps naturels' (see Vachon, 1972). 'The word *espèce* and its synonyms are not used, not once.' This last statement is not exact. Lamarck often utilises the word *espèce* in his *Flore françoise*, in his *Philosophie zoologique*, his *Histoire naturelle des Animaux sans Vertèbres* and other writings. Certainly, there are many different species concepts. That of Lamarck does not exactly correspond to that of Darwin; the species of a modern population geneticist is again different from that of Darwin. The species concept is also quite different for an ornithologist, a bacteriologist or a botanist studying apogamous plants.

Lamarck did not treat the problem of the origin of species in the sense of Darwin, and in this respect he was not a precursor of Darwin. But we must admit his outstanding merit, to have understood and said when it required an exceptional courage to do so, against the heavy power of a Cuvier, that individuals of the same species are not alike in different environments and that living organisms are a product of transformation, by natural laws, beginning with very simple first organisms up to the most complex mammals. He has also the great merit to have included man without hesitation in his transformation series as the most complex product of evolution. The fundamental evolutionary idea of the progressive complexification of organisms, taken over later by Teilhard de Chardin, Vandel and many others, is one of the main original concepts of Lamarck.

It has sometimes been noticed that Lamarck did not use the word evolution, but rather transformism. I would certainly not try now to abolish the term 'evolution'. But 'evolution' is derived from *evolutio* and *evolvere* which had for some older philosophers and naturalists, as Leibnitz, Swammerdam, Redi, Malpighi, Bonnet, etc., the sense of a preformation, which means simply to evolve, to unfold, to produce an organism. In opposition to this preformationism Lamarck chose 'transformism', and he had good reasons to do so.

Lamarck quite often made philosophical speculations about nature, without trying to present facts as proofs for his hypotheses. The title of his famous *Philosophie zoologique* has nothing unusual, and fits well with the habits of his time. Between 1809 and 1859 there took place an important shift in scientific methods. It is Darwin's great merit to have utilised with his great efficiency the new experimental approach to the study of evolution.

But Lamarck did not present only speculations. Those who have read the *Philosophie zoologique* know that one of his important and original achievements was precisely the introduction of a very new method in zoology. He emphasised the study of large numbers of the individuals of a species. He was the very first who pointed out clearly that the typological approach in taxonomy is wrong, because it does not correspond to the reality which exists in nature. The description of a species on the basis of one individual was nonsense for Lamarck; studying all available specimens he showed the enormous variability of organisms. Lamarck's method of taxomomy and zoology was a very profound modification of the older attitudes and has only been adopted in the middle of our century.

Lamarck states clearly many times that this fundamental observation of the great variability of organisms is at the origin of his conviction that species do not have the absolute constancy which was claimed at his time. He concludes that species have been and still are under transformation. That is to say, Lamarck cannot be simply contrasted to the experimental naturalist Darwin, as an eighteenth-century philosopher, since he introduced a very new experimental method in zoology, if we consider as we ought the observation of large samples as an experimental method.

We have of course to avoid the pitfall of putting in Lamarck's writings our present ideas and knowledge in population genetics. But there is no doubt that Lamarck, as an excellent naturalist and zoologist, did see the heterogeneity of populations concretely, and that his concept of the high intraspecific variability is closer to the ideas we have now in this field than to those which prevailed as recently as the thirties of this century. His explanation of the mechanisms producing this variability, by volition, and the idea of evolution by hereditary transmission of acquired characters, are of course wrong. But let us not forget that the latter theory was generally accepted even much later, since it had also the favour of Darwin.

This idea of adaptation by volition is very widespread and even popular. It is appropriate to quote Jean Giono, who has a peasant of the mountains of Provence saying:

> Tu sais: l'orage couche le blé; bon, une fois. Faut pas croire que la plante raisonne pas? Ça se dit: bon, on va se renforcer, et, petit à petit, ça se durcit la tige et ça tient debout à la fin malgré les orages. Ça s'est mis au pas.

Another important observation of Lamarck concerns the differences between individuals of the same species living in different conditions. This is the main basis of his theory of transformism. Lamarck observed this fact in natural populations and in plants cultivated in particular conditions by man. In the *Philosophie zoologique* he states:

> Ceux qui ont beaucoup observé et qui ont consulté les grandes collections ont pu se convaincre qu'à mesure que les circonstances d'habitation, d'exposition, de climat, de nourriture, d'habitude de vivre, etc. viennent à changer, les caractères de taille, de forme, de proportion entre les parties, de couleur, de consistance, d'agilité et d'industrie, pour les animaux, changent proportionnellement.

Lamarck insisted also on the fact that in a given geographical region conditions change but not with constant rhythms. These modifications of climate and other conditions of life are often so slow that we cannot notice them. That makes some people believe that conditions are stable.

It is well known that Lamarck thought that the evolution of organisms is driven by change of the conditions of life. But he did not advance the hypothesis that physical and other conditions of the environment act directly on the organisms. On the contrary, he rejects this hypothesis, declaring explicitly that it is an error. He adds that circumstances do not operate directly on the morphology and organisation of animals. He believes that important modifications of the conditions of life induce modifications in their needs (*besoins*) and then in turn animals act necessarily in a different way. Hence, transformation by volition.

Let us repeat that this is not an observation but an erroneous speculation. Weismann (1893) made a useful effort to disprove the theory of heredity of acquired characters. He says: 'Nicht das Rennen hat die Pferde in 200 Jahren zu Rennpferden gemacht, sondern die Auswahl der für das Rennen vortheilhaftesten Variationen unter den Nachkommen ausgezeichneter Schnelläufer.'

In spite of suggesting the wrong explanatory mechanism, Lamarck has to be considered as one of the earliest evolutionists. Others also had some vague evolutionary ideas, as for example Maupertuis in his *Vénus physique*. But Lamarck was the first to develop and present, as a conclusion from his studies of large invertebrate collections, a coherent theory of evolution, general for earth and specific for the organisms. He states clearly that a good observer knows that nothing on the planet is always maintained in the same state. Everything changes according to the circumstances. He utilises the term 'mutation' for the description of these transformations, as Spallanzani did before him.

Mayr (1972) points out the importance of Lamarck in the history of evolutionary theories:

> It would seem to me that Lamarck has a much better claim to be designated 'the father of the theory of evolution', as indeed he was by several French historians [e.g. Landrieu, 1909]. No author before him had devoted an entire book exclusively to the presentation of a theory of organic evolution. No one before had presented the entire system of animals as the product of evolution.

Mayr (1972) emphasises, as I did (Boesiger, 1971), besides other achievements of Lamarck, one that seems especially important and new:

> No writer prior to Lamarck appreciated as clearly the adaptive nature of much of the structure of animals, particularly the characteristics of families and classes. Even though he reiterated forever his statements on 'growing perfection' and 'increasing complexity', he nevertheless realised clearly that the nature of evolution was dual and that an irregular phenomenon of *ad hoc* adaptation was clearly superimposed on the linear trend of increasing perfection.

And:

> The importance of Lamarck's emphasis on adaptation is usually ignored, and all that is mentioned, is the fact that his explanatory mechanisms are invalid. This misplaced emphasis

fails to appreciate that Lamarck was far ahead of the contemporary essentialist morphologists in recognising that the particular conformation of the morphology of a kind of animals was neither an accident of nature (*Lusus naturae*) nor the result of the inscrutable design of the creator but the product of an analysable interaction between structure and environment. In that sense Lamarck quite clearly *was* a forerunner of Darwin, who based much of his argument on that same relationship between structure and (selective factors of) the environment.

Lamarck insisted on the fact that the evolution of organisms is a consequence of the natural forces and mechanisms which are still at work at present, so that it is possible to observe evolution going on. Man also is a product of the natural evolution of animals. Evolution is a very slow and gradual process; but that does not mean that evolution progresses at a constant rate. As long as there is no change in the condition of the environment, as long as there is no need, the organisms are not under evolutionary pressure. Lamarck discusses this problem after having seen the mummified animals of ancient Egypt, which were brought to Paris, and were not morphologically different from the corresponding species actually living in Egypt.

Neo-Lamarckism

It is not possible to present a coherent account of the neo-Lamarckian theory of evolution, simply because no coherent neo-Lamarckian theory exists. But even so, many evolutionists have been classified as neo-Lamarckians. For most of them the label neo-Lamarckism is not correct, especially when we think about the obvious fact that many so-called neo-Lamarckians diverge so considerably with respect to the true ideas of Lamarck. Evolutionary theories have often been presented under the name of Lamarck, although they are distortions of his theory. The situation is even more complicated by the fact that Lamarck's philosophy evolved considerably between his eighteenth-century book *Flore françoise* and his last books *Animaux sans Vertèbres* and *Système analytique des Croissances positives de l'Homme restreintes à celles qui proviennent directement ou indirectement de l'Observation.*

The following is a brief account of some of the theories which are often falsely considered to be of Lamarckian inspiration.

Haeckel followed partially Lamarck, when he proposed the hypothesis of the existence of complex molecules, the plastidules, in living organisms. These plastidules are not fundamentally different from other chemical substances, but they have a memory which explains the recapitulation of the phylogeny of an organism during its ontogeny. External influences, ecological conditions, modify the movements and associations of the plastidules. This explains the adaptations of organisms.

Herbert Spencer, vigorously attacked by Weismann, advanced at about the same time a similar hypothesis. He assumed, besides the ordinary mineral and organic molecules, the existence of specific groups of molecules forming what he called physiological units, which circulate in the organisms. When an organism is modified under the influence of external conditions, the physiological units record the change. Since they also enter in the constitution of the sex cells, acquired characters can be transmitted by these units. Haeckel's and Spencer's hypotheses are close to the theory of pangenesis proposed by Darwin. The physiological units of Spencer, the plastidules of Haeckel and the gemmules of Darwin have about the same properties as Maupertuis' *particules*, able to be modified under the influence of external conditions, circulating in the organism and passing through the germinal tract into the sex cells. Maupertuis conceived in this way the possibility of evolution. Lamarck, and later Darwin and others, followed this widespread belief.

Edward Cope (1896) admitted some influence of natural selection in evolution, but he thought that the will and the habits are mainly responsible for the evolutionary changes of structures. He postulates a consciousness in all living matter. All vital manifestations were in the beginning conscious and passed then to an unconscious state, becoming inscribed in the germ plasm. In this way acquired characters are transmitted to the next generation by heredity. The unconscious memory is based on particular arrangements of the germ plasm.

Packard, Hyatt, Osborn, Eimer, Naegeli and many others may be considered as neo-Lamarckians, but it is not our purpose to present a complete historical survey. In France, the great majority of evolutionists can be considered, and in some cases consider themselves, as neo-Lamarckians. Some of their ideas are really Lamarckian, others are not. Let us use a few examples only.

Cuénot (1911, 1951) fears to be treated as finalist. He says that mechanism has always replaced the spontaneous finalistic explanation of biological phenomena. Many global finalistic appearances are only fortuitous coincidences; the neo-Darwinian explanation of evolution is acceptable as a whole. Cuénot does not feel the need of a distinction between micro and macroevolution. He has himself introduced Mendelian genetics in France and is convinced that the modern evolutionary theory has to be based on the knowledge of Mendelian genetics. But Cuénot is nevertheless reluctant to accept Darwinism. His book about biological evolution has the sub-title *The facts, the uncertainties.* He formulates hesitations and criticisms, often advanced by neo-Lamarckians. His questions about the structure of the cell nuclei have of course been answered since the publication of his book by the discovery of the genetic code.

Cuénot has a preference for some palæontological intuitions, in spite of their metaphysical character. He believes that the founders of new groups of organisms possess potentially the plan of the future species. In the same sense in which ontogeny is preparing the future of the individual, orthogenesis is preparing the future of evolutionary lines.

Cuénot has also doubts about the Darwinistic interpretation of mimetic colorations or patterns. He discusses at length the inheritance of acquired characters, concluding that nature has a tendency to replace adaptations (*accomodats*) by corresponding germinal mutations. He says that it is not possible, in the actual state of our knowledge, to be sure of the absolute impossibility of a hereditary transmission of acquired characters.

A main point on Cuénot's uncertainty is the opposition between chance and finality. This actually is an insoluble question; acceptance of neo-Darwinism and the unlimited power of chance, and of finality applied to the tools of man and animals, means a refusal of discussion and research.

Contemplating the extraordinary achievements of nature, its aesthetical creations, harmony of colours, beautiful patterns of feathers and astonishing adaptations in animals, Cuénot (1951) raises the question of the religious attitude of man. Man has made gods. Nature is a deity for him, and the appropriate worship is pantheism. This pantheism is simple, has no theology, is similar to the materialistic pantheism of d'Holbach and Diderot. His priests are scientists, its prayer is research.

Cuénot was certainly not simply a neo-Lamarckian. But his partial acceptance of finalism, and his question—who had the idea of all these realisations of nature? who made the plan?—are characteristic of many neo-Lamarckians. So is his partial acceptance of orthogenesis. But the profound scientific culture of Cuénot and his personal contribution to Mendelian genetics at the beginning of this century led him to accept partially also the basic concepts of the biological theory of evolution.

Wintrebert (1962, 1963) gave at the end of his life a synthesis of his ideas on evolution. His two last books are a passionate and highly partisan defence of Lamarck's ideas. He frankly admires Lamarck's ingenious intuition. For him, Lamarck is the founder of transformism, who was right since he considered the intrinsic adaptability of living matter to be the essential evolutionary mechanism. Wintrebert claims that the actual progress of physiology, of biochemistry and of ultramicroscopic cytology are confirming the genial intuition of Lamarck.

Wintrebert declares that mutations cannot explain evolution, that they prove only that it is easier to destroy than to construct. The new genotypes obtained by mutation are regressive and teratological. Mutations represent only intraspecific variations which cannot produce evolution. The creation of a new species is the consequence of the general reaction of organisms, and not of the localised mutation of genes. Progressive evolution of sense organs cannot be explained by a series of happy random events; that is even more so for the brain. Wintrebert refuses to admit that evolutionary genetics could explain evolutionary trends. The protoplasm creates the needed adaptations which are then transmitted by the genes to the offspring. The gene is nothing else than a tool of the cytoplasm, receiving from it by delegation the hereditary adaptations created by the protoplasm. Living matter does not need any special regulatory mechanisms added to its normal properties.

Conscious and unconscious intelligence are innate parts of the macromolecular structure of living matter, which creates its own destiny. Wintrebert attacks biologists who explain evolution by Providence, as well as those who admit only random factors. He thinks that these two tendencies are closely related, since both arrive at the wrong conclusion that living matter is not able in itself to create, to guide its own evolution, and to produce the hereditary genes for the transmission to the offspring. Living matter has escaped from the physical milieu by individualisation, and is always defying it. Adaptive and immunity reactions are responses to the hostile environment. However, living matter is strictly subject to the universal laws of matter. The secrets of life and evolution cannot be explained by any preorganisation. They can only be understood by an analysis of the physiological and biochemical functions of living matter. Wintrebert postulates a kind of biochemical Lamarckism. He accuses the neo-Darwinists of finalism, and claims that Lamarckism is not finalistic. For Lamarck living matter is always actual and clearly determinated in response to the conditions of the environment. Living matter has no final goal; it only tries to live and to conserve life.

Vandel (1938) admits up to a certain limit the role of mutation and selection in evolution. His studies of the genetics and evolution of terrestrial and cave-dwelling isopodes, the Trichoniscidae, permit him to conclude that the specific adaptations of these cave animals are the result of mutations, responsible for the disappearance of the body pigments, regression of the eyes and other characters affecting size, ornamentations and the secondary sexual characters. The appearance of albinos in laboratory systems, where all individuals are living in the same conditions, proves that albinism cannot be explained by a Lamarckian mechanism. The existence of normally pigmented forms in caves, with normal eyes, and on the other hand the existence of partially or completely depigmented forms living on the surface in normal light, is for Vandel irrefutable proof that albinism and blindness in cave animals are not induced by the cave conditions. Animals which had by genetical preadaptation—that is, by mutation—characteristics which are useful for life in cave conditions have penetrated in the caves and continued then by selection their specific evolution.

In another very interesting study of an isopode, *Trichoniscus elisabethae*, Vandel (1940) shows that the triploid race is more resistant than the normal diploid race against two unfavourable ecological conditions, cold and aridity. He was one of the first to postulate that this kind of mutation of the karyotype can be a factor of evolution and can be responsible for the geographical distribution of the varieties of a species. He claims that this evolutionary process may lead, by geographic isolation, to speciation.

It is obvious that Vandel can be considered neither a Lamarckist nor a neo-Darwinist. In a brilliant synthesis of his concepts, Vandel (1968) presents a subtle intermediate and independent standpoint. Like Lamarck, he rejects vitalism and sees no fundamental difference between living and non-living matter. The difference between the two types of matter is not a chemical one, but given by the different organisation. The actual state of the world, and the evolution which produced it, are not the work of a supreme engineer, creator of the world, nor can it be the result of lucky random events as Epicur and the modern neo-Darwinists think. Evolution is a highly complex problem. The evolutionary process is not the same for bacteria and for man. Vandel admits the sequence mutation–selection for bacteria but not for metazoans, in which the cytoplasm is as important as the nuclear genome. Mutations are furnishing the material for evolution, but in metazoans they cannot be assimilated directly. Accumulation of mutations, produced at random, has to take place for the establishment of organised structures. Only autoregulatory processes can achieve the necessary integration of the mutants within the previous organisation of an organism; these regulations take place during the embryonic development. The underlying genetic mechanisms have not the same signification in bacteria and metazoans. In the latter the transmission of genetic information is less direct; there exist steps and structural levels. Vandel points out that the initial influence of the genes is diluted, reduced by substances such as hormones and organisers, by psychological factors such as imprinting, and by multiple interactions.

Grassé has in several papers thoroughly discussed the problems of evolution and takes also an intermediate position. In a lecture given in 1947 he admits the reality of natural selection, but he asks if the raw materials for evolution by natural selection, the useful mutations, present themselves always just at the right moment. He admits that useless characters can be maintained by natural selection, when they are pleiotropic effects of a physiologically or otherwise useful gene, and that their presence does not contradict the effectiveness of selection. Anyway, an individual has positive and negative characters. The essential point is that the organism obtains on the whole a favourable balance between positive factors and deficiencies. Under these circumstances natural selection has an explanatory value. Grassé admits that neo-Darwinism offers a logical and mechanistic solution of the problem of evolution, and accepts large parts of it. But he is not convinced of the universality of natural selection, and of its omnipotence in the process of evolution, which has many mechanisms and causes. Grassé adds that the fundamental principle of Lamarckism has not been given experimental test, that neo-Darwinism has on several points an explanatory value, but that evolution has not been achieved by only a single mechanism.

In a recent address, Grassé (1971*b*) speaks of the evolution of man who emerged from animality, stressing the complexity of the structures and phenomena which acted in our evolution. The hominisation affected several organs, which must have been modified simultaneously and in coordination. Grassé concludes that we are still ignorant of the physicochemical laws of evolution.

The main point which hinders many evolutionists from accepting the biological theory of evolution is the importance given to random events: the chance of the occurrence of mutations, of Mendelian segregation and of recombination. They cannot believe that a harmoniously built organism, with complex organs as the brain and the eyes, well adapted to the environment, with useful behaviour patterns, is simply a product of random events. And they find it even more difficult to conceive how evolution has maintained adaptation for long periods under changing conditions, showing at least in some cases continuous, directional evolutionary trends, which have been called orthogenesis.

Some biologists tried to solve the paradox of chance and determinism in evolution by postulating special capacities in living matter, which distinguish it fundamentally from non-living matter. The attribution of such vitalistic theories to Lamarck is an error, since he declared unambiguously that there is no fundamental difference between the living and the nonliving, that life is based on strictly physical phenomena, and that even the 'spirit' is an expression of physically defined matter.

A famous example of a vitalistic model is that of the great philosopher Bergson (1907), postulating a specific vital agent, the *élan vital*, responsible for creative evolution.

A quite different solution of the problem is to believe in a supreme coordinator of evolution and of the whole world. Lecomte du Noüy (1947) explains evolution by the intervention of a transcendent principle, acting towards a known and preestablished goal. For him, evolution is not the result of random events but an expression of the will of God, whose final aim is the creation of man.

Teilhard de Chardin (1959) takes an original and interesting position. No doubt his evolutionary theory is speculative, but it is also based on a background of solid palæontological and biological knowledge. He does not separate his religious and his scientific positions; he declares the need to unite both. Teilhard de Chardin does not postulate predetermination. Evolution is not preplanned since it proceeds by *tâtonnements*. Teilhard is much more interested in general evolution, which is converging to the superior final stage of point Omega, than in particular evolutions of separate groups or organisms. The goal of evolution is the communion of humanity with its Creator. Mankind may fall out of love with its destiny, but Teilhard trusts that this will not happen. Lamarck's rule of the general tendency of evolution to increase complexity and consciousness, to increase the degree of organisation and differentiation, and to create new properties, is further developed by Teilhard.

For Teilhard de Chardin, evolution is not a product of chance, and yet the progression towards the final goal proceeds by *tâtonnements*. Evolution is not directed in the details but is directional in advancing to the point Omega. Man has reached the level of self-consciousness and he can, to a certain degree, choose the orientation of his future evolution. The meaning of life for man comes from his participation in the evolution towards the point Omega. The final success of hominisation has not yet been attained, but the neogenic forces are constantly acting towards the achievement of the evolution by convergence in the point Omega.

Most people admit nowadays the fact of biological evolution. Some curious exceptions nevertheless exist. Vialleton (1929), for example, assumes the existence of a partial orthogenetic evolution in some groups, for example in Equidae and Ammonites. But he denies the general theory of evolution of organisms, from the simplest to the most complex. Evolution cannot attack and modify the fundamental characters of a group of organisms. All higher taxa have been created separately and are not products of evolution. Lemoine (1937) goes even further and declares that there is no evolution at all.

Some people maintain this archaic position even now, and they have some success in schools and universities. The Creation Research Society in the United States declares in its publications that the description of the creation of the world and of the organisms, as it is given in the original text of the Bible, is literally true in all details, and that man and all other organisms have really been created by God in six days. This might simply be considered a curiosity had it not succeeded, in 1969, in passing a recommendation prepared by the California State Board of Education that the origin of life by creation according to the Bible be taught together with Darwinism. State funds for biology books to be distributed in schools are only given if this recommendation is respected. Biology teachers in California fear that the teaching of evolution and of biblical genesis as equivalent explanations might become obligatory. In the State of Tennessee, it was prohibited to teach the theory of evolution, and in South Carolina it was prohibited until recently to mention in official schoolbooks the term 'evolution' and the name of Darwin.

Darwin

Darwin's theory of evolution is well known. To avoid misunderstandings, I have to say clearly that Darwin's theory of natural selection, together with Mendelian genetics and the discovery of mutations, is of a paramount importance in the modern theory of evolution. The following remarks do by no means reduce the explanatory value of natural selection for biological evolution.

Darwin happened to be misunderstood or misinterpreted by followers as well as by enemies. The slogan 'struggle for life' or 'struggle for existence' has often been utilised wrongly, in spite of Darwin's careful presentation in *The Origin of Species:*

> I should premise that I use the term Struggle for Existence in a large and metaphorical sense, including dependence of one being on another, and including (which is more important) not only the life of the individual, but success in leaving progeny. Two canine animals in a time of dearth, may be truly said to struggle with each other which shall get food and live. But a plant on the edge of a desert is said to struggle for life against the drought, though more properly it should be said to be dependent on the moisture.

On the important point of the effect of random events in evolution, Darwin declares in *The Origin of Species:*

> I have hitherto sometimes spoken as if the variations—so common and multiform in organic beings under domestication, and in a lesser degree in those in a state of nature—had been due to chance. This, of course, is a wholly incorrect expression, but it serves to acknowledge plainly our ignorance of the cause of each particular variation.

It may be useful to say again that Darwin was in full agreement with the most contradicted part of Lamarck's theory, the hereditary transmission of acquired characters. Discussing the effects of use and disuse of organs, Darwin states:

> From the facts alluded to in the first chapter, I think there can be little doubt that use in our domestic animals strengthens and enlarges certain parts, and disuse diminishes them; and that such modifications are inherited. Under free nature we can have no standard or comparison, by which to judge of the effects of long-continued use or disuse, for we know not the parent-forms, but many animals have structures which can be explained by the effects of disuse.

Darwin gives the example of animals of different classes, inhabiting the caves of Styria and of Kentucky, which are blind: 'As it is difficult to imagine that eyes, though useless, could be in any way injurious to animals living in darkness, I attribute their loss wholly to disuse'.

Darwin was well aware of the impossibility to explain evolution without hereditary transmission of the selected characters. He created speculatively his own theory of heredity, pangenesis, which is in agreement with Lamarck's ideas about heredity. On a point which became very important in the modern theory of evolution, Darwin disagreed considerably, although not totally, with Lamarck. He did not think that climate and other conditions of the environment have a considerable influence on evolution. Darwin, like Lamarck, was sometimes very speculative, for example dealing with the hereditary transmission of acquired characters. Both had a solid knowledge of zoology and botany. In Lamarck's time this was very new, in Darwin's it became a current attitude of naturalists. Darwin's knowledge was so vast that his positions are in most cases very circumspect and differentiated. This is so when he discusses the influence of environmental conditions on evolution. He did not think that this influence was important, but he was aware of the complexity of the situation:

> Owing to this struggle for life, any variation, however slight, and from whatever cause proceeding, if it be in any degree profitable to an individual of any species, in its infinitely complex relations to other organic beings and to external nature, will tend to the preservation of that individual, and will generally be inherited by its offspring.

Darwin means that selection acts and adapts organisms mainly with respect to other organisms, to competitors, and not to external conditions:

> . . . the structure of every organic being is related, in the most essential, yet often hidden manner, to that of all other organic beings, with which it comes into competition for food or for residence, or from which it has to escape, or on which it preys.

He does not absolutely exclude selection by and for ecological conditions, but this selection seems unimportant to him. Darwin gives a theoretical example:

Look at a plant in the midst of its range, why does it not double or quadruple its numbers? We know that it can perfectly well withstand a little more heat or cold, dampness or dryness. . . . In this case we can clearly see that if we wished in imagination to give the plant the power of increasing in number, we should have to give it some advantage over its competitors. . . . On the confines of its geographical range a change of constitution with respect to climate would clearly be an advantage to our plant, but we have reason to believe that only a few plants or animals range so far, that they are destroyed by the rigour of the climate alone.

Darwin makes this clear by another example speaking of the adaptations of an animal or a plant to a new country:

If we wished to increase its average numbers in its new home, we should have to modify it in a different way to what we should have done in its native country for we should have to give it some advantage over a different set of competitors or enemies.

That is certainly right. Organisms have to be fit for survival. But it would seem that Darwin failed to see the great importance of adaptation to the ecological conditions by natural selection.

In his *The Variation of Animals and Plants under Domestication*, Darwin discusses many aspects of the causes of variation:

Under nature, the individuals of the same species are exposed to nearly uniform conditions for they are rigorously kept to their proper places by a host of competiting animals and plants . . . but it cannot be said that they are subject to quite uniform conditions, and they are liable to a certain amount of variation . . . These facts alone, and innumerable others could be added, indicate that a change of almost any kind in the conditions of life suffices to cause variability.

But then Darwin declares once more:

Moreover, it does not appear that a change of climate, whether more or less genial, is one of the most potent causes of variability. . . . It is doubtful whether a change in the nature of the food is a potent cause of variability. . . . In some few cases, however, plants have become habituated to new conditions. . . . It is certainly a remarkable fact that changed conditions should at first produce, as far as we can see, absolutely no effect; but that they should subsequently cause the character of the species to change.

Darwin speaks about Lamarck's ideas, without direct reference:

Many naturalists, especially of the French school, attribute every modification to the *monde ambiant*, that is to changed climate, with all its diversities of heat and cold, dampness and dryness, light and electricity, to the nature of the soil, and to varied kinds and amount of food. . . . A new sub-variety would thus be produced without the aid of selection.

That means clearly that Darwin postulated alternative ways of evolutionary modifications: either by natural selection or by conditions. He did not yet have the now very common idea of ecological genetics: natural selection and evolution not by ecological conditions but for specific conditions; he did not see the 'challenge and response' situation. This is very clear when Darwin says:

I will first give in detail all the facts which I have been able to collect rendering it probable that climate, food, etc., have acted so definitely and powerfully on the organisation of our domesticated productions, that they have sufficed to form new subvarieties or races, without the aid of selection by man or of natural selection. I will then give the facts and considerations opposed to this conclusion, and finally we will weigh, as fairly as we can, the evidence on both sides.

After having said that a great 'number of animals and plants which range widely and have been exposed to great diversities of conditions, yet remain nearly uniform in characters', Darwin presents the balance. 'We are thus driven to conclude that in most cases the conditions of life play a subordinate part in causing any particular modification.' In opposition to Lamarck, who rejected explicitly this hypothesis, Darwin thinks that 'some variations are induced by direct action of the surrounding conditions . . .'. But he is included 'to lay very little weight on the direct actions of the conditions of life'.

It is well known that Darwin was convinced that use and disuse have played a considerable part in the modification of organisms and that these changes are transmitted to the offspring. The opposition between the evolutionary concepts of Darwin and Lamarck was less profound than that between neo-Lamarckians and neo-Darwinists. Both Darwin and Lamarck thought that use and disuse are quite important in evolution; both believed in hereditary transmission of acquired characters; both postulated a very similar theory of heredity; both were convinced that all organisms are a product of evolution, including man, that evolution continues and that the principle of actualism permits its explanation. Both rejected finalism and vitalism.

Darwin disagreed with Lamarck on the importance of the ecological conditions in evolution and was wrong on this point. But he postulated, simultaneously with Wallace, the principle of natural selection, which has a paramount importance in evolution. Lamarck did not consider selective elimination a factor of evolution and attached no importance to it, although he made the following vague remark in his *Philosophie zoologique*:

Par suite de l'extrême multiplication des petites espèces et surtout des animaux les plus imparfaits, la multiplicité des individus pouvait nuire à la conservation des races . . . si la nature n'eut pris des précautions pour restreindre cette multiplication dans des limites qu'elle ne peut jamais franchir. . . . On sait que ce sont les plus forts et les mieux armés qui mangent les plus faibles et que les grandes espèces devorent les petites.

Modern evolutionary biology is actually establishing a new synthesis, integrating the two fundamental ideas: evolution and geographically differentiated adaptation of organisms through natural selection to particular ecological conditions. This synthesis is new, compared with neo-Darwinism and with the synthetic theory as it was presented in the thirties.

Neo-Darwinism

Weismann is the founder of neo-Darwinism. The relations between neo-Darwinism and Darwin have some analogy with the relations between neo-

Lamarckism and Lamarck. Attacking Spencer and other neo-Lamarckians, Weismann rejects vigorously the hypothesis of the hereditary transmission of acquired characters, and the influence of the use and disuse of organs. He also proposes against Darwin's genetical theory of gemmules his own model based on the strict separation of the mortal somatic cells and the potentially immortal germ cells bearing the hereditary material. Later, Weismann's speculations on the hereditary mechanisms have been, at least partially, justified by the chromosomal theory of heredity. The rediscovery of Mendel's laws and the establishment of a solid basis of formal Mendelian genetics by Morgan and his collaborators contributed largely to the present general rejection of the theory of the hereditary transmission of acquired characters.

The new theory of evolution, created in the thirties on the basis of natural selection and spontaneous mutation as well as on the knowledge of formal and population genetics, has been classified under the name of Weismann's neo-Darwinism. It is preferable to utilise the term 'synthetic theory of evolution' or 'biological theory of evolution'. At the beginning, the well-known synthetic theory of evolution was based on the chance occurrence of mutations, the sieving effect of natural selection and on speciation by divergent evolution in isolated populations. It was accepted at that time that selection acts separately and independently on each couple of alleles. Natural selection sorted out 'good' from 'bad' alleles. This model of microevolution implied necessarily a tendency to form homogenous populations with a high degree of homozygosity of individuals. Once such models were established, their authors and others became prisoners of the scheme. Some criticisms have nevertheless been rightly addressed to this original form of the synthetic theory of evolution. This model did not offer enough evolutionary flexibility and potentiality. It was indeed difficult to accept that the maintenance of adaptation to ecological conditions, the directional trends of selection and evolution and the formation of complex organs could be furnished by this sieve mechanism.

Development of the biological theory of evolution during the last forty years

Since the establishment of the synthetic theory of evolution in the thirties, its founders and followers have made most important contributions to its development. Although this theory is still based on mutation, natural selection and isolation mechanisms, it is now quite different from what it was forty years ago. The most important impulse came from the study of the genetic constitution of natural populations and from research on the modification of gene pools in experimental and natural populations, and of the underlying mechanisms.

Fisher presented in 1927 the fundamental concept of the overdominance of heterozygotes with a selective advantage over both homozygotes. At the same time he established the hypothesis that gene effects can be modified in natural conditions by selective pressures on the genetic background. This led him to formulate the hypothesis that the influence of natural selection acting on the genetic background modifies the state of recessivity and dominance.

Haldane (1956) developed the idea that heterozygosity in individuals and heterogeneity in populations have an *a priori* advantage since they increase the number of biochemical tools and consequently the adaptability and microevolutionary potentialities of populations. He proposed the hypothesis that, in human populations, the lethal gene causing sickle cell anæmia has, in the heterozygous state, a considerable advantage because it confers resistance against malaria.

Teissier and L'Héritier showed that, in experimental populations of *Drosophila melanogaster*, natural selection does not eliminate one of two alleles which are in competition but favours their coexistence at characteristic equilibrium frequencies which may be modified when the ecological conditions or the genetic background change. They showed that the selective value of an allele is not constant but depends on the conditions and on the frequency of each allele. Teissier (1945) introduced also the idea of two different modalities of natural selection: 'Sélection conservatrice et sélection novatrice'.

Timofeeff-Ressovsky, Tchetverikov, Dubinin and others studied the heterogeneity of natural populations of Drosophilids. Ford (1940, 1965, 1971) introduced the term genetic polymorphism, and made important contributions to the study of genetic polymorphisms in Lepidoptera. He showed that natural selection can produce evolution much faster than it was formerly admitted, that an allele can pass from the recessive to the dominant state when it has a selective advantage and that natural selection has a tendency to organise groups of genes with coordinated functions in supergenes. Ford made also an important contribution to the study of ecological genetics. Lerner (1954) developed the idea of developmental homeostasis, presented earlier in a different way by Mather and Waddington, and postulated the important new idea of genetic homeostasis.

The most important contributions to the modern biological theory of evolution have been accomplished by Dobzhansky (1970). His personal work, together with that of his students and collaborators, on the heterogeneity of natural populations of Drosophilids, as well as the new knowledge and understanding of the genetic structure of natural populations, the mechanisms of evolution and particularly of the modalities and effects of natural selection, has had a great influence on the progress of population genetics. If the biological theory of evolution is nowadays much more sophisticated, subtle, differentiated and supported by many strong observational and experimental data, this is in good part due to Dobzhansky. His observations on the frequencies of lethal and other deleterious genes, his extensive work on the polymorphisms of structural types of chromosomes as well as the more recent research accomplished with his collaborators on the biochemical polymorphisms in natural populations of Drosophilids, together with other investigations which he stimulated, have considerably changed ideas about the structure of natural populations.

All this proves that natural populations are extremely heterogeneous. Each individual has an unique genotype. The degree of heterozygosity is much higher than it had been imagined before. In most organisms, natural selection does not simply act as a sieve and eliminate the less valuable alleles in favour of the best available allele at each locus. Natural selection maintains many deleterious and even lethal genes in a heterozygous state. That changes the concept of the genetic load. Dobzhansky shows that natural selection is complex and acts in different ways: by elimination of deleterious alleles as normalising selection, by maintaining different alleles at a locus as balancing selection, and by producing evolutionary changes as directional selection.

Dobzhansky has taken advantage of all this work and of his exceptionally broad knowledge to give on several occasions brilliant new syntheses of the actual state of the theory of evolution. This was the case for example in the lecture he gave in Paris in 1966 about *L'Evolution créatrice* (1967), in his *Genetics of the Evolutionary Process* (1970) and again in the paper delivered at the present symposium. In relation to our subject, I would like to quote four sentences:

> Natural selection constitutes a bond between the gene pool of a species and the environment. It may be compared to a servomechanism in a cybernetic system formed by the species and its environment. Somewhat metaphorically, it can be said that the information about the states of the environment is passed to and stored in the gene pool as a whole and in particular genes. Yet the environment does not ordain the changes that occur in the genes of its habitants.

Dobzhansky's synthesis postulates natural selection as the anti-chance factor in evolution, as the agent of creativity of evolution, working in an opportunistic manner. Natural selection and evolution permit the response of the organisms to the challenges of the environment. Dobzhansky's synthesis answers many of the questions and reduces the weight of the doubts of some neo-Lamarckians. Dobzhansky shows that there is no paradoxical conflict between chance and order in evolution. Mutations are, within biochemical limits, chance events. But natural selection is an ordering factor, an anti-chance agent, maintaining a meaningful and necessary correlation between the organisms and their environment. That does not mean that there exists a preestablished programme. Natural selection is often capable of maintaining indispensable adaptations; but many species disappear as a consequence of the lack of an evolutionary solution, while others evolve by anagenesis. Dobzhansky's concept of biological evolution is an important step in the development of the synthetic theory of the thirties. He presents a new synthesis, integrating the importance of the ecological conditions and the different modalities of adaptation to these conditions by natural selection.

Modern tendencies in genetics opposed to natural selection

Besides neo-Lamarckians, several groups of geneticists are now in different degrees opposed to the theory of natural selection. One of these tendencies is not very recent in reality; it is the hypothesis of evolution by random genetic drift. The proponents of this hypothesis do not exclude the action of natural selection, but they assume alongside it a more or less important effect of random drift on the evolution of populations. This factor plays indeed a role in small laboratory populations, but the efficiency of random drift in large natural populations has not been demonstrated. The proponents of the theory admit that 'most of the mutations which are important in evolution have much smaller selection coefficients than it is practicable to demonstrate in the laboratory', but nevertheless consider the possibility that different inversions in Drosophilids, or blood groups in man, may be adaptively neutral, that is to say have no adaptive value. A more extreme version of the same theory is the theory of so-called non-Darwinian evolution, in which natural selection as a directional factor plays little or no role and in which evolution is primarily the product of random events.

Another problem is whether there are great numbers of neutral alleles in enzyme polymorphisms. An allele which is in competition with another allele can indeed be neutral under certain ecological conditions, at a certain frequency of the allele or in a certain genetic background. But it is highly probable that the same allele is not neutral in other situations. It is in reality a false problem. Most of those who espouse the hypothesis of neutral alleles postulate that each allele has its specific and intrinsic selective value, independent of the genetic background and the ecological or other conditions. This oversimplified hypothesis is seriously misleading.

A third modern evolutionary hypothesis, which is at least partially anti-selectionistic, is that of the prevalence of biochemical evolution over other forms of evolution. This, in the extreme form, proposes that phylogeny can be inferred exclusively from the biochemistry of the organisms and that modifications of the genetic code, and consequently the production of new molecules, are the *only* cause of evolution. In a more subtle form Monod (1970) states that the theory of evolution by natural selection got all its sense, precision and certitude only since less than twenty years ago. He means, after the discovery of the genetic code. This underestimation of the primordial value of studies on organisms and populations is characteristic of many molecular biologists. Monod (1970) ascribes to proteins almost a kind of teleonomic intelligence, permitting them an oriented, coherent and constructive activity. On this point he is, curiously enough, close to Wintrebert's idea of the intelligence of living matter.

In the most extreme form, this modern 'non-Darwinian' theory assumes that evolution is largely a series of chance events in the modification of DNA molecules by replacement of codons. Crow and Kimura 'would not be surprised if it turns out that an appreciable fraction of nucleotide replacement in evolution is of this type . . .', that is to say, by neutral substitutions.

François Jacob (1970), one of the great molecular biologists, defends a much more sensible standpoint. What is evolving is not matter compounded with energy, but the organisation, the unity of emergence. In organisms the effects of chance are immediately corrected by the necessity of adaptation, of reproduction, of natural selection, says Jacob. Of course, we do not deny that biochemistry and molecular aspects of phylogeny are of a great help for the theory of evolution, as Florkin (1966) and Zuckerkandl and Pauling (1965) have rightly shown.

Still another modern tendency, which is in its extreme form antiselectionist, is numerical taxonomy, trying to measure phylogenetic differences simply by biometrical identity tests (Cole, 1969).

The place of Lamarck and Darwin in the new synthesis

Almost all founders and actual adherents of the biological theory of evolution completely reject Lamarck and accept Darwin. On the other side, the neo-Lamarckians who partially accept the biological theory of evolution are opposed to the rigid form of 'neo-Darwinism' of the thirties. They are attached to some arguments attributed to Lamarck. Both Lamarck's and Darwin's evolutionary theories can hardly be considered valid in their original forms. This statement certainly does not diminish the merit of these two great evolutionists. It is a general rule and a necessity for scientific progress that theories evolve.

Lamarck was right when he stressed the heterogeneity of natural populations. Darwin also studied carefully the intraspecific variability essential for his theory. But some neo-Darwinists thought later that the heterogeneity of populations has to be very limited, since they postulated that natural selection eliminates almost all mutant alleles. They have retarded the establishment of a correct knowledge of the genetic structure of natural populations. Lamarck was right when he insisted on the great importance of the ecological conditions for the diversification of organisms and for their evolution. Darwin did not pay enough attention to the influence of the environment on evolution. This may have retarded the creation of ecological genetics, which is now a main pillar of the modern theory of evolution.

The distribution of animals and plants and the frequencies of different genotypes of a species in a locality represent at any given moment a more or less well-established equilibrium between the organisms and their physical milieu. The elements of this equilibrium are constantly changing. These changes may be slow or rapid, feeble or important. Evolution is on the one side the expression of the constant readaptation of the organisms by natural selection to the changing conditions, but also increase of possibilities of action. In this sense evolution is also progress.

For the sake of evolutionary biology it is fortunately not true that we are still more or less at the point of the evolutionary model of the synthetic theory of the thirties. An important progress in knowledge and in concepts has been made. There is no need for a new name for this theory of evolution. 'Biological theory of evolution' fits well. But it seems useful to admit that we need a new synthesis, which is indeed on the way to being established at present.

The modern biological theory is not, and cannot be, an eclectic puzzle of pieces of Lamarck, Darwin, Mendel and de Vries. The most important elements of the biological theory of evolution of our days are doubtless:

(1) emergence of biological variants by mutations (de Vries);

(2) hereditary transmission of genetic variants according to the laws of Mendel;

(3) adaptation of organisms and populations to particular ecological conditions (Lamarck);

(4) adaptation and evolution of organisms and populations by natural selection (Darwin);

(5) maintenance of heterogeneity, evolutionary potential and adaptations to multiple challenges by the advantage of heterozygotes and by coadaptation of chromosomes (Dobzhansky);

(6) equilibrium between the indispensable renewal of heterogeneity of populations and the obligatory constant maintenance of the correlation between populations and the conditions of their environment by genetical homeostasis (Lerner).

Dobzhansky, Simpson, Mayr, Ford, Lerner and others have changed the evolutionary conceptions. From older typological and analytical standpoints they shifted to a population view of biology and evolution. They took the main steps towards experimental proofs, and towards comprehension of the importance of factors of integration, of interaction between opposed forces and of the widespread situation of balanced equilibrium.

We always need, and we all constantly utilise, the method of reduction in biological research. But it is important to say very clearly that we need even more, especially in the field of evolutionary biology, the methods and the concepts of the integration between environment and organisms, and between the different organisms.

Some biologists have in recent years denied the usefulness of this populational, organismic and synthetic views and research trends. This symposium is a good occasion to insist once more on the importance and necessity of the organismic point of view in biology.

References

Boesiger, E. (1971). Evolution des tendances néolamarckiennes et néodarwinistes au vingtième siècle en France. *13th Int. Congr. Hist. Sci., Moscow*. Ed. Naouka, Moscow.

Bergson, H. (1907). *L'Evolution créatrice*. Alcan, Paris.

Cole, A. J., ed. (1969). *Numerical Taxonomy*. London.

Cope, E. (1896). *The Primary Factors of Organic Evolution*. Chicago.

Crow, J. and Kimura, M. (1970). *An Introduction to Population Genetics Theory*. Harper and Row, New York.

Cuénot, L. (1911). *La Génèse des Espèces animales*. Paris.
Cuénot, L. and Tetry, A. (1951). *L'Evolution biologique: les faits, les incertitudes*. Masson, Paris.
Dobzhansky, Th. (1967). L'évolution créatrice. *Diogène*, no. 58, Paris.
Dobzhansky, Th. (1970). *Genetics of the Evolutionary Process*. New York.
Dobzhansky, Th. and Epling, C. (1944). *Contributions to the genetics, taxonomy and ecology of* Drosophila pseudoobscura *and its relatives*. Carnegie Inst., Washington, Publ., 554.
Fisher, R. A. (1927). On some objections to mimicry theory. *Trans. R. Ent. Soc. London*, **75**, 269–78.
Florkin, M. (1966). *Aspects moléculaires de l'Adaptation et de la Phylogénie*. Paris.
Ford, E. B. (1940). Polymorphism and taxonomy. In *The New Systematics*. Oxford.
Ford, E. B. (1965). *Genetic Polymorphism*. London.
Ford, E. B. (1971). *Ecological Genetics*. London.
Grassé, P. P. (1947). Les mécanismes de l'évolution. *Coll. Paléont. Génét. et Evolution*, CNRS, 201–19.
Grassé, P. P. (1971*a*). Discours d'ouverture. In *Colloque Intern. 'Lamarck'*. Libr. Sci. et Tech. Blanchard, Paris.
Grassé, P. P. (1971*b*). L'homme et son ADN, l'inné et l'acquis. *Institut de France*, **17**er, 1–11.
Haldane, J. B. S. (1956). *Biochimie et génétique*, Paris.
Hodge, M. J. S. (1971). Species in Lamarck. In *Coll. Intern. 'Lamarck'*. Paris, 31–46.
Jacob, F. (1970). *La Logique du vivant*. Gallimard, Paris.
Lamarck, J. (1778). *Flore françoise, ou description succincte de toutes les plantes qui croissent naturellement en France*. Paris.
Lamarck, J. (1802). *Recherches sur l'Organisation des Corps vivants*. Paris.
Lamarck, J. (1809). *Philosophie zoologique*. Paris.
Lamarck, J. (1820). *Système analytique des Connaissances positives de l'Homme restreintes à celles qui proviennent directement ou indirectement de l'Observation*. Paris.
Landrieu (1909). Lamarck, le fondateur du transformisme. *Mém. Soc. Zool. de France*, **21**.
Lecomte du Noüy (1947). *Human Destiny*. New York.
Lemoine, P. (1937). *Encyclopédie française*, vol. 5. Paris.
Lerner, M. (1954). *Genetic Homeostasis*. Edinburgh.
Mayr, E. (1972). Lamarck revisited. *J. Hist. Biol.*
Monod, J. (1970). *Le Hasard et la Nécessité*. Seuil, Paris.
Rabaud, E. (1911). *Le Transformisme et l'Expérience*. Paris.
Simpson, G. G. (1964). *This View of Life*. New York.
Teilhard de Chardin, P. (1959). *The Phenomenon of Man*. Harper & Row, New York.
Teissier, G. (1945). Mécanismes de l'évolution. *La Pensée*, **2**, 5–19, and **3**, 15–31.
Vachon, M., Rousseau, G. and Laissus, Y. (1972). *Inédits de Lamarck: d'après les manuscrits conservés a la bibliothèque du Muséum National d'Histoire Naturelle de Paris*. Masson, Paris.
Vandel, A. (1938). Contribution à la génétique des isopodes du genre Trichoniscus. *Bull. Biol. Fr. Belg.*, **72**, 121–46.
Vandel, A. (1940). La parthénogénèse géographique. *Bull. Biol. Fr. Belg.*, **74**, 94–100.
Vandel, A. (1968). *La Génèse du vivant*. Masson, Paris.
Vialleton, L. (1929). *L'Origine des Etres vivants: l'illusion transformiste*. Plon, Paris.
Weismann, A. (1893). *Die Allmacht der Naturzüchtung: eine Erwiderung*. Fischer, Jena.
Wintrebert, P. (1962). *Le Vivant créateur de son évolution*. Paris.
Wintrebert, P. (1963). *Le Développement du vivant par lui-même*. Paris.
Zuckerkandl, E. and Pauling, L. (1965). Molecules as documents of evolutionary history. *J. Theor. Biol.*, **8**, 357–66.

Discussion

Rensch

Lamarck's idea that the *besoins*, the needs, of a species may cause phylogenetical alterations can perhaps be accepted partly, in those cases in which the behaviour has influenced evolution. The shape and colour of many flowers, for instance, has been determined to some extent by the behaviour of pollinating insects and birds. The need to become pollinated guided the special evolutionary development. This is particularly conspicuous in the morphology of the flowers of the genus *Ophrys*. They became so insect-like that certain species of bees try to copulate with the lip of the flower and really fertilise the plant by this behaviour, although of course with the pollen of another specimen of the same species of *Ophrys*. We must also take into consideration that all animals have certain ecological needs and that most of them actively search for the suitable ecological conditions. This is particularly important when the offspring spreads over the borderline of the habitat of their parents. This can lead to new adaptations mainly if some varieties are preadapted.

4. The Problem of Molecular Recognition by a Selective System

GERALD M. EDELMAN

Immunology has been profoundly altered in the last decade by two major developments: the theory of clonal selection and the chemical analysis of antibody structure. As a result of these developments, it has become clear that the central problem of immunology is to understand the mechanisms of selective molecular recognition in a quantitative fashion. The molecules and cells mediating selection in the immune response are known or can be known and, above all, the time scale of the selective events is well within that required for direct observation and experimentation. Aside from evolution itself, there are few such well-analysed examples of selective systems in biology, or in other fields for that matter. For this reason, the immune system provides a unique opportunity to analyse the problem of selection under defined and experimentally measurable conditions which have so far been hard to achieve elsewhere.

In this paper, I shall briefly outline the idea of clonal selection as it operates in immunology, compare some of its features to those of natural selection and attempt to show how progress in molecular immunology (the reductionist approach) has clarified and sharpened the notion of clonal selection and its relation to the specificity of molecular recognition. My purpose is not epistemological; nor shall I attempt an analysis of the reductionist or holist position, for I believe that in their most extreme forms there is a fruitless opposition implied by these terms. Nevertheless, a description of the interplay between the phenomenological approach of the clonal selection theory and the mechanistic approach implicit in the analysis of antibody structure may provide a useful example of the importance and limitations of reductionism in a particular biological system.

The work of the author was supported by USPHS Grants AM 04256 and AI 09273.

Antibodies and the idea of molecular recognition

The most striking feature of the immune response is not just its specificity but its specificity over such an enormous range of chemical structures. Antibodies may be generated to proteins, carbohydrates, nucleic acids and various artificially synthesised organic compounds or haptens. Indeed, antibodies may be generated to antibodies themselves in both the same and different animal species. It is the combination of the range of recognition with the exquisiteness of the specificity that gave rise to the early paradoxes in immunological thinking and led to an important idea as well as an erroneous theory.

The important idea, which emerged from Ehrlich's early studies, from Landsteiner's epoch-making experiments and from Pauling's analysis of chemical bonding, was that of molecular complementarity (Landsteiner, 1945). A thermodynamic analysis of antigen–antibody interactions indicated that the free energies of binding were of the order of 5–10 kcal/mole. From a knowledge of the chemistry of non-covalent interactions and a comparison of cross-reacting haptenic homologues which differed slightly in shape, it was concluded that antigens and antibodies interacted over very short distances. This implied that there was a very close fit between some portion of the antigenic determinant and the antigen-combining site of the antibody molecule.

Given this notion, it was reasonable for Pauling to clarify earlier ideas of Breinl, Mudd and Haurowitz (Jerne, 1967) into an instructive theory of antibody formation. This theory states that the complementarity is imposed upon the antibody by the antigen which acts as a template for the combining site. At first glance, this seems to be the only reasonable solution to the problem of molecular recognition, for the range of antigens is so great and some of them are so obviously unable to act during evolution of the immune system, that only a direct information transfer from antigen to antibody would seem to account for the specificity.[1]

This theory has turned out to be incorrect. It is not appropriate here to trace the origin of the error, but it is important to realise that it was a reductionist theory and that its failure rested not in its reductionism but in approaching the analysis of the system at only one level, that of the molecule rather than that of the cell. It was clear, for example, quite early that the instructionist theory could not easily explain immunological memory; nonetheless, the instructionist position held sway for a long time.

The paper that finally opened the way to modern theories of selection was that of Jerne entitled 'The natural selection theory of antibody formation' (Jerne, 1955). Jerne approached the problem from an entirely different point of view. He assumed that there was (1) a *random* mechanism for ensuring the limited synthesis of antibody molecules possessing all possible combining

[1] It is interesting to note the analogy to Lamarckian thinking: the environment (antigen) actually alters the biological information system (antibody-forming apparatus).

sites *in the absence of antigen*, (2) a *purging* mechanism for repressing antibody synthesis against autoantigens, (3) a *selective* mechanism for promoting synthesis of those antibodies that fit a particular antigen best.

This selective theory (the origins of which have been charmingly described by Jerne, 1966) still made the curious assumption that it was the antibodies that were circulating in the blood that brought the antigen to the appropriate cell. This failure to analyse the site at which recognition takes place led to a variety of objections. But it must be stated that even though Ehrlich (1900) had formulated a 'selective' theory very early, Jerne was the first to see truly the fundamental nature of the problem as one of selection.

Burnet (1959) modified Jerne's theory and shifted the object of selection from free antibody molecules to cells which were committed to making antibodies of one specificity. These cells were supposed to generate the types by *somatic mutation* and were then selected by antigen to divide and produce progeny which could synthesise more of the same kind of antibody (Figure 1). This theory accounted for a variety of biological phenomena in immunity including memory, the increase in avidity of antibodies with time of immunisation, and the problem of self-recognition. This last difficulty was fully recognised by Jerne, Burnet, and particularly by Medawar (1957). According to the theory, self-recognition could be stated in terms of repression or death of the appropriate clones of cells. Lederberg made a clear and incisive analysis of the genetic requirements of clonal selection which included a consideration of the problem of tolerance and self-recognition (Lederberg, 1959).

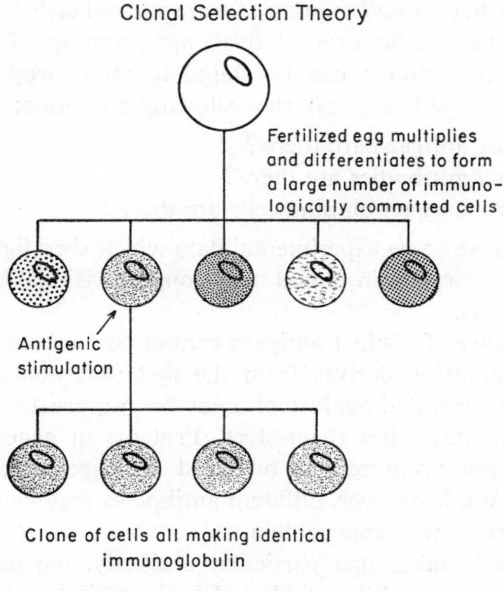

Figure 1. The clonal selection theory.

The appearance of Burnet's theory was pivotal in that it shifted the level of reduction from antibodies *per se* to cells and spoke in terms that biologists could understand. It thus was a clear challenge to instructive theories and, as shown by the subsequent facts, it was pointed in the right direction. On one point the theory was triumphantly clear: one cell makes only one kind of antibody, and antigen in the appropriate form can interact with that antibody and stimulate cell maturation and division. Both points have been amply confirmed since by experiment.

It must not be assumed, however, that the molecular approach was useless. Indeed, the analysis of antibody chain structure (Edelman, 1959), the finding that different antibodies appeared to have different amino acid compositions and the finding that no antigen could be demonstrated in cells actively synthesising antibody (Nossal, 1965) all quite independently dealt severe blows to instructive theories. The mortal blow came when it was shown that antibodies could be denatured and renatured *in vitro* in the absence of antigen and yet regain their specificity (Haber, 1964).

The incompleteness of the theory of clonal selection

The difficulties of the theory of clonal selection as originally formulated centred around its failure to analyse exactly how antibody specificity was generated over a sufficient range. It also failed to present a framework for the quantitative analysis of the number of different antibodies required, of the thresholds of cellular response and of the population dynamics of lymphoid cells. Thus, for example, the theory came under strong attack when Simonsen (1967) showed that a *small* number of transplanted cells in a graft *versus* host reaction could be specific for a relatively *large* number of antigens.

The specificity problem can be stated in many ways, but perhaps the simplest way to pose it is to ask the following questions:

(1) how many antigens are there?
(2) how many antibodies are there?
(3) how many antigen-binding cells are there?

Before considering some experimental data which shed light on the problems of specificity and range in clonal selection, provisional answers may serve to set some limits.

(1) The number of distinct antigens cannot be measured but must be very large. This conclusion derives from the fact that non-cross-reacting antibodies may be generated against proteins from a great variety of species as well as against antibodies themselves. Proteins in general have aperiodic structures and possess more than one kind of antigenic determinant on each polypeptide chain. Moreover, different antibodies may be made to denatured and native forms of the same protein.

(2) If one cell makes one particular antibody, the number of different antibodies cannot exceed the number of antigen-binding cells. As I will show

later, however, structural analysis indicates that the number of possible antibodies is legion.

(3) The total number of lymphoid cells varies in different organisms capable of producing antibodies. It can be as small as 100 000 in a tadpole or as large as 10^{11} or 10^{12} in a man.

The main point deserving analysis is this: the clonal selection theory implies that there must be very many more ways in which an antigen can be bound by a variety of antibodies than there are antigens to be recognised. Another way of putting this is to assert that the antibody forming system must be *degenerate* with regard to a given antigen—that is, many *different* antibodies must be capable of binding that antigen *more or less well.* Were this not the case, and were there a best or even a unique antibody for each antigen, then either each antibody must have been selected specifically by its antigen during evolution, or the clonal selection theory would have to assume a miraculous 1:1 correspondence between each antibody and its corresponding antigen. In either case, the theory would be vitiated.

The facts are in accord with the idea of degeneracy: each antigen in general elicits the synthesis of a heterogeneous collection of antibodies. But this degeneracy points up the problem quite sharply, for it must imply that at some level there is a lack of specificity or an extensive cross-reactivity among different antibodies. At what level does this occur, and how is it that one does not see more evidence of it in circulating antibodies?

To answer this question even provisionally we must examine the requirements of the clonal selection theory in greater detail, particularly in the light of modern findings on antibody structure.

The requirements of clonal selection

The theory has three main requirements:

(1) A source of diversity amongst different antibodies and a means of committing a cell or phenotypically restricting it to one of each possible kind of antibody (see figure 1).

(2) A means of trapping antigen or favouring encounter with antigen binding cells.

(3) A means by which this encounter is *amplified* by triggering the cells to divide and produce more antibodies of the same type.

From analyses of the structure of antibodies (Edelman and Gall, 1969; Edelman, 1970*a*; Gally and Edelman, 1972), it has become clear that the basis for antibody diversity is variation in the amino acid sequences of approximately the first 110 residues (variable regions) of the light and heavy polypeptide chains of which the antibody is made (figure 2). In addition, the variability is increased by the possibility of random assortment among different light and heavy chains. The number of different sequences and combinations possible is of the order of 10^{14} or greater. There are 10^{11} or so lymphoid cells, and it is obvious that the diversity *possible* in the system is

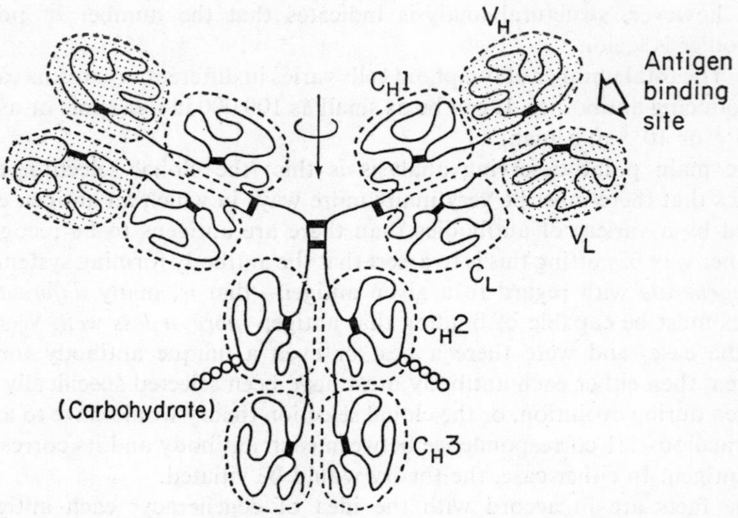

Figure 2. A model of the structure of a human IgG antibody molecule. The variable regions of heavy and light chains (V_H and V_L), the constant region of the light chain (C_L) and the homology regions in the constant region of the heavy chain (C_H1, C_H2 and C_H3) are thought to fold into compact domains (delineated by dotted lines), but the exact conformation of the polypeptide chains has not been determined. The vertical arrow represents the twofold rotation axis through the two disulphide bonds linking the heavy chains. A single interchain disulphide bond is present in each domain. Carbohydrate prosthetic groups are attached to the C_H2 regions.

sufficient for the binding of a large number of antigens of different shape. The intriguing problem of the genetic origin of sequence diversity remains unsolved. It has been critically examined elsewhere (Edelman and Gall, 1969; Gally and Edelman, 1972), and for this reason it will only be considered here as it bears upon the problem of specificity.

Commitment of a cell to one of the possible light–heavy chain pairs is not fully understood but may be related to the unusual arrangement of the genes that specify antibody chains in a series of clusters. Each of these clusters contains separate genes for variable regions (V genes) and for constant regions (C genes) of antibody molecules. This arrangement has been called a translocon (Gally and Edelman, 1972), to emphasise the requirement that information from a V gene must be translocated and linked with that from a C gene to make a single VC gene for an antibody chain. Whatever the mechanism of such an event, it would serve to restrict the cell so that it would make only one kind of light chain (say $V_{L22}C_L$) and one heavy chain (say $V_{H89}C_H$). Although additional mechanisms of gene control must be invoked (as in any differentiated cell), it is not difficult to see how the requirement for phenotypic restriction might be met by such an arrangement.

It appears that the second requirement of clonal selection theories, that for antigen trapping, is mediated by a subtle interaction between two types of cells *both* with antibody receptors. These are respectively the thymus-derived lymphocyte (T cell) and the bone marrow-derived lymphocyte (B cell). Primary immunisation by protein antigens requires cooperation between a T cell and a B cell; one of the more likely hypotheses is that T cells locally concentrate the antigen for presentation to B cells (Gowans, Humphrey and Mitchison, 1971), possibly by attachment of the antigen–antibody complex from a T cell to a third cell which then presents the complex to the B cell. In this way, the B cells are triggered to mature and divide (Feldman, 1972).

This triggering phenomenon is the third requirement of clonal selection and here practically nothing is known of either the threshold or the mechanism of stimulation. This may be the major factor, however, in providing a solution to the central dilemma of specificity in clonal selection. As originally proposed, the clonal selection theory was noncommittal about the strength of binding of an antigen required for triggering or about whether triggering was a function of the avidity for antigen of the receptor antibody. Because an analysis of this problem is central to the relationship between specificity and selection, and because the problem can be attacked on logical as well as experimental grounds, it deserves detailed analysis.

A two-factor theory for the specificity of clonal selection

It has been observed that Darwin's theory of evolution differed from previous theories in that they considered only one factor as the driving force for evolution. Darwin's theory and its more sophisticated successors considered the interplay between variation among organisms and natural selection. A similar situation may obtain in clonal selection. Diversification of antibodies may to a certain extent be under selective pressure by groups or classes of chemical substances during evolution. But according to the theory of clonal selection, it is *impossible* for *each* cell receptor to have been selected for or against during evolution. Instead, a great number of antibody variants have been generated (by whatever process; see Edelman and Gall, 1969; Gally and Edelman, 1972; Edelman, 1970*b*), many of which will never be selected during the lifetime of the organism. The central question with which this paper has been concerned is: under these circumstances alone, how specific can the system be?

In order to approach this question, it is important to distinguish several levels (Gally and Edelman, 1972) of potential diversity in the clonal system (figure 3). Almost all antibody-producing cells whose products have been characterised are derived from precursor cells that have been stimulated to proliferate (by antigen or otherwise) to form a clone of cells, each committed to the production of the same antibody. The phenotypic expression of the antibody genes is therefore mediated in the animal by the somatic division of

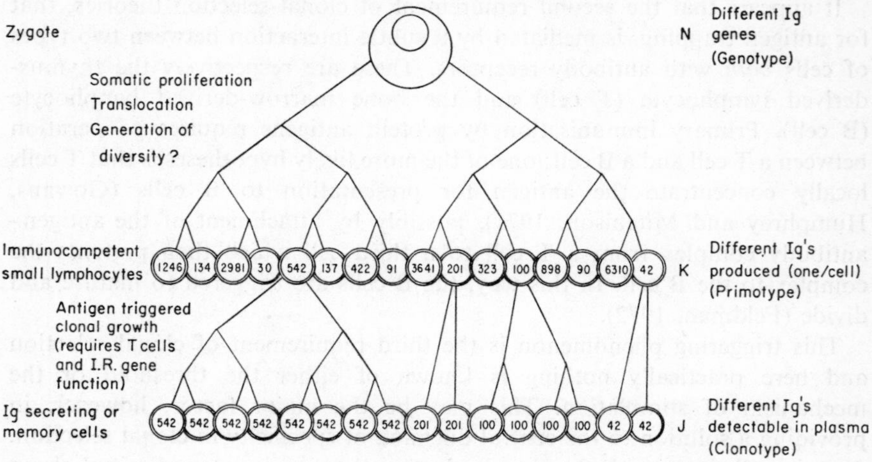

Figure 3. A model of the somatic differentiation of antibody-producing cells according to the clonal selection theory. The number of Ig genes, which equals N in the zygote, may increase during somatic growth so that in the immunologically mature animal K different cells are formed, each committed to the synthesis of a structurally distinct Ig (indicated by the arabic number). A small proportion of these cells proliferate upon antigenic stimulation to form J different clones of cells, each clone producing a different antibody. $N < J < K$.

precommitted cells. Since an animal is capable of responding specifically to an enormous number of antigens to which it is usually never exposed, it follows that the animal must contain genetic information for synthesising a much larger number of different antibody molecules than actually appear in detectable amounts in the bloodstream. The clonal selection theory suggests that large numbers of cells commit themselves to the production of antibody molecules with binding sites that are not complementary to any known antigen. In other words, the antibody molecules whose properties we can examine may represent only a minor fraction of those for which genetic information is available. Specific terms to designate the levels of expression of antibody genes may be useful in attacking the problem of specificity:

(1) The genotype: the genetic information stored within the zygote which can be passed on to future generations, specifically translocons containing V and C genes and their regulatory genes.

(2) The primotype: the sum total of different, structurally distinct antibody molecules generated within an organism during its lifetime. While the number of different antibody molecules in the primotype may be much less than that which is *potentially* present in the genotype, it is probably orders of magnitude greater than the number of effective antigenic determinants to which the animal is ever exposed.

(3) The clonotype: the different antibody molecules an animal actually

produces which can be detected and classified according to antigen-binding specificity, class, antigenic determinants, primary structure, allotype, or any other experimentally measurable molecular property. This definition takes explicit account of the fact that, in general, for an antibody to be examined in terms of these properties, a large number of identical molecules must be present, usually far more than can be obtained from a single cell. As a class, the clonotype is smaller than the primotype and wholly contained within it (figure 3).

So far, we are unable to examine *directly* those antibody molecules in the primotype whose production has not been clonally amplified, and can only assume without evidence that they are similar to those which we can examine. In any case, the question about specificity can be now posed in the following terms: is the probability of cross-reactivity with various antigens the same in the primotype and the clonotype? If the degree of specificity were the same in the primotype and the clonotype, one would expect that considerable cross-reactivity exists in both populations.

There is now some evidence, however, to suggest that this is not the case and that the degree of cross-reactivity may be much greater in the primotype. If this is the case, selection for specificity cannot merely be the result of antigen-binding but must also depend on a second factor. The most likely candidate is the triggering threshold of the cell carrying the antibody receptor. Thus, if two cell populations contain cells which bind two different antigens but also contain a subpopulation which can bind both, specificity would be lost. It would be preserved, however, if a particular cell capable of binding both antigens is more likely to be triggered by only one of the two. The trigger threshold might depend on the state of the cell but might also be a function of steric factors reflected in the free energy of binding for each particular antigen and particularly of the *surface density* of the antigen which increases the *avidity* of the binding.

Whatever the detailed mechanism of triggering, the implication of this two-factor theory is that variation at the level of the primotype leads to a relatively non-specific set of molecules capable of binding various antigens with low specificity. The selective forces that yield specificity are a product of both the probability of binding and the probability of triggering above a certain triggering threshold.

This model may now be tested by specific methods of cell fractionation (Edelman *et al.*, 1971) which promise to provide a detailed picture of the population dynamics of lymphocytes and thus permit formulation of a quantitative theory of clonal selection. The preliminary results suggest that as many as 1 per cent of the lymphoid cells from the spleen of an unimmunised mouse (Rutishauser *et al.*, 1972) can specifically bind a given antigen over a wide range of affinities. Only cells with receptors of higher affinities appear to be stimulated after immunisation, however, and the number of progeny cells appears, *above a certain threshold*, to be proportional to the affinity of the antibody receptors on the cells. This result may help to explain several of the outstanding objections to clonal selection in its original form.

Clonal selection and natural selection

The system of clonal selection bears resemblances to natural selection during the evolution of the vertebrate organisms in which immunity occurs. It should be emphasised, however, that these processes differ in a number of fundamental features, some obvious and some subtle. The most obvious is the time scale over which selection takes place, for in this respect they are truly disparate—days and months *vs* millions of years. It is also clear that the clonal selection system is simpler: only one kind of target molecule and only a few cell types are involved and the selection occurs within the somatic cells of an organism rather than among organisms.

There are some more subtle differences that are worthy of mention. In the immune system, degeneracy prevails in the primotype and many variants are not pertinent. In the natural selection of organisms, there are probably few or no neutral mutations. Clonal selection results in a set of individuals which do not reproduce beyond their clonal expansion, which do not show evidence of sex, and most of which are useless; in natural selection most surviving genomes are adaptively useful and viable. Clonal selection is a fixed system with variations on a theme; in natural selection completely new themes are possible (although relatively rare) and the system is open-ended. Above all, at least for the basic protein structure of antibodies, natural selection is obviously required before clonal selection can operate.

Despite these differences, it is important to examine clonal selection as a tractable experimental model and as the only known example of a selective system with a short time scale. Indeed, one of the most interesting theories of the origin of antibody diversity is the somatic mutation theory (Jerne, 1971; Edelman, 1971). This theory in effect proposes that a certain number of germ line V genes are subject to somatic mutation during embryogenesis. They are then either selected for by antigen or not selected against during embryogenesis (Jerne, 1971). In the mouse this would leave no greater than fifteen days for all of the necessary mutants to be selected. If this theory is correct, the selective immune response would be the result of a process resembling neo-Darwinian evolution *in utero*.

A significant test of this theory would be to score the rate of appearance of antigen-binding cells in embryogenesis. If a significantly large number of cells capable of binding many different antigens appears within only four or five division times, the somatic mutation theory would be greatly weakened. This experiment is now in progress and my prediction is that this theory will not be supported. Nonetheless, there would still remain the task of distinguishing whether each V region is specified by a single V gene or whether a small number of V genes with evolutionarily selected mutations generate antibody diversity by somatic recombination.

The role of reductionism in formulating the idea of a selective immune system

Rather than enter the debate concerning reductionism and holism, it is perhaps more useful to distinguish among the heuristic value, the logical completeness and the predictiveness of an idea. To make this point succinctly, I shall state rather dogmatically that, in its initial form, the idea of clonal selection had enormous heuristic value but was logically incomplete and indeed vague. A good deal of that vagueness was removed by the analysis of antibodies at the molecular level (Edelman, 1971), and as a result the questions of immunology could be formulated in a more restricted and clearer fashion (Jerne, 1967).

In describing these developments, it is helpful to distinguish the hierarchical levels at which various ideas operate. For example, the idea of specificity and molecular recognition was prior to, and was necessary but not sufficient for, the idea of clonal selection which operated at the level of cells. In turn, the particular picture of variability and constancy which emerged at the level of the molecular analysis of antibodies could be applied to refine the idea of clonal selection at those points where it seemed most dubious. It now seems likely that a further refinement will come from a chemical analysis of the statistical effects of multipoint binding (avidity rather than affinity), from analysis of the mechanism and threshold of lymphocyte triggering in relation to specificity and, finally, from the quantitative analysis of the gains and feedbacks in the clonal system that promises to emerge from the application of new methods of specific cell fractionation.

References

Burnet, F. M. (1959). *The Clonal Selection Theory of Acquired Immunity*. Cambridge University Press.

Edelman, G. M. (1959). Dissociation of gamma globulin. *J. Amer. Chem. Soc.*, **81**, 3155.

Edelman, G. M. (1970*a*). The covalent structure of a human γG-immunoglobulin. XI. Functional implications. *Biochemistry*, **9**, 3197.

Edelman, G. M. (1970*b*). The structure and function of antibodies. *Sci. Amer.*, **223**, no. 2, 34.

Edelman, G. M. (1971). Antibody structure and molecular immunology. *Ann. NY Acad. Sci.*, **190**, 5.

Edelman, G. M. and Gall, W. E. (1969). The antibody problem. *Ann. Rev. Biochem.*, **38**, 415.

Edelman, G. M., Rutishauser, U. and Millette, C. F. (1971). Cell fractionation and arrangement on fibers, beads and surfaces. *Proc. Natl. Acad. Sci. US*, **68**, 2153.

Ehrlich, P. (1900). On immunity with special reference to cell life. *Proc. Roy. Soc. London B*, **66**, 424.

Feldman, M. (1972). Specific collaboration between T and B lymphocytes across a cell impermeable membrane *in vitro*. *Nature New Biol.*, **237**, 13.

Gally, J. A. and Edelman, G. M. (1972). The genetic control of immunoglobulin synthesis. *Ann. Rev. Genet.*

Gowans, J. L., Humphrey, J. H. and Mitchison, N. A. (1971). A discussion on cooperation between lymphocytes in the immune response. *Proc. Roy. Soc. London B*, **176**, no. 1045, 369–481.

Haber, E. (1964). Recovery of antigenic specificity after denaturation and complete reduction of disulfides in a papain fragment of antibody. *Proc. Natl. Acad. Sci. US*, **52**, 1099.

Jerne, N. K. (1955). The natural selection theory of antibody formation. *Proc. Natl. Acad. Sci. US*, **41**, 849.
Jerne, N. K. (1966). The natural selection theory of antibody formation: ten years later. In *Phage and the Origins of Molecular Biology* (ed. J. Cairns, G. S. Stent and J. D. Watson). Cold Spring Harbor Laboratory of Quantitative Biology.
Jerne, N. K. (1967). Antibodies and learning: selection *vs* instruction. In *The Neurosciences: A Study Program* (ed. G. C. Quarton, T. Melnechuk and F. O. Schmitt). Rockefeller University Press, NY, 200,
Jerne, N. K. (1967) Waiting for the end. *Cold Spring Harbor Symp. Quant. Biol.*, **32**, 591.
Jerne, N. K. (1971). The somatic generation of immune recognition. *Eur. J. Immunol.*, **1**, 1.
Landsteiner, K. (1945). *The Specificity of Serological Reactions*, revised edition. Harvard University Press, Cambridge, Mass.
Lederberg, J. (1959). Genes and antibodies. *Science*, **129**, 1649.
Medawar, P. (1957). *The Uniqueness of the Individual*. Methuen, London, 143.
Nossal, G. J. V. (1965). How cells make antibodies. *Sci. Amer.*, **211**, no. 6, 106.
Rutishauser, U., Millette, C. F. and Edelman, G. M. (1972). Specific fractionation of immune cell populations. *Proc. Natl. Acad. Sci. US.*
Simonsen, M. (1967). The clonal selection hypothesis evaluated by grafted cells reacting against their hosts. *Cold Spring Harbor Symp. Quant. Biol.*, **32**, 517.

5. A Geometric Model of Reduction and Emergence

PETER MEDAWAR

My special purpose in this contribution is to use the hierarchy of geometries as Felix Klein conceived it to illustrate the relationship between the various tiers of the hierarchy of natural sciences. I want to try to clarify the concept of 'emergence', by which I mean the emergence at each tier of the hierarchy of concepts peculiar to and distinctive of that tier, and not obviously reducible to the notions of the level immediately above or higher still. One such notion, as we shall see, is that of a 'circle' in Euclidean geometry. The contextually distinctive notions of biology at the organism level include, for example, 'heredity', 'infection', 'immunity', 'sexuality' and 'fear'.

These notions are not reducible to physics and chemistry, if only in the naive sense that they simply do not appear in the syllabuses of university courses in physics and chemistry. Alternatively, if they did so appear, then we should want to invent a special name to distinguish such unusual and specialised branches of physics or chemistry. The name we should be most likely to hit upon would be none other than 'biology'—and this, I think, is the point, or will appear to be so in the course of this contribution.

You may of course say that to speak of a 'hierarchy' of the natural sciences is rather to beg the question, but I think we all believe in the existence of such a hierarchy as I can illustrate by the following imaginary lantern slide:

1. Physics
2. Chemistry
3. Ecology/sociology
4. Organismic biology

Every scientist who looks at this list (unless he is being deliberately obtuse) will see that there is something wrong with it, namely that items 3 and 4 should be interchanged, whereupon the list makes sense. But the concept in terms of which it thereupon makes sense is the concept of a hierarchy of the empirical sciences.

The geometric parallel

In propounding this parallel I propose for simplicity's sake to adopt the old-fashioned or 'common sense' view that space is a vessel containing points or other geometric objects like lines, circles or squares. I shall also make use of the idea of a 'transformation' in a sense not far removed from the intuitively obvious one. But I shall also need the idea of a transformation formula or 'mapping function'. A 'mapping function' is the formula which enables one to define a transformation or change of shape precisely. As the name signifies, the most familiar everyday context of transformation is cartography.

Suppose

$$\text{Circle } (x, y) = 0 \qquad (1)$$

is the conventional formula or equation for a circle when the points $[x, y]$ are plotted on ordinary square graph paper, and suppose we want to transform this circle into an ellipse.

This can be done in alternative ways: either (*a*) by keeping the old formula Circle $(x, y) = 0$ and plotting out the points $[x, y]$ on a special graph paper which is elongated in one dimension—that is, by plotting the old points on transformed coordinates; or (*b*) by changing the old points $[x, y]$ into new points $[x', y']$ which are precisely related to the ones they replace and then plotting out the new points $[x', y']$ on the old square graph paper. In either case we shall have to specify the mapping function or substitution rule for the change of points or change of coordinates, namely

$$\begin{aligned} x' &= \mathrm{f}_1\,(x, y) \\ y' &= \mathrm{f}_2\,(x, y) \end{aligned} \qquad (2)$$

These mapping functions (2) are the key to the transformation.

Another notion I shall have to introduce is that of 'invariance', because it plays an important part in the argument later on. In any process of transformation some properties will change and others will remain unchanged. In the transformation of a circle to an ellipse the figures remain unchanged in so far as they remain closed lines which divide the plane into an inside and an outside. On the other hand circularity, as such, changes, for in the new figure points on the perimeter are no longer equidistant from a point in the middle. In a world subject to transformations of the kind I have just described the word 'circle' would have no meaning.

Klein's conception of the geometries

Most people who study logic or philosophy are taught to believe that Euclid's geometry is the paradigm of all deductive systems, and so it came to be thought that all geometries could be defined like Euclid's, in the sense in which Euclid's geometry consists of axioms and postulates (and definitions, of course) together with all the theorems that issue from them by mathematical, that is, deductive, reasoning. If one alters the axioms, one gets a different geometry—a non-Euclidean geometry, perhaps—or a geometry of some quite different kind like projective geometry.

Somewhat after the middle of the nineteenth century an entirely new concept was introduced into geometry by the man Coolidge calls 'the greatest synthesist that geometry has ever known'—Felix Klein.

In this conception of geometry any given geometry—for example, metrical geometry—may be thought of as the ensemble of properties of geometrical objects that remain unchanged, or invariant under the transformations of a given group. Metrical geometry is then the invariant theory of a certain 'metrical' group of transformations. (The exact meaning of the word 'group' in this context is explained in the appendix, but for the moment we may simply read 'set'.)

Metrical geometry

Suppose we confine the group of operations to those which may be represented geometrically as translations of figures on a plane, rotations or inversions or any combination of any of them. The ordinary properties of figures that we associate with shape obviously do not change under this group of operations. Figures related to each other by metrical transformations can be said to be 'similar' and are indeed superimposable one upon the other. Another geometric property that is conserved under transformations of this group is 'distance'. It becomes possible to say of two points A, B, that A is a certain distance from B, and if it is so it will remain so under the transformations to the metric group.

Euclidean geometry

The group of operations that defines Euclidean geometry allows for symmetrical magnification, that is, for an enlargement of geometric objects to the same degree in all three dimensions of space. Figures related to each other by Euclidean transformations can be said to be 'of the same shape'. In Euclidean geometry we can clearly speak of circles and squares and right-angled triangles. Distance, however, is not an invariant property, so theorems depending entirely upon precise measurements of distance do not appear in Euclidean geometry; that is, nothing turns upon a point A's being precisely 2 or 3 units of distance from point B.

Affine geometry

When the new points or new coordinates are related to the old points they replace by linear integral functions—that is, when the mapping functions are of the form

$$\begin{pmatrix} x' = a_1x + b_1y + c_1z + d_1 \\ y' = a_2x + b_2y + c_2z + d_2 \\ z' = a_3x + b_3y + c_3z + d_3 \end{pmatrix}$$

—then, geometrically speaking, the transformations involved are much the same as those of Euclidean geometry except that the degree of extension or diminution in the three different dimensions of space is not necessarily the same. In affine geometry, therefore, it will hardly be meaningful to speak of a 'square' or of a 'circle', but we may speak more generally of an 'ellipse' (of which a circle is a special case) or a 'rectangle' (of which a square is a special case). Geometric objects related to each other by affine transformations are sometimes referred to as 'homographic'.

Projective transformations

The mapping functions which specify projective transformations are those in which the new points or coordinates are *fractional* linear functions of the points or coordinates they replace. The mapping functions run:

$$x' = \frac{a_1x + b_1y + c_1z + d_1}{a_4x + b_4y + c_4z + d_4}$$

$$y' = \frac{a_2x + b_2y + c_2z + d_2}{a_4x + b_4y + c_4z + d_4}$$

$$z' = \frac{a_3x + b_3y + c_3z + d_3}{a_4x + b_4y + c_4z + d_4}$$

The denominator, you will notice, is the same throughout. If the value of this denominator should tend to zero, then certain points under projective transformation will be 'carried to infinity'. Among the properties invariant under projective transformation will be linearity, for straight lines are transformed into straight lines, certain ratios like the anharmonic ratio and the property being a conic section—that is, a circle, an ellipse, hyperbola or parabola, as the case may be. Parallelism is not an invariant, however.

Topology

Lastly we come to the most general relationship imaginable between the new points and the old ones—that in which all we ask of the mapping functions is (*a*) that they should be single-valued both ways and so bring the new points into a one–one correspondence with the old points, and (*b*) that they should be continuous functions. As a model of continuous one–one transformations we may consider a drawing of any kind upon a sheet of very flexible rubber. The rubber may be stretched and distorted in any way we please, but it must not be torn, for otherwise the transformations would not be continuous. Figures which can be brought into correspondence by topological transformations of this kind are said to be 'homeomorphic'. In topology, which is the invariant theory of continuous one–one transformations, all ordinary geometrical properties of figures cease to mean very much, for clearly there are no straight lines in this geometry and certainly no such properties as parallelism. Nevertheless some very general properties of space remain—for example, a closed figure still divides a plane into an inside and an outside, the order of points in a line is not upset. The 'sidedness' of a surface remains unchanged. The knotty quality of a certain kind of knot is also invariant under transformation, as anyone may verify by tying a knot in a piece of elastic. Evidently topology deals with the most elementary properties of space—simple properties like insideness, outsideness and sidedness.

The hierarchy of geometries

If we write down the geometries in the following order, following F. Enriques:

1. Topology
2. Projective geometry
3. Affine geometry
4. Euclidean–metrical geometry

we can see that each geometry is a special case of the one above it, for in each case the group of operations that defines it is a subgroup of the group relating to the one above, and that as we go down the line the theorems and the concepts of the geometries become progressively richer and more particular. This progressive enrichment occurs not *in spite of* the fact that we are progressively restricting the range of transformations, but precisely *because* we are doing so. Topology deals with the most general and fundamental properties of space where metrical geometry deals with ordinary homely familiar properties like rectangularity, circularity and so forth. It will also be clear that every statement that is true in one geometry is also true in the geometries entered below it. Every topological theorem must also be a theorem in projective geometry and is also 'true in' Euclidean geometry. But as the geometries become more and more restricted, so new concepts 'emerge', concepts that simply do not appear in the geometry preceding it—for example, we cannot retain the notion of a 'circle' in projective geometry but have to content ourselves with the more general notion of a 'conic section'. In Euclidean geometry the notion of a circle emerges for the first time.

The analogy with the empirical sciences

If we write down a list of empirical sciences in the order:

1. Physics
2. Chemistry
3. Biology
4. Ecology/sociology

somewhat similar considerations apply. As we go down the line, the sciences become richer and richer in their empirical content and new concepts emerge at each level which simply do not appear in the preceding science. Furthermore, it seems to be arguable that each science is a special case of the one that precedes it. Only a limited class of all the possible interactions between molecules constitutes the subject matter of biology, and a limited class of the possible interactions between human beings constitutes the subject matter of sociology. It is because of this progressive restriction and limitation that we get the enormous conceptual enrichment of sciences like biology and ecology. Just as a concept like that of a 'circle' simply cannot be envisaged in the world of topology, so a contextually distinctive notion like the 'foreign exchange deficit' cannot be envisaged in the world of physics.

To pursue the analogy still further let me point out that, as we go down the list in the hierarchy of empirical sciences, every statement which is true in physics is true also in chemistry and biology and ecology and sociology. Likewise any statement that is true in biology and 'belongs' to biology is true also in sociology. Thus a characteristically physical proposition like $E = mc^2$ is true also in all the sciences below it in the list. More usually, however, a physical or chemical statement such as 'the atomic weight of potassium is 39' is simply not interesting in a subject like sociology, and does not bear at all upon its distinctive problems. Many rather simple-minded scientists believe that the special statements they are accustomed to in their own special branches of science *ought* to be regarded as very interesting and important by people practising in the lower sciences of the hierarchy like biology or sociology.

Conclusions

The entire notion of reducibility arouses a great deal of resentment among people who feel that their proprietary rights in a given science are being usurped, but if the parallel as outlined above were to be accepted, I think it would purge the idea of 'reducibility' of the connotation of diminishment or depreciation.

Suppose we accept John Stuart Mill's pronouncement that

> the laws of the phenomena of society are and can be, nothing but the laws of the actions and passions of human beings gathered together in the social state. . . . Human beings in society have no properties but those which are derived from and may be resolved into, the laws of the nature of individual men.

Even so, the 'reduction' of sociology to biology would have to take the form of asking what special qualifications and what restrictive clauses must we envisage in the interactions between human beings if the interactions between them are to constitute such and such a sociological phenomenon? There is reduction here but no diminishment. Biology is not 'just' physics and chemistry, but a very limited, very special and profoundly interesting part of them. So with ecology and sociology.

The seriousness of the sense of diminishment in analytic reduction is shown by Keats's well-known denunciation of Newton for destroying all the poetry of the rainbow by reducing it to the prismatic colours. This seems rather comic to us now, but I must say that denunciations of the use of analytic reduction in biology seem to me to be equally comic, for experience shows that analytic reduction provides us with an immensely powerful methodological weapon which enables us not merely to interpret the world but also, if need be, to change it.

Appendix: bibliography and additional notes

For Felix Klein's conception of geometry as embodied in the Erlanger Programme of 1872, see F. Klein, *Elementary Mathematics from an Advanced Standpoint* (*Geometry*), Macmillan, London (1939). The best account known to me, however, is in L. Godeaux, *Les Géométries*, Armand Colin, Paris (1937). See also J. L. Coolidge, *A History of Geometrical Methods*, Oxford (1940). The argument outlined in this paper was first adumbrated in P. B. Medawar, *Induction and Intuition in Scientific Thought*, London and Philadelphia (1969).

Groups

A *group* of operations is a set of operations S_1, S_2, S_3, S_4, etc., having the following properties which unite them into a sort of family circle: (1) the product of any two operations of the group is itself an operation of the group (the product of two operations is the operation made up of performing the two operations successively); (2) each operation has an inverse, and the group contains the 'Identity' operation which is necessarily the product of any operation with its inverse, and which restores the *status quo*; (3) the operations of a group obey the associative law but are not necessarily commutative, of course—that is, the operation $S_1 \times S_2$ may well be different from the operation $S_2 \times S_1$.

The anecdote about Keats is recounted in Benjamin Robert Haydon's autobiography, a book to which my attention was drawn by Dr J. R. Philip of A.N.U. Canberra.

6. Changing Strategies: A Comparison of Reductionist Attitudes in Biological and Medical Research in the Nineteenth and Twentieth Centuries

JUNE GOODFIELD

I am overpowered by a feeling of *déjà vu* verging at times on the very edge of intellectual impotence. 'Reductionism'; 'antireductionism'; 'beyond reductionism'; 'holism'. We have seen these words and heard the accompanying arguments so many times before. The issue is a very old one, recurring in various forms with unfailing regularity throughout biological history, and the feeling of impotence arises because, after all this time, the issue never seems to get any clearer. The arguments remain fuzzy round the edges while the actual progress of biological and medical research continues regardless of cries of 'nothing buttery' as opposed to 'greater than the sum of the individual parts'. David Newth's (1969) question 'Need a scientist's philosophical linen be as clean as his laboratory glassware?' is apparently well taken. The arguments for reductionism and antireductionism seem irrelevant to what is actually done in the laboratory, mere echoes from the sidelines whose impact and influence are effectively nil.

So what relevance, or point, can there be to our series of discussions at this conference if, as one is by now tempted to say, there seems little left which would be either true or new? Perhaps one way round frustration is to put the issue under a microscope and examine in some detail not only the antireductionists' utterances but their actual work too, trying to see what are the theoretical and experimental consequences—if any—of their standpoint, tackling the issue in concrete terms rather than in overall abstractions. I shall, therefore, be referring to Chaptal, Bichat, Magendie, Liebig and

Bernard in the nineteenth century rather than the *Natur*-philosophers, and in the twentieth to Loeb, Weiss, Spemann, Crick, Watson and Lettvin, rather than Teilhard de Chardin, Bertalanffy, Woodger and Koestler.

In the first part of this paper I shall remind us of the early nineteenth-century situation, and most particularly at the point of puzzle and paradox, when a man like Bernard tries, at one and the same time, both to preserve the methodological attitudes which formed the basis of his work and yet do justice to the biological phenomena as he found them. How did Bernard manage to hold his conviction that physicochemical laws did apply in living organisms, which must be studied by physicochemical methods, yet at the same time not only to recognise the uniqueness of organisms but also to specify this uniqueness? Secondly, I shall examine the experiments and writings of Paul Weiss and Jerome Lettvin, by comparison with those of practising reductionists of our century, in an attempt to focus upon certain questions. I make no apology for confining my attention to physiology rather than to evolutionary theory. It is the area that I have studied most closely, and since its laboratory methods are intimately linked with those of physics and chemistry, a comparison of reductionist and antireductionist attitudes amongst physiologists should be correspondingly more instructive.

The questions that I want us to bear in mind as an agenda for subsequent discussion arise in this way. Philosophers of science have long tended to assume that the preconceptions a man holds about the nature of the phenomena with which he is dealing affect his work in a variety of ways. To a certain extent, we believe, it affects what a scientist is prepared to accept as a genuine problem in the first place. This in turn affects the questions he then goes on to ask, which in turn affect the form of experiment he designs. His subsequent concept and theory formation—the general synthesis he makes of his data—are themselves affected by the attitudes with which he began his work, and though this has been most extensively documented within the physical sciences, we might expect the same general pattern to hold true within the biological sciences. So looking at the antireductionists of two centuries, one would want to know:

(1) Given their antireductionist attitude, how far does this affect the selection of the problems they choose to study? That is, do they focus quite deliberately on certain phenomena as posing genuine problems which a twentieth-century reductionist either feels safe in ignoring or never does recognise in the first place?

(2) Do these people—or could these people—ever agree, even in principle, with a twentieth-century reductionist as to what constitutes a genuinely biological problem?

(3) To what extent do antireductionists continue to work quite deliberately within certain conceptual frameworks, and thus at the outset render

down their problems to forms which differ markedly from a reductionist's problems?

(4) To what extent, if at all, do these people employ different laboratory procedures towards the solution of their problems?

(5) Given the same piece of empirical data, do their explanations tend to be different, and if so, where does this difference lie?

(6) Will, perhaps, the difference between antireductionists and reductionists in the twentieth century turn out only to be the following? Though they may agree on what constitutes a problem in the first place, and though there may be no obvious difference in laboratory procedure and methodology, they do not agree on what constitutes the *resolution* of a particular problem, and—the corollary of this—they do not agree on what does constitute, or should constitute, a biological explanation.

In the very beginning—in the era when a scientific physiology was just beginning to emerge—the issue took on a deceptive clarity, and the questions acquired a precision and a simplicity which in the long run were to prove misleading. It is a paradoxical fact in the history of biology, that, at the very moment when—through the work of people like Black and Lavoisier—it seemed one could draw an absolute distinction between inorganic and living matter with the problems of rationality and the mind outstanding, Crawford and Lavoisier were also demonstrating that chemical processes were going on in living organisms, which were at least analogous with, if not absolutely identical to, those of the inorganic world. On the one hand there were inert substances. These could be weighed, measured and contained, and these chemical substances, if left exposed to the air, were often oxidised away. The laws governing their behaviour and properties were those of physics and chemistry. On the other hand, one had living organisms, which breathed, respired and reproduced—and which apparently did not conform to the laws of physics and chemistry. For until death, they withstood the oxidising and destructive effects of the atmosphere and, in the case of warm-blooded animals, they maintained their body temperature at a constant level irrespective of the fluctuations of the environment. (The whole of this particular issue has been extensively documented elsewhere (see Goodfield, 1960, 1969; Mendelsohn, 1964).) But by the turn of the nineteenth century, enough work had been done by Hales, Black and Lavoisier, Priestley and Spallanzani, to demonstrate conclusively that *chemical* reactions, involving gaseous exchanges, went on in living organisms, and certainly so far as respiratory processes were concerned the reactions could be quantatively assessed. So several questions immediately presented themselves. Given the obviously different properties of organisms and inorganic substances, what was the significance and importance of the physicochemical reactions going on within living things? Were these similar or only analogous in the two realms? What was the value—if any—of quantitative methods when applied to living organisms, and what was the relevance of the results obtained? What in fact did the physicochemical experiments tell us? Did organisms, or did they not, conform to the laws of physics and chemistry? Should they, or should they not, be studied with physicochemical techniques? And if the answers to the last two questions were to be 'yes', then how could one *account for those observable differences of properties which were the starting point for the very recognition of a living organism in the first place*? This last question is the essential one: whether one is a reductionist or an antireductionist, in the nineteenth century or twentieth century, this is the question which has to be answered.

It is not at all surprising that, in the early years of the nineteenth century, various forms of answer were presented, and it took some fifty years before a deterministic methodology was accepted. In fact one finds every possible variation of opinion presented. At the extreme, there are men like the chemist Chaptal, and the physiologist Bichat, who take an uncompromising stand on the methodological issue. (Both men had connections with the University of Montpellier.) Bichat's (1805, p. 81) views are classic:

> One calculates the return of a comet, the speed of a projectile; but to calculate with Borelli the strength of a muscle, with Keill the speed of blood, with Lavoisier the quantity of air entering the lung, is to build on shifting sand an edifice solid itself but which soon falls for lack of an assured base. This instability of the vital forces marks all vital phenomena with an irregularity which distinguishes them from physical phenomena remarkable for their uniformity. It is easy to see that the science of organised bodies should be treated in a manner quite different from those which have unorganised bodies for object.

Elsewhere (Bichat, 1801) he said:

> The laws of natural philosophy are constant and invariable; they admit neither of diminution nor increase . . . on the contrary the vital properties are at every instant undergoing some change in degree and kind; they are scarcely ever the same. . . .
>
> . . . to apply the science of natural philosophy to physiology would be to explain the phenomena of living bodies by the laws of an inert body. Here . . . is a false principle.

Chaptal (1791, p. 279) expresses it in another way:

> In order to direct with propriety the applications of chemistry to the human body, proper view must be adopted relating to the animal economy itself, together with accurate notions of chemistry itself. The results of the laboratory must be regarded as subordinate to physiological observations. It is in consequence of a departure from these principles that the human body has been considered as a lifeless and passive substance. . . . In the mineral kingdom everything is subject to the invariable laws of affinities. No internal principle modifies the action of natural agents. . . .
>
> In animals functions are much less dependent on external causes and nature has concealed the principal organs in the internal parts of their bodies as if to withdraw them from the influence of foreign powers. But the more the functions of the individual are connected with its organisation, the less is the empire of chemistry over them, and it becomes us to be cautious in the application of this science to all phenomena which depend essentially on the principles of life.

In both cases their precepts limit, even prohibit, the use of physicochemical methods and techniques in the study of living organisms. The justification for this prohibition is based upon two different arguments. In Bichat's case it is his quite understandable preoccupation with the plasticity of behaviour that as a doctor he has observed in the recuperating patient, and he is quite as justified in emphasising the importance of this as a phenomenon as Lavoisier and Crawford were in emphasising that, in the respiration of an animal, a measurable quantity of carbon is oxidised by a measurable volume of oxygen to release a measurable quantity of carbon dioxide and heat.

Chaptal, the chemist, was undoubtedly more influenced by sheer *chemical* facts. The striking thing for him was the way in which the organisms held at bay the destructive chemical effects of the outside atmosphere. Inside the animal the laws of vitality apply, whatever form they may be found to take. Outside the animal is the realm of physics and chemistry, of destruction and absolute determinism. And the boundary layer between the two worlds is the critical layer.

In the years between Lavoisier and Laplace's classic paper (1780) and Claude Bernard's classic book *An Introduction to the Study of Experimental Medicine*, published in 1865, physicochemical studies were extended, being given an impetus in England by the foundation of a Society for the Promotion of Animal Chemistry (founded 1802 or 1803), and while in these earlier years there were many people who unhesitatingly used physicochemical techniques to study living organisms, the era of the total reductionist had yet to come. It was, as I have indicated elsewhere (Goodfield, 1965), a time of methodological heart searching; a time of paradox and uncertainty for those physiologists who were really concerned with what form a physiological explanation should take. Extremists like Magendie were initially few. Having 'left all hypotheses outside the door of his laboratory together with his overcoat', Magendie proceeded to tackle physiological problems with physicochemical methods, announced the empirical results, and left it at that. His physician—or doctor—counterpart like Bichat does no experiments at all. By far the majority of physiologists, however, admitted the need for the experiments, admitted the physicochemical nature of much that was going on in living organisms, but worried a great deal about the exact relevance of their results and the nature of the explanation they should be formulating. One finds the whole gamut of possible types of explanation, from the anticipation of our twentieth-century views—that life phenomena can be interpreted only in terms of a complex of interrelated processes—right through to the very simple, indeed 'dead-end', explanations of vital principles and vital forces. Once again, it is by no means the chemists who are the reductionists, by no means the biologists who are the antireductionists. The balanced views of Sir William Lawrence (1816), the surgeon, are a remarkable anticipation of the equally balanced views of Claude Bernard (1865). Life cannot be constrained into one term, within one definition, formula or phrase, such as the proponents of the vital principle and life-force might try to do. For what we are faced with in the phenomena of life, says Lawrence, is not immaterial abstractions, but a set of mutual interactions; the very structures and processes within organisms are in constant flux, by virtue of the exchanges between the organism and the external world. 'Organisation is the instrument. Vital properties are the acting power; function, the mode of action; *and life is the result*' (Lawrence, 1816).

By contrast, take Liebig's (1842) views. Liebig was a chemist who, single-handedly, did almost more than anyone to analyse and give a definitive chemical account of many living processes, such as digestion and respiration. And his was not just a chemical analysis; it was a quantitative chemical analysis at that. Liebig added an extra sophistication to the problem posed by Bichat, because he did try to find some physiological reason for the inevitably varied nature of the quantitative results obtained in his day. Yet Liebig was not a reductionist. To be more exact: if the word implies a belief in the deterministic nature of individual physicochemical processes going on in living organisms, then he was one. If the word implies a conviction that only by the application of physicochemical techniques can we really come to understand the nature of the organisms, then again he was a reductionist. But faced with the phenomena as he saw them, with the properties as he had analysed them, Liebig felt no alternative but to opt for the introduction of a special type of force—along with electrical, magnetic and chemical forces—to explain the capacity of the organism to hold destructive chemical influences of the environment at bay. (Echoes of Chaptal, once more.) Here we have a situation in which the term 'vital force' is introduced *not as an* a priori *statement of faith about the nature of organisms, which effectively excludes them from experimental study, but as an explanation, after experiment, for the results he has obtained.*

But successful methodology is one thing; dealing with one's results quite another. By the mid-century, as Everett Mendelsohn (1963) points out, the German biophysicists Carl Ludwig, Herman von Helmholtz, Ernst von Bruke and Emile Dubois-Reymond could give a frank assessment of the situation in these terms:

> We four imagined that we should constitute physiology on a chemico-physical foundation, and give it equal scientific rank with physics . . . but the task turned out to be much more difficult than we had anticipated.

One suspects that the difficulty lay not so much in establishing the physicochemical methodology for physiology, nor in giving, as we increasingly can, a physicochemical account of the detailed individual processes within living organisms or individual cells, but in specifying *why* the vital properties we observe are, in fact, a *consequence* of the structures and processes we have analysed.

Undoubtedly the first successful attempt to tackle this problem came in Claude Bernard's *Introduction to the Study of Experimental Medicine*, published in 1865. From the very outset it is clear for Bernard that physiological explanation lies in seeing 'physics and chemistry worked out in the special field of life' (Bernard, 1865). And without ever allowing himself to lapse into tautology, what Bernard does for us is to try to cash in that phrase 'special field of life'. This means determining the conditions under which these important physicochemical processes are taking place, which themselves result in the vital properties as we see them.

. . . I should agree with the vitalists if they would simply recognise that living beings exhibit phenomena peculiar to themselves and unknown in inorganic nature. I admit, indeed, that manifestations of life cannot be wholly elucidated by the physico-chemical phenomena known in inorganic nature. . . . I will simply say that if vital phenomena differ from those or inorganic bodies in complexity and appearance, this difference obtains only by virtue of determined or determinable conditions proper to themselves. So if the science of life must differ from all others in explanation and in special laws, they are not set apart by scientific method (Bernard, 1865).

To make any progress in our understanding of living things, physiology must be scientific, but to be scientific it must be deterministic. What Bernard sets out to demonstrate is, as he says in the *Cahier Rouge* (1850–1860, p. 97), that 'in animals which have an apparent independence, this independence is illusory; that the law is always the same, that is to say that the organic properties have not changed; only the conditions of their actions have altered'. The apparent resistance of the organism to the destructive forces of oxidisation that had so impressed Bichat and Liebig must find their explanation in causes other than a unique vital force, quite distinct from all others and in essential nature unanalysable according to the methods of inorganic chemistry. Because otherwise:

We should either have to recognise that determinism is impossible in the phenomena of life, and this would be simply denying biological science; or else we should have to acknowledge that vital force must be studied by special methods and that the science of life must rest on different principles from the science of inorganic bodies. . . . I propose, therefore, to prove that the science of vital phenomena must have the same foundations as the science of the phenomena of inorganic bodies, and that there is no difference in this respect between the principles of biological science and those of physico-chemical science. Indeed, as we have already said, the goal which the experimental method sets itself is everywhere the same; it consists in connecting natural phenomena with their necessary conditions or their immediate causes (Bernard, 1865, p. 60).

It is only that, as Bernard says, 'the physical and chemical phenomena of the organism have in the living being conditions which they do not have elsewhere' (Bernard, 1850–1860, p. 112). And it is these conditions which we must study; these conditions which together add up to the animal's internal environment, that self-sustaining liquid, with constant pH and temperature, whose composition and whose constancy enables the higher functions of the animal to be maintained. Without this constantly maintained medium around the cells, the vital processes of the animal would indeed be influenced and affected by variations of the external environment, as they are in the simplest forms of life. For then the appearance of independence is not found, and without the correct environment conditions vital properties cannot be manifested. And at this level the influences which produce or accelerate or alter these vital phenomena are *exactly the same* as those that produce, accelerate or alter physicochemical phenomena in the organic world. And Bernard (1865, p. 62) concludes:

. . . so that instead of following the example of the vitalists in seeing a kind of opposition or incompatibility between the conditions of vital manifestations, we must note, on the contrary, in these two orders of phenomena a complete parallism and a direct and necessary relation. Only in warm-blooded animals do the conditions of the organism and those of the surrounding environment seem to be independent; . . . an inner force seems to join to combat with these influences and in spite of them to maintain the vital forces in equilibrium.

But fundamentally it is nothing of the sort; and the semblance depends simply on the fact that, by the more completely protective mechanism which we shall have occasion to study, the warm-blooded animals' internal environment comes less easily into equilibrium with the external cosmic environment. External influences, therefore, bring about changes and disturbances in the intensity of organic functions only in so far as the protective system of the organism's internal environment becomes insufficient in given conditions.

Bernard, of course, had a profound influence. He left to his predecessors not only a definitive statement and justification of his methodology, not only an analysis of the internal environment and a detailed account of the regulatory mechanisms which helped to maintain it, but also a concept of regulation that was to find its epitome in the work of Henderson (1913), Cannon (1929, 1932), Wiener (1948) and the later school of cyberneticists. And in the twentieth century, the very facts of homeostasis and self-regulatory control were used as one starting point for the whole systems theory of biological explanation—a whole new antireductionist approach. And if one looks at Bernard and at the practising antireductionists of the twentieth century, we will find that most of them are speaking to us in the same kind of terms. What Bernard did was to force us to focus on rather different kinds of questions: to look not so much at structures as at regulatory mechanisms, to study not only the physicochemical nature of the physiological process but the special conditions under which it exists and because of which it has a quite special character.

After Bernard, things were never quite the same again. No one could really dispute the value and importance of physicochemical methods for the study of organisms, or the deterministic nature of the laws, even though many people would continue to question whether biological explanation could be completely reduced to, or should be subordinate to, physicochemical explanation. It was not only biologists who were sceptical; once again, we find physicists and chemists brought up sharply against the immense complexities of the living cells. Maxwell, particularly, deplored the way in which biologists cut and pared at the facts, he said, in order to bring the phenomena of life into the range of the dynamics of the nineteenth century. He calculated that the smallest units of life had not more than about 1 000 000 molecules, and this number he considered totally inadequate to embody, 'even in a code', the character and complexities of the full-grown animal. How could the same number of molecules contain in capsule form all the varieties of structure? The complexity of organised form would baffle scientists for a long time.

But if we look at Bernard—that self-styled 'physical-vitalist'—and his work in relation to the questions I posed at the beginning of this paper, we can see that in this case he did indeed focus our attention on problems which his contemporaries never saw as problems or even, perhaps, never saw as phenomena. It was not only that he recognised the structure and the functions of the living cellular tissues as being the real important area for detailed study—Bichat did so too;[1] but it was Bernard's emphasis on the process and its relationship, on the mechanisms for control and integration as the essentially biological phenomena which was fundamental. This integration, this control and apparent direction, Bernard could handle at the level of cellular physiology, but the embryological field presented him with phenomena which he could not see his way past. Later in the twentieth century, this was to prove one of the taking-off points for a whole school of antireductionists.

The twentieth-century situation

One feature among others distinguishes the situation in the early twentieth century from its counterpart in the nineteenth: namely, there is no debate about the validity and use of physicochemical techniques in the laboratory. The methodological battle is won and an irritated reductionist could well wonder why the contestants will not quit. But this victory only served to sharpen up the problem of explanation, the questions about what really constitute genuine biological problems and their resolution. Looking over the history of the two centuries, one is continually intrigued by the persistence of the antireductionists' views. And amongst the antireductionists of the twentieth century we can find one example of a biologist whose ideas stand out as brightly coloured threads in the fabric of theoretical and philosophical biology. For as a first-class experimentalist, whose laboratory 'hardware' both in terms of empirical results and conceptual innovations is widely recognised and appreciated, Paul Weiss stands apart from the majority of antireductionists in the organismic school of biology. For above all he *is* an experimentalist, whose philosophical statements have been backed up by a wealth of empirical detail, and when he is referred to by organismic biologists such as Bertalanffy, Woodger and Agar it is more often than not his experimental work on motor units, developmental biology, tissue regeneration, that they point to. When Weiss himself, in his classic polemic against Jacques Loeb, took on the most formidable reductionist of the twentieth century, he chose to meet his adversary in the laboratory rather than the armchair, presenting *experimental* evidence to counter the naive theories of a man whose ideas were based on the simple cause-and-effect machine principles of

1 Bernard often acknowledged the importance and significance of Bichat's work on tissues as the source of the vital properties and metabolic processes.

ITHACA COLLEGE LIBRARY

classical physics. Weiss's first paper—a directed polemic against Jacques Loeb—was not a philosophical treatise but a formal statement, in precise logical and polemical language; his opinions and arguments—unlike those of most organismic biologists who rarely went into a laboratory and whose ideas never really filtered down to the experimental level—could not only be grasped by analytical scientific minds—since the arguments were written in their terms—but could be tested and verified by any working scientist.

The Mechanistic Conception of Life is Jacques Loeb's most famous book. As a religious tract for biological methodology it is quoted over and over again, and the sweep of ambition revealed in this book is breathtaking. As the publisher's blurb on the most recent edition tells us, Loeb took the title piece for the book from

> . . . an address in which he reduced life to a physicochemical phenomenon, free will to an illusion generated by tropistic causes, and religious faith to an absurdity. He proclaimed the total validity of mechanistic principles and derived from them a system of human ethics based on instincts whose unobstructed expression would rejuvenate world society.

Paul Weiss was content to tackle him on the first of these ambitions.

Jacques Loeb was not the first biologist in history to turn from philosophy to biology (Peter Medawar is another distinguished, and rather more happy, example), nor will he be the last. Hans Driesch went in the opposite direction. But both Loeb and Driesch ended up in positions so extreme that one turns to Medawar's and Weiss's balanced viewpoints with relief. Freedom of the will was Jacques Loeb's motivating problem; failing to find any satisfactory answer from philosophers to his questions about free will, he turned to biological experimentation to answer it. By the time he addressed the first international congress of the Monists, in September 1911, his empirical findings on artificial fertilisation, on development and on animal tropisms put him in a position to assert with a wealth of empirical backing that the nature of the will was rapidly coming within the scope of physicochemical explanation. Man's wishes and hopes, disappointments and sufferings, found their basis in instincts 'comparable to the life instinct of the heliotropic animals'; and for many of these instincts the chemical base was so clearly understood as to make it 'only a question of time' before they were to be fully accounted for on mechanistic lines.

Loeb began with the familiar phenomena of plants orientating towards light and gravity. Having then observed similar behaviour in sessile animals, also during the growth process, he went on to study directional growth reactions, in animals which could move freely; movements which were influenced and seemingly directed, by external factors. As Weiss (1959) pointed out in his paper, the total construction of Loeb's theories rests on the tacit assumption that similarity of movements in all these cases *must* be due to an identity of underlying mechanisms.

For a student, to have the self-confidence to tackle in his PhD thesis a man of Loeb's stature, and on his most famous topic, is akin to a graduate student demolishing the central dogma of Crick and Watson and demonstrating empirically that they had taken a wildly simplistic view of the gene. When Weiss did his first experiments, he was interested in the behaviour of a certain species of butterfly, *Vanessa*, and he noticed that the animals would all go to sleep in a head-down position. He observed them in light, and the animals crawled around but would always turn their heads away from the light source. Now an interpretation of these observations in the strict manner of Loeb would be given in terms of the animals being 'negatively phototropic'. But, as Weiss recalls, his mind rebelled against this, for he noticed that none of the animals got into this position by any fixed machinery, mechanisms or route, and this he proceeded to demonstrate. It was merely that the terminal result was always the same. According to Loeb the animal was *forced* to orientate itself towards the light; in other cases, the animal was *forced* to cling to a surface. And in the very first page of his thesis, Weiss (1959, p. 2) lays clearly before us a view of reductionism and antireductionism, from which he was never to shift.

No matter whether or not one will succeed in an ultimate reduction of biological concepts to chemical or physical ones, the attempt to do so will always have to start out with the basic elements of biology and will thus be a reduction *in toto,* destined not to replace the laws of the more complex field, but to coexist with them.

To put it another way, Weiss, unlike Loeb, was prepared to be critical about the *range* of certain laws. Whereas Loeb believed that the lowest order of magnitude to be found in a phenomenon is the fundamental one for its explanation—it was almost *a priori*, self-evident to Loeb—this was by no means so obvious to Weiss, who believed that the laws might be valid only within the range of given magnitudes.

Weiss pinpointed the difficulty quite precisely. In fact, by making suitable substitutions one can make Weiss's formulation of the difficulty sound like Claude Bernard.

As Weiss saw it, the problem once again arises because the two separate sets of facts had to be kept in mind.

First, one cannot escape the conclusions that the organism (in its development and function) reacts as a unitary whole.

Secondly, one cannot ignore the fact that in spite of this recognised unitary wholeness, a great number of processes in the organism can be explained as a fixed linear sequence of component processes, *i.e.*, through mechanisms. These two facts have to be reconciled now (Weiss, 1959, p. 2).

Weiss finds his way out of this paradox in the same way that Claude Bernard had earlier found it, by an extension of the idea of a homeostatic system. The organism had 'the faculty of responding to an exogenous alteration of the equilibrium state with reactions from within, itself of such direction that it tends to attain a new unequivocal systemic state of equilibrium'.

And by a series of extremely exact and detailed experiments which tested the butterfly's—both adults and the newly hatched—behaviour to gravity, to light, to light and gravity combined, Weiss was able to present a complex systemic account of this animal's behaviour. Once again the individual reactions and processes retained their importance, but the constancy of the system was preserved by compensatory internal reactions arising in response to external environmental changes. As Claude Bernard had earlier seen the external phenomenon of warm-blooded animals in terms of a variety of physicochemical processes, linked to a system of feedback and control, so arranged and integrated to restore any inbalance imposed by fluctuations of the external environment, so now Weiss was to see beyond the apparent simplicity of the tropic movements described by Jacques Loeb. These were a series of individual, cause-and-effect responses, but so coordinated and integrated that they tend to restore the system to its original condition after external environmental disturbances. In the case of animal heat, it was the simplest observation—that the warm-blooded animal disobeyed the laws of physics and chemistry, defying Newton's law of cooling—that gave rise to a vitalistic interpretation of animal physiology, which could be countered only by a detailed analysis of the individual physicochemical processes acting in a coordinated manner. So now, again, the apparent simple capacity of an animal to respond to light and gravity stimuli, had led in its turn to an equally naive reductionist view: namely, that animal behaviour was nothing more than simply physicochemical responses. It is interesting to compare the two situations, for both the naively vitalistic view in the nineteenth century and the naively mechanistic view of the twentieth century are wrong. Both phenomena in question, heat maintenance and tropic movements, find their explanation in the same kind of terms; Weiss, in most of his papers, is echoing Claude Bernard over and over again but in relation to different phenomena. Physicochemical methods, physicochemical laws, physicochemical processes are one thing; what makes them uniquely biological is the special conditions under which they manifest themselves. *It is the nature of these special conditions that we must study*. Claude Bernard brings himself up sharply against the facts of development and, in the absence of any general theory or detailed analysis of developmental mechanics, is unable to move past the wall of incomprehension at that point but remains a 'physical vitalist'. Decades of research culminating in molecular biological theory enable Weiss and his colleagues to see developmental mechanics in the detail of which Claude Bernard was completely ignorant, but Weiss too, I suspect, remains a 'physical vitalist', and is so, I would want to argue, for reasons very similar to those of Bernard.

The problem of organisation

A century after the appearance of Theodore Schwann's *Microscopical Researches* (1838), at a symposium on cell theory, Paul Weiss (1940) focused on the biological mystery of organisation of living organisms. The problem which biologists had puzzled over, ever since Schwann first proposed the theory of the cell, was the problem of cell individuality and the development of a complex metazoan organism from a single primordial cell. 'At the end of development', wrote Weiss,

> ... we are confronted with a unitary organised system, called an organism, which, at the same time, is a collective of cells. At the beginning of development we find just one primordial cell—the egg. We call a system 'organized' when its multiple elements appear in typical diversity, typical spatial distribution and typical temporal order. The elements are subordinated to this order and their freedom is restricted by it; hence, the order is a supra-elemental property of the system. In the developed system, 'organism', the cells represent the elements; hence, organisation is a supra-cellular property. But the primordium of the organism—the egg—does not consist of cells. Now, there arises a dilemma.

Weiss observed that faced with this dilemma, biologists were forced to choose one of two courses of interpretation. 'Either the egg already possesses supra-cellular organisation of the same order as the later body—then it is not just another cell, but an uncellulated organism; or it is merely a cell like others—then it cannot (*sic*) be at the same level of organisation as the later body.' In the latter case, development would give rise to a higher order of organisation, and it is to this view that the cell theory committed itself. 'The individual cells', according to Schwann, 'so operate together in a manner unknown to us as to *produce* a harmonious whole' (italics mine). In Weiss's words, 'the organism would be synthesised by progressive integration of cells into higher units, tissues, organs and the body as a whole. Cells would form the organism.'

There were some, however, who vigorously challenged this view of progressive organisation and charged that the cell theory of development was inadequate. If organisation was to be accepted as arising *de novo* in every ontogeny, then some principle was thought necessary to mould order out of chaos, and vitalistic agents such as Semon's 'mneme' and Driesch's 'entelechy' were evoked. But faced with *this* alternative, others found it more comfortable to assume that the powers of organisation were inherent in the egg itself. The cells would not form the organism in this case, but the organism would simply break up into cells.

These two opposing views clearly represented a new version (applied to organisation) of the old antithesis: epigenesis *versus* preformation. ' "Epigenesis" of organisation was the claim of the "egg-equals-cell" theory, while preformation of organisation was the tenet of the "egg-equals-organism" doctrine.' The extremists on one end urged an 'elementarian' view, professing that the activity of individual cells was responsible for all that happens in an organism as a whole, while the advocates of the other extreme fought for a 'totalitarian' concept of development, arguing that the cells were subjected from the beginning to the unchallenged control of the organism as a whole. Much of the fight between 'elementarians' and 'holists' took place 'on philosophical grounds rather than on the factual grounds of observation and evidence' (Weiss, 1940, p. 34). But, as Weiss points out, experimental embryology managed, on the whole, to steer clear of both extremes.

The revelation of the multiplicity of developmental processes and mechanisms has been a sad disappointment; for it has removed all hope of a general, comprehensive and universal formula of development. . . . We no longer ask: 'Is development epigenetic or preformed?' but focus on a single contributory phrase, asking: 'How much of it is due to epigenetic and preformed conditions?', only to find that the answer varies with the object. It is this abandonment of the unitarian claim which has rendered us immune to both the strictly elementarian and the strictly totalitarian view, and which has steadied our picture of the relative role of cell and organism in development.

At the end of this article on cell theory and cell development, Weiss concluded that:

Practically every step in development reveals the cell in a double light; partly as an active worker and partly as a passive subordinate to powers which lie entirely outside of its own competence and control, *i.e.*, supra-cellular powers. . . .

And he implies that biologists and embryologists cannot help paying particular attention to the interaction and interdependence of organismic elements, rather than concentrating solely on isolated elements without regard for the whole organism, or *vice versa.*

Two decades later, however, he would feel compelled to reemphasise his argument for considering the cell as a unit.

Lest our necessary and highly successful preoccupation with cell fragments and fractions obscure the fact that the cell is not just an inert playground for a few almighty masterminding molecules, but is a *system*, a hierarchically *ordered* system, of mutually interdependent species of molecules, molecular groupings, and supra-molecular entities; and that life, through cell life, depends on the *order* of their intractions; it may be well to restate at the outset the case for the cell as a *unit* (Weiss, 1962).

In terms of its material composition a cell *is* no more than the sum of its parts. But a living cell is different from its homogenate because in the degrading process from cell to nonliving constituents the cell is deprived of the interrelations that had existed between the formerly united parts, and the hierarchical organisation of the living state is destroyed. Operationally the cell is an open system which can exist only in its entirety. During the progressive decomposition of a cell, 'information content' is irretrievably lost on the way down, and the possibility of resynthesising the system from its decomposed elements seems unlikely. This suggests, he argues, that when we mentally retrace such synthetic courses, we rely on verbal crutches—for instance, 'reconstitution', 'reintegration'—which may only be abstractions without concrete counterparts in our experience with living things (Weiss, 1963).

Weiss considers the resynthesis of a unitary system (such as a cell) from a disordered pile of its constituent elements to be the true test of a consistent theory of reductionism. Why, we may ask, does he not consider *synthesis de novo*? But Weiss believes that despite recent success in synthesising higher-order systems from lower-order elements, one cannot extrapolate this trend and expect to synthesise a living cell, because ' "synthetic" success at one level does not automatically spell success for any other level'. There is nothing in our experience to justify the expectation of a 'synthetic cell', says Weiss, and 'to predict whether or not cells will ever be synthesised from scrambled molecules is in that generality not only an idle but a logically unsound undertaking with more emotional and cultural overtones than scientific foundations'.

The crucial test for a consistent theory of reductionism also has a corollary for Weiss, in the test for the 'cell-as-a-unit' theory. 'Can such interlocking (interdependent) systems', asks Weiss,

> be taken apart and put together again *stepwise*, like a machine or jigsaw puzzle, by adding one piece at a time, or is the very existence of the system as a whole predicated on the *simultaneous* presence and operation of all components? In the former instance, an eventual 'synthesis' of artificial cells could be envisaged; in the latter case, it could not (Weiss, 1963).

For Weiss the answer to this question is merely an empirical problem which can be solved by surveying the evidence. The evidence in this case, and Weiss's interpretation of it, is reviewed in two earlier articles: 'The compounding of complex macromolecular and cellular units into tissue fabrics' (1956) and 'From cell to molecule' (1962). Briefly, he concludes that all so-called syntheses of higher-order complexes from simpler elements, such as the integration of collagen into a basal lamina, require an intimate connection with living cells. That is, 'elements endowed for such ordered group performance have always been prefitted for it by properties *previously imparted to them as members of just such an organised group unit*'.

This brings us to a fundamental tenet of Weiss's systems doctrine: *omnis organisatio ex organisatione*, which he first proposed in 1940 as a supplement to Virchow's *omnis cellula e cellula*. If the latter denies spontaneous generation of living matter, Weiss declared, the former denies spontaneous generation of organisation. This categorical statement poses a serious problem: namely, how does Weiss account for the origin of life? Obviously 'all organisation from organisation' excludes the possibility of life originating from non-living matter. Weiss scrupulously avoids this topic in his writings.

After deriving his argument from the record of evidence, Weiss finds that his empirical studies really have not yielded any useful information about the actual mechanisms of organisation of cellular structures and coordination of cellular processes. Accordingly, he leaves the firm ground of observation for conjecture.

> Now, it is intriguing to speculate in the interest of consistency that perhaps the structured portion of the cell might itself also subserve this function of coordinating the unstructured fraction of the cell content by establishing and maintaining differential topographic distributions within the otherwise unsegregated molecular populations of the intracellular pools (Weiss, 1963).

In another instance, he suggests that a 'structured' environment may be responsible for imparting organised structure to subcellular components

(Weiss, 1968). But in any case he assumes that cellular (organismic) activity is ultimately directed by a supramolecular (supracellular) 'coordinating principle'. Organised components are the operational expression of a 'field'. Thus formed lamellae, for instance, 'become the visible traces of the configuration of an invisible field of interactions, figuratively comparable to the iron filings which trace lines of force of a magnetic field'.

Could this invisible coordinating principle be of the same categorical order as individual physical reactions themselves—only one more of them? Weiss's argument precludes the possibility of compounding interaction systems. He likens the problem of the synthesis of a cell from its constituent elements to the 'many-body' problem in physics, and echoes his PhD thesis as he does so. The symbols used in the mental reconstruction of a higher-order system from 'symbolic' elements and operations do not have true counterparts in the physical reconstruction of such a system from separate components.

> By reasons of logic and scientific honesty, the problem of *coordinated unity* of the cell must therefore be acknowledged as a real one. It cannot be hedged by assuming that starting from gene reduplication and the first steps of protein synthesis, all further developments would run off collaterally and uninterrelated; for this would imply that once having been mapped out micro-precisely down to the most minute details, they would then actually be capable of pursuing with absolute rigidity their individual courses so predesigned as to yield blindly, but unfailingly, a viable product—a modern version of Leibnitz' 'pre-stabilised harmony'. The unpredictability of the vicissitudes of the environments in which those courses materialise rules out any such concept of absolute predetermination as utterly unrealistic and absurd . . . (Weiss, 1963).

The key ideas here are predictability and variability. The legacy of classical physics to biology was the notion of predictability based on linear causal trains—a notion which found full expression in Jacques Loeb's *Mechanistic Conception of Life*. The wheel comes full circle, for in his polemic against Loeb, Weiss categorically denounced the 'mechanistic' approach to biology and demonstrated the fallacy of considering the reaction of the whole simply as a product of a string of component reactions. Over the years, Weiss has continued to refer to predictability in the classical physical sense of that word, employing phrases like 'microprecise automatic assembly' and 'micro-precisely programmed linear chain reactions' (Weiss, 1969). For Weiss, the observed variability of living systems thus precludes a deterministic description of those systems.[2]

But though organisation poses a problem, and most especially with regard to the organisation of life, as Weiss himself says in the epilogue to his article 'From cell to molecule'

[2] Any current writings on molecular biology reveal that the problem of order and organisation in living things is now one central problem of biological research. Indeed, the language of molecular biologists mimics that of organismic biologists, *e.g.*, 'integrated system of macromolecular structures and functions', 'open system', 'hierarchy of organisation', 'the only source of biological order is biological order' (see Andre Lwoff, *Biological Order*, 3–6, 87–97; Francis Crick, *Of Molecules and Men*, 10–16; John Kendrew, *The Thread of Life*, 13–16).

There are now available practical and constructive approaches to the gradual replacement of symbolic reference to 'organisation' by a true insight into the dynamics involved. Our knowledge of the dynamics is rudimentary and spotty, but it is consistent for us to realise that every mechanism in a living matter system employs a combination of dynamic principles, rather than must a single kind.

How often do we hear in the discussions of biological problems pronouncement to the effect that this or that event is 'biochemical'. Such statements are platitudinous unless they are accompanied by indications of how the particular reaction is conditioned by the physical setting in which it occurred, and how its effects, in turn, modify the physical settings for subsequent reactions. In this broader perspective, ordering process and ordered structure become a single continuum, determining and limiting each other in endless sequences of activities, so that as a given chemical event may control (that is, condition) the appearance of a particular physical array or 'structure', the latter will then go on to control (condition) the next chemical transaction, which in further consequence may again alter the prior structure, and so forth, almost *ad infinitum* (Weiss, 1962).

To sum up: when one looks at the detail of Weiss's experiments, what he actually does in the laboratory is, methodologically speaking, no different from what was actually done by Magendie or Bernard in the last century, and by Loeb, Crick and Watson in this. So that what men like Bernard or Weiss do for the subject is to force biologists to turn their attention to certain phenomena which otherwise might be ignored, and to see them as posing genuine problems. In the same way as nineteenth-century classical physicists were more concerned with the nature of atoms than with the things that *happen* to these atoms, and Cartesian physicists were more concerned with particles rather than with what they *did*, so too reductionist biologists have been more concerned with biochemical substances and biological structures, rather than *interactions* and the special conditions under which structures and processes operate. I suspect that Weiss, who came to biology from engineering, would want to say like many others who have come the same way, 'Only if I could build it *and make it function* would I understand it', rather than 'If I can build it, then I'll understand it'. Giving a detailed account of each individual process in the organism, or coming to understand the complex chemical substances that make it up, is by itself not enough; if something '*more*' is needed, then this 'more' is not a super-added principle or a vague word like 'organisation' but, firstly, a greater understanding of the two-way relationship of the cell and its environment, and the organism and its environment; and, secondly, a recognition that this relationship may actually *impose* upon the organism not only both a pattern of behaviour and response, but—because all organisms are a product of an evolutionary process—even a pattern of structure and form.

I would like to illustrate this final point with regard to some remarkable work on neurophysiology and perception. This comes from Massachusetts Institute of Technology, born out of Norbert Wiener and Warren McCulloch, bred through Pitts, Maturana and Jerome Lettvin. Looked at from the point of physicochemical methodology, again there is nothing to choose between these men and any other biophysicist or neurophysiologist. Their methods and procedures are impeccable; their experiments remarkable. Their paper

'What the frog's eye tells the frog's brain' (Lettvin *et al.*, 1959) is one of the classics in the field. From the point of view of the historian of biology, what is exciting is not so much the results they produced but how the formulation of a new kind of question led to those very results. In contrast to the naivety with which Jacques Loeb approached questions of animal behaviour, simple psychophysical parallelism as a starting point for their studies was just not on. Showing the frog's eye a beam of light, or giving it an electrical stimulus, or analysing the chemical changes which went on in the retinal cells, would seem to them to be a self-limiting type of procedure. Substituting instead the ethological question 'What is it important for a frog to see?' and then proceeding to start from that basis, showing the frog a series of bug-like objects, had a fantastic pay-off. For the first time we began to appreciate just how extensive is the perceptual sorting and sifting which goes on in the retina, and how this seems to be related in a functional relationship to the anatomy of the retinal cells.

Lettvin and Robert Gesteland were to try again with the frog's nose, asking this time how the frog smells as opposed to how it perceives. They found the olfactory tissue infinitely more difficult to handle than the eye, and that the responses of single nerve fibres were not simply related to the stimuli presented. What Lettvin and Gesteland (1965) then wanted to do was to have 'accepted the combinational responses as given', and then made inductions on the nature of the olfactory code. But in the end what happened was that though they developed good methods for looking at olfactory tissue and getting data, they had no idea how to handle the results. If visual perception revealed a form–function relationship in a sense organ which is related very closely both to the external environment on the one hand and the central nervous system on the other, the olfactory system presented this even more strikingly. Smells are, of course, infinitely more difficult to handle than light and light spectra with their well-defined wavelengths. And in their paper 'Speculations on smell', that refreshing 'cloud of opinion, prejudice and hunch, that we try not to show when offering data for', once again we get a remarkable insight into a biological engineering approach to a problem in a way which the more formal and restrictive writings of most scientists never reveal.

The nature of the complication became apparent from the very beginning of their work. Lettvin and Gesteland assumed that there were probably a large number of different kinds of olfactory receptors, and all the combination qualities which obviously go to make up our appreciation of smells might well be conferred by 'the second order and later olfactory systems in the brain'. But as they say:

> The response of a single axon in the olfactory nerve is not odor-specific. That is, it does not discharge only when a particular chemical and its related compounds are wafted into the nose. While specific receptors of this sort exist elsewhere . . . we have found no analogous elements in the frog's nose. Instead, almost every odor seems to affect almost every receptor one way or another (Lettvin and Gesteland, 1965).

Lettvin and Gesteland found that the response to a mixture of odours was totally unpredictable, and wondered whether this might be accounted for by the interdependence of olfactory sites on the same receptor; that continuous 'traps' might affect each other so that 'the response to substance A is contingent on whether a substance B or even another molecule of A has been clasped elsewhere on the same membrane'. And they go on:

Such a picture, if even partly correct, makes a temporarily helpless object for the biophysicist. The raw coding which evoked this picture also baffled the decipherer. This sort of coding can be called holistic. It is as if every axon expresses a point of view with respect to all compounds and combination of compounds, and each axon has a separate point of view. Considered in the limit, such a system was first described by Leibnitz in his *Monadology*. A restricted case is now found in the now popular optical hologram, in which every point in the representation of the scene expresses an integral function with respect to the whole scene. But integral functions need not be simple and linear like Fourier transforms: if only they are regular, holistic coding preserves information about relations between elements represented. So, for example, in studying the frog's eye one finds certain contextual matters around a point more significant in determining the firing of a ganglion cell than the light value at that point. One of the advantages of holistic codes is that certain relations between elements are encoded together with the elements themselves, so that, in so far as form inheres in relation between elements, resolution of forms is what increases as the number of points of view increase.

This kind of coding implied by the very idea of 'receptive field' as first voiced by Kuffler, occurs everywhere in physiology. When it is possible to guess at the nature of the transformation, as we did with the Frog's eye, then all is golden; one has invariance by the tail and one's colleagues murmur approval. True, the results cannot be handled analytically, but then it is only last grey lackeys of positivism who still hope for the quick and dirty algorithm to plug into a computer. Real nervous systems are above such low cunning. Yet we had not bargained for complete anarchy such as we seem to have in olfactory receptors. If we hold to the consequences of our studies, that every receptor differs in its ordering of odors from every other, we have a result worthy of the Royal Academy of Laputa. But the lie is given us by our gross records. How is it possible that there should be such consistent changes in the electro-osmogram with pyrrole, ethanol, methanol, that we can pick out by proved signs, one from the other, if there were no underlying irregularities? That, on the level of the single fibre, we have missed these regularities is beyond doubt; and it is to them that we must go in order to characterise *the kind of transformation that the whole receptor makes on the world of odors* (Lettvin and Gesteland, 1965).

And whatever the coding procedure is, it preserves the relations between odours in combination. And Lettvin and Gesteland concluded:

The initial transductive process is not adducible from receptor action and the psychological laws for smell cannot be synthesised from knowing what we have said about receptors, even if we suppose that all we have said can be confirmed. All we can say is that such a receptor language is of the same form as other languages of the nervous system, has the same kind of provenance, and resists reading for the same reasons. That is not saying much.

As with the internal self-regulatory mechanisms of the nineteenth century, and the developmental mechanics and simple tropisms of the earlier twentieth century, so now too with the nervous system. We know what our methodology has to be. There is no alternative. Donald Fleming (Loeb, 1912), in the concluding paragraph to his introduction to Jacques Loeb's book, said:

To act *as if* the organism is a machine, and to act *because* it is are operationally identical; and it is the only operation which is open to an experimentalist. At the moment of closing with the organism in an experiment—of coming alive as an investigator—every biologist recovers the posture of Loeb. It is the only posture that will make biology go at the moment of truth.

I wonder. If this means no more than that our laboratory tools have to be physicochemical ones, then I agree. But if it means that in the formulation of our concepts, in the analysis of our biological problems, in the synthesising of our results, in short in the whole way in which we approach organisms and their study, we are committed to a naive view of the organism, as a machine in Loeb's terms, then the examples of Weiss and Lettvin and the greatest of them, Claude Bernard, give us the lie. For what makes biology 'go at the moment of truth'? Not merely our experiments, but our questions and our concepts too; in fact, our whole attitude to the organism. And, as break-throughs in biology have come just as much from people who call themselves antireductionists as from reductionists, the difference in strategies lies now, not at the level of experiment, but in men's theoretical approach.

References

Bernard, C. (1850–1860). *The Cahier Rouge* (trans. H. H. Hoff, L. Guillemin, R. Guillemin). Cambridge, Mass. (1967).

Bernard, C. (1865). *Introduction a l'Etude de la Médecine expérimentale.* Paris. English trans. C. Greene. Dover, New York (1957).

Bichat, M. F. X. (1801). *Anatomie générale appliquée à la Physiologie et Médecine.* Paris.

Bichat, M. F. X. (1805). *Recherches physiologiques sur la Vie et la Mort*, third edition. Paris.

Cannon, W. B. (1929). Organization for physiological homeostasis. *Physiological Reviews*, **9**, 397. Norton, New York.

Cannon, W. B. (1932). *The Wisdom of the Body.* Norton, New York.

Chaptal, J. A. (1791). *Elements of Chemistry* (trans. W. Nicholson). London.

Goodfield, G. J. (1960). *The Growth of Scientific Physiology.* Hutchinson, London.

Goodfield, G. J. (1965). Review of E. Mendelsohn, *Heat and Life. Isis*, **56**, 465.

Goodfield, G. J. (1969). Some aspects of English physiology, 1780–1840. *J. Hist. Biol.*, **2**, 283–320.

Henderson, J. L. (1913). *The Fitness of the Environment.* Macmillan, New York.

Lavoisier, A. L. and Laplace, P. S. (1780). Mémoire sur la chaleur. *Mémoires de l'Académie des Sciences*, 355.

Lawrence, W. (1816). On life. In *An Introduction to Comparative Anatomy and Physiology.* London.

Lettvin, J. R. and Gesteland, R. C. (1965). Speculations on smell. *Cold Spring Harbor Symposia on Quantitative Biology*, **30**, 217–25.

Lettvin, J. Y., Maturana, H. R., McCulloch, W. S. and Pitts, W. H. (1959). What the frog's eye tells the frog's brain. *Proc. Inst. Radio Engineers*, **47**, 1940–51.

Liebig, J. von (1842). *Animal Chemistry.* London.

Loeb, J. (1912). *The Mechanistic Conception of Life* (ed. and introduced by D. Fleming). Belknap Press, Cambridge (1964).

Mendelsohn, E. (1963). Cell theory and the development of general physiology. *Archives d'Histoire des Sciences*, **65**, 421.

Mendelsohn, E. (1964). *Heat and Life.* Harvard University Press, Cambridge, Mass.

Newth, D. (1969). A critique of the Koestler clique—review of *Beyond Reductionism* (ed. A. Koestler). *New Scientist*, 2 October.

Weiss, P. (1940). The problem of cell individuality in development. *The American Naturalist*, **74**, 34–46.

Weiss, P. (1959). Animal behaviour as system reaction: the orientation towards light and gravity in the resting postures of butterflies (Vanessa). In *General Systems: Yearbook of the Society for General Systems Research*, **4**, 1–44. English trans. Gudrun S. Johnson, of a paper originally in *Biologia Generalis*, **1**, 167–248 (1925).

Weiss, P. (1959). The compounding of complex macromolecular and cellular units into tissue fabric. *Proceedings of the US National Academy of Sciences*, **42**, 819.

Weiss, P. (1962). From cell to molecule. In *The Molecular Control of Cellular Activity* (ed. J. M. Allen). McGraw-Hill, New York.
Weiss, P. (1963). The cell as a unit. *Journal of Theoretical Biology*, **5**, 389–97.
Weiss, P. (1968). One plus one does not equal two. In *The Neurosciences* (ed. Schmidt). Rockefeller University Press, New York.
Weiss, P. (1969). The living system: determinism stratified. In *Beyond Reductionism* (ed. Koestler). Hutchinson, London.
Wiener, N. (1948). *Cybernetics*. MIT Press, Cambridge.

Postscript in the light of discussions at Serbelloni

The conclusion of my paper seemed lame and rather trite; after the discussion I was even deprived of that! I had hoped to show both that there was some genuine difference in outlook between men like Bernard and Magendie, Weiss and Crick, and that this difference was reflected in their science. The first was easy enough to demonstrate; all it took was a comparison of the men's writings. But the other point proved elusive. While acknowledging that, methodologically speaking, there was nothing to choose between these people, I felt that if reductionist or antireductionist attitudes had *any* relevance or point at all, then this should have been expected to show in experimental or theoretical payoffs. And I made the suggestion (above) that the attitude with which a man approached biology might help him to focus on certain problems that otherwise might be ignored.

In the discussion, I put this question to Professors Medawar and Edelman particularly, asking whether in their opinion and so far as laboratory and theory were concerned they saw and recognised any difference between the people mentioned. The answer was a flat negative. Both Medawar and Edelman felt that it made no difference at all. A man was a good scientist or he was not. Problems as biological problems *eventually* will force themselves on to the attention of a scientist irrespective of the general philosophy with which he approaches the subject. When it arrives it may be only a question of 'ripe' or 'unripe' time. To this Karl Popper added his rider in a form of criticism of my paper, that since reductionism was essentially a problem of a 'world-view' we should not expect it to have direct consequences within the laboratory, nor at the level of a theory, where the continuation of a theory is dependent on laboratory findings.

If all this is true, then what follows? Firstly, there has been a great 'much-ado-about-nothing'. The reductionist/antireductionist arguments among biologists may have as little relevance and impact on the direction of biology as similar arguments conducted in the abstract by philosophers. A salutary conclusion, perhaps? Secondly, if reductionism, etc., *is* only a matter of world-view, then it becomes like a belief in God, or a cynicism about free will: namely, a factor that may be only a motivating force, propelling a man into science in the first place and keeping him there when he arrives. But so far as the course of *science* goes, it becomes as irrelevant as whether or not he regularly beats his wife on a Saturday night. He subscribes to the communal methods, the intellectual ambitions, the problems and theories accepted by scientists at that point in time. He cannot escape this, and personal considerations—his aesthetic predispositions which in fact may be all that antireductionism amounts to—become subjugated to the demands of the collective enterprise.

So the question 'What makes a particular science go at any moment and what influences the direction and pace of scientific theory?' becomes very much more subtle in form and much more difficult to answer.

The other two points that emerged from the discussion were corrective, though minor. I drew attention to the similarity in attitudes and views of Bernard and Weiss, showing how they specified the form of the biological problem in essentially the same terms. I would want to add Medawar to the list. Medawar's paper, with its emphasis on 'How progressive conceptual enrichment occurs because we are progressively restricting the range of transformations', is another and rather more precise way of expressing Bernard's views, that vital phenomena differ from inorganic ones by virtue of 'determined or determinable conditions proper to themselves'. This, I kept emphasising, was the crux of the problem. One question occurred to me as a natural consequence, and it took the following form. I wanted to argue that a biologist should want to know why the vital properties exist only given these special conditions. One could ask, *why*, given certain conditions, is DNA replication a consequence of DNA structure? But the question in this form was not accepted by Edelman, who felt that molecular biologists would simply rephrase it to 'What kind of structure do we have to have in order to achieve replication?' Phrased like this, it is both a much more limiting type of question and raises the possibility of *more* than one answer. But is the *why* element always likely to be ruled out as an impermissible form of question for science?

Lastly, whatever conclusions we draw about the attitudes of reductionists and antireductionists and the value for biology of their approaches and philosophies, Shapere did remind us of one important point. Reductionist methodology may have been extremely successful, but the history of science abounds with examples where forms of explanation, successful in one field, have turned out to be disastrous when imported into another. Shapere gave the example of the failure that followed when Newton's theories were applied as a basis for chemical explanation, at the time of John Dalton. I can add another: in the early nineteenth century attempts were made to explain living properties with a theory of 'The Vital Principle'. At every point of pressure this form of theory was justified by an appeal to the examples of Newton and his theory of the gravitational principle. Success in one field is no guarantee of applicability of explanation in another, for a new phenomenon remains a new phenomenon and must be dealt with as such. The trouble is that it may be extremely difficult to recognise when one is dealing with a genuine one.

7. Cerebral Activity and Consciousness

JOHN C. ECCLES

Introduction

By consciousness I mean conscious experience, which each of us has privately for himself. It is the primary reality for each of us, as I have argued in my book (Eccles, 1970). I try to avoid the words 'mind' and 'mental' because they have been so indiscriminately misused that they now are devoid of precise meaning. For example, mental attributes have been postulated for matter in some ordered state. It has been stated: 'We arrive at another concept of order in matter in which events analogous to mental events in man, maintain the order, respond to previous events and anticipate immediate future events, for that is what we mean by mental events' (Birch, 1974); and Polten (1973) states that 'a rock is subject to mind in that it is ruled by law. . . . I maintain that a rock is held together by substantial binding energies that are mental in nature'. As I have stated earlier (Eccles, 1970):

> In order to preserve a continuity in the evolutionary process and to avoid a special and unique emergence or discontinuity, many eminent thinkers [Sherrington, 1940; Teilhard de Chardin, 1959; Huxley, 1962] have taken refuge in the vague generalisation that there is a mental attribute in all matter. As the organisation of matter gradually became perfected in the evolutionary process, there was a parallel development of the mental attribute from its extremely primordial state in inorganic matter, or in the simplest living forms, through successive stages until it reached full fruition in the human brain.

On the contrary, Dobzhansky (1967) has stated that there are two exceptions to this continuity in the evolutionary process—the origin of life and the origin of man.

> The origin of life and the origin of man were evolutionary crises, turning points, actualisations of novel forms of being. These radical innovations can be described as transcendences in the evolutionary process. Human mind did not arise from some kind of rudimentary 'minds' of molecules and atoms. Evolution is not simply unpacking of what was there in a hidden state from the beginning. It is a source of novelty, of forms of being which did not occur at all in the ancestral states.

For my part I think that there is no meaning or scientific value in ascribing mental properties to systems that exhibit order or apparent purpose or memory, or even intelligent action. I fail to distinguish between these modern philosophic views and the ancient Aristotelian concepts of vegetative soul, sensitive soul and rational soul to which we could also now add mineral soul! Even when we come to the apparently intelligent actions of higher animals with their remarkable abilities to learn and remember, I have not found any reason to go beyond the purely mechanistic neurophysiology in explaining their brain performances, which of course was the position of Descartes. I am a reductionist in that I know of no compelling reason to postulate some mental basis for any of the reactions that neurophysiologists have observed in animal brains or any of the animal behaviours observed by psychologists. And the same could be said for human brains, except when scientific investigations are carried out under very special circumstances on conscious human subjects that are expressing or reporting their conscious experiences or the lack of them. In fact one can state that the programme of the neurobiologist is strictly mechanistic and reductionist because we attempt to explain the neural events and behaviour of experimental animals solely in terms of physics and chemistry and their various developments in biophysics, biochemistry, neuropharmacology and neurocommunications.

In agreement with Polten (1973) one may consider conscious experience at three levels, as diagrammed in figure 1. There is firstly outer sensing, which is the perceptual experience due to input from sense organs, not only from the external world by exteroceptors, such as the organs of sight, hearing, smell, taste and touch, but also from body states, for example by proprioceptors from muscles, joints, fascia, etc., and by receptors for pain, hunger, thirst, etc. Secondly there is inner sensing, which is not directly derived from sense data, though it is often triggered by these data, and has many derivatives from the data. It includes the experiences of thinking, emotions, dispositional intentions, memories, dreams and creative imagination. Thirdly there is the pure ego or self that recognises itself by apperception (Kant) and is central to all experience. It transcends immediate experiences and gives each of us the sense of continuity and identity throughout a lifetime. This sense of continuity bridges periods of unconsciousness in sleep and in other less pleasant ways, but it is essential to the concept each of us has as being a self, and in the religious sense it corresponds to the soul.

Conscious experiences and brain states

From these basic considerations we now raise the question: 'How are these various levels of conscious experience related to brain states?' The hard-core materialist and behaviourist answer was that this question was meaningless—there were only brain states and the rest was introspective fantasy and not worthy of philosophical or scientific enquiry. We can reject this crude materialistic dogma not only because it ignores or distorts the facts of experience, but also because it is self-stultifying: how can brain states describe themselves? As a philosophy it is discredited. The less radical

WORLD OF CONSCIOUSNESS

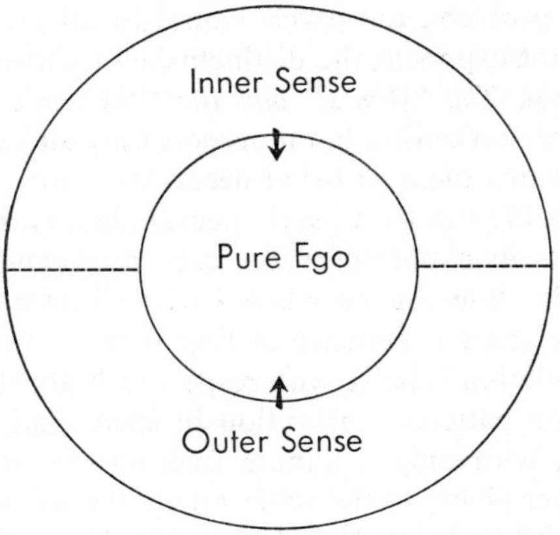

Outer Sense	Inner Sense	Pure Ego
Light	Thoughts	The self
Colour	Feelings	The soul
Sound	Memories	
Smell	Dreams	
Taste	Imaginings	
Pain	Intentions	
Touch		

Figure 1. World of consciousness. Diagram of the three postulated components in the world of consciousness, together with a tabulated list of their components.

behaviourism of the Skinner (1971) variety, for example, does not deny conscious experiences but relegates them to a meaningless role with respect to the behaviour of man and animals. The behaviouristic dogma of Skinner is that investigations on the pattern of stimulus–response–reinforcement will eventually lead to a complete explanation not only of animal but also of human behaviour, and to its complete control by operant conditioning, as witness his recent book *Beyond Freedom and Dignity*. I reject this philosophy because it resolves the brain–mind problem by ignoring both brain and mind; the former is safely and inviolably enclosed in the black box, and the latter is as ineffective as a fantasy. This type of behaviourism leads to a caricature of man—beyond freedom and dignity—that ignores the personal experiences that for each of us is the primary reality. It can appeal only to the philosophically naive and to those seeking the power that devolves from the absolute control of man.

Behaviourism has been replaced by a much more sophisticated solution of the brain–mind problem, the psychoneural identity hypothesis, that has as its most important exponent the distinguished philosopher Herbert Feigl (1967) with his book *The 'Mental' and the 'Physical'*. This philosophy is essentially a materialist monism, but it accepts fully all varieties of conscious experience and explains them as being necessary components or aspects of brain states, there being strictly a psychoneural identity. It is postulated that every brain state has its counterpart in a conscious experience, the analogy being that the brain state can be recognised by external observation, and consciousness is the inner experience of that same state. Unfortunately the philosophical formulation is naive with respect to brain states. These are not recognised as having patterned operation in space and time of an almost infinite complexity, with only a minute fraction—less than 1 per cent of cortical activity—ever giving to the subject a conscious experience (Moruzzi, 1966; Jung, 1970). Nevertheless the identity hypothesis has won acceptance amongst most neuroscientists. Undoubtedly it has been a great relief to them to have a respectable philosophical umbrella sheltering from further bother about how the mind may interfere in their neurophysiological investigations!

I will not embark on a philosophical disputation, but recently there has been a most critical appraisal of the psychophysical identity hypothesis by Polten (1973), who has demonstrated that it leads to paradoxes and contradictions and so stands refuted. My attack on the hypothesis is based on a consideration of the brain events and of the manner in which the identity hypothesis relates them to consciousness. What is to my mind a variant of the identity hypothesis has recently been published by Laszlo (1972) in his book *Introduction to Systems Philosophy*. He criticises the identity theory, but it seems to me that subsequently he arrives at an almost identical conclusion in his theory of biperspectivism, namely that there is an identity, the mental experiences being the inner aspect of the neural events that could be observed by an investigator armed with a suitable instrument such as the mythical cerebroscope of Feigl. Laszlo states:

> But rather than reducing mental events into physical ones through some stratagem such as declaring them to be contingently identical, or doing away with the world of scientific entities by means of an essentially Berkeleyan type of argument, we can recognise the fundamental character of both types of events and still produce an internally neat and consistent ontology.
>
> Cognitive systems constitute the *mind*; and natural systems range over the full extent of the microhierarchy on the surface of the earth: they include *homo sapiens*, *i.e.*, the *body*. Now, mind is not a public observable; rather than being observed, it is 'lived'. But a system that is 'lived' from one point of view does not mean that it cannot be 'observed' from another. In other words, an introspectively lived system of mind-events (*i.e.*, a cognitive system) may well be an externally observed system of physical-events, if the viewpoint of the observer filing the report is correspondingly shifted.

Another variant of the identity theory has recently been proposed by Globus (1974).

Since psychoneural identity hypothesis should specify that *phenomenal experience is identical with events in the unobserved brain,* the hypothesis is irrevocably untestable by empirical means. Corollary to this hypothesis is that *phenomenal experience is not identical with but informationally equivalent to observed neural events.*

I would agree in general if the last section were changed to: phenomenal experience is not identical with but informationally coherent with observed neural events.

Experiments on the cerebral cortex and conscious perception

Let us now consider problems of conscious perception in relation to the remarkable investigations by Sperry and associates (Sperry, 1968*a*, 1968*b*, 1969, 1970*a*, 1970*b*) on human subjects in which communication between the two cerebral hemispheres *via* the great cerebral commissure (the corpus callosum) has been interrupted for therapeutic reasons. This dominant hemisphere is the speech hemisphere, which is almost always (98 per cent) the left hemisphere (Penfield and Roberts, 1959), and was so in all of Sperry's cases. It has been a remarkable finding that only the input from receptor organs to the dominant hemisphere gives conscious experiences to the subject. For example, he knows nothing of flashed signs on the left visual field or touch and movement of the left hand, for these are exclusively projected to the right (minor) hemisphere. Despite the subject's ignorance of all the perceptual inputs to the minor hemisphere, this hemisphere is able to carry out with the left hand appropriate and skilled actions deriving from these perceptual inputs. When it is pointed out to the subject that his left hand has carried out appropriate and intelligent actions, all he can reply is that it must have been a guess or an unconscious accident, but never that it is a result of his own experience and understanding. In fact he has no voluntary control of the left hand, which of course is programmed from the minor hemisphere. He makes various statements such as 'I cannot work with that hand', that the hand 'is numb', that I 'just can't feel anything or can't do anything with it', or that I 'don't get the message from that hand'. The whole perceptual side of the minor hemisphere remains impenetrable when an attempt is made to discover whether there are some mental attributes associated with all the skill and correct responses programmed from the minor hemisphere.

We can conclude that the rigorous testing of the subjects who have been subjected to section of the corpus callosum has revealed that conscious experiences arise only in relationship to neural activities in the dominant hemisphere. It was tentatively suggested (Eccles, 1965) that this exclusive relationship is normally present but unrecognisable until revealed by the callosal section. Popper (1970) stresses particularly the association with the linguistic areas of the dominant hemisphere. This relationship is shown in figure 2 where arrows lead from the linguistic and ideational areas of the dominant hemisphere to the conscious self that is represented by the circular area above. It must be recognised that figure 2 is an information flow diagram and that the location of the conscious self is for diagrammatic convenience. It is of course not meant to imply that the conscious self is hovering in space above the dominant hemisphere! It is postulated that in normal subjects activities in the minor hemisphere reach consciousness only after transmission to the dominant hemisphere, which very effectively occurs *via* the immense impulse traffic in the corpus callosum, as is illustrated in figure 2 by the numerous arrows. Complementarily, as will be discussed in full later, it is postulated that the neural activities responsible for voluntary actions mediated by the pyramidal tracts normally are generated in the dominant hemisphere by some willed action of the conscious self (see downward arrows in figure 2). When destined for the left side, there is transmission to the minor hemisphere by the corpus callosum and so to the motor cortex of that hemisphere.

It must be recognised that this transmission in the corpus callosum is not a simple one-way transmission. The 200 million fibres must carry a fantastic wealth of impulse traffic in both directions. In the normal operation of the cerebral hemispheres, activity of any part of a hemisphere is as effectively and rapidly transmitted to the other hemisphere as to another lobe of the same hemisphere. The whole cerebrum thus achieves a most effective unity. It will be appreciated from figure 2 that section of the corpus callosum gives a unique and complete cleavage of this unity. The neural activities of the minor hemisphere are isolated from those cerebral areas that give and receive from the conscious self. The conscious subject is recognisably the same subject or person that existed before the brain splitting operation and retains the unity of self-consciousness or the mental singleness that he experienced before the operation. However, this unity is at the expense of unconsciousness of all the happenings in the minor (right) hemisphere.

We can regard the minor hemisphere as having the status of a very superior animal brain. It displays intelligent reactions and primitive learning responses and it has a great many skills, particularly in the spatial and auditory domains, but it gives no conscious experience to the subject. Moreover, there is no evidence that this brain has some residual consciousness of its own. Sperry postulates that there is another mind in this brain, but it is prevented from communicating to us because it has no speech. I would agree with this statement if it be linked with the further statement that in this respect the minor hemisphere resembles an animal brain, though its performance is superior to that of the brains of the highest anthropoids. In both of these cases we lack communication except at an extremely poor linguistic level, so it is not possible to test for the possibility of some consciously experiencing being. We therefore must be agnostic about the question of mental activities and consciousness in the manner in which I have defined it at the beginning. The superiority of the minor hemisphere over

MODES OF INTERACTION BETWEEN WORLD 1 : WORLD 2 : WORLD 3

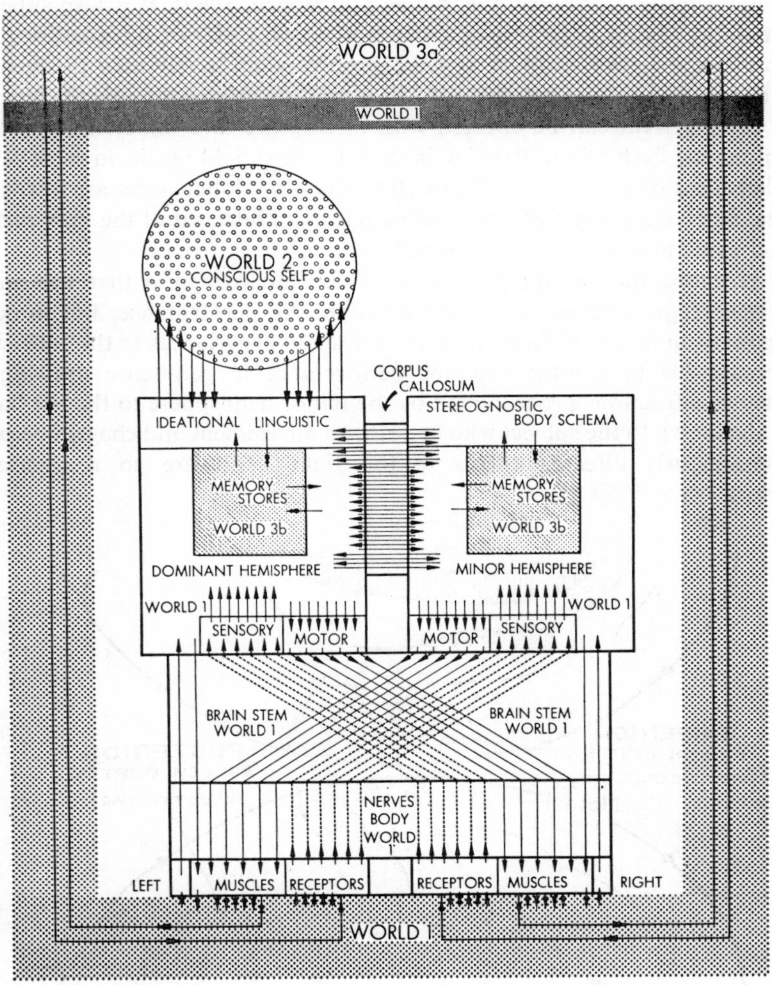

Figure 2. Communications to and from the brain and within the brain. Diagram to show the principal lines of communication from peripheral receptors to the sensory cortices and so to the cerebral hemispheres. Similarly, the diagram shows the output from the cerebral hemispheres *via* the motor cortex and so to muscles. Both these systems of pathways are largely crossed as illustrated, but minor uncrossed pathways are also shown. The dominant left hemisphere and minor right hemisphere are labelled, together with some of the properties of these hemispheres that are found displayed in figure 5. The corpus callosum is shown as a powerful cross linking of the two hemispheres and, in addition, the diagram displays the modes of interaction between Worlds 1, 2 and 3, as described in the text.

subhuman primate brains is demonstrated, for example, by the time of many minutes during which an initial signal can be held in its memory before a successful retrieval (Sperry, Gazzaniga and Bogen, 1969). It is also superior to an animal brain in respect of cross-modality transfer of information. For example, a visual or auditory signal can be very effectively used to signal to the minor hemisphere an object to be retrieved using kinaesthetic sensing, and this retrieval can be effected with intelligence and understanding. For example, the flash of a dollar sign in the left visual field results in retrieval by the left hand of some coin—25c or 10c—when no dollar notes are available, or the flash of a picture of a wall clock results in retrieval of the only related object available—a child's toy watch!

It is important for our present arguments to recognise that the minor hemisphere programmes a refined stereognostic performance, but none of the goings-on in that hemisphere gives conscious experiences to the subject. It is remarkable to see the superior performance in geometric drawing or mosaic constructions programmed by the minor hemisphere to the left hand, all unbeknown to the subject who sees it with amazement and chagrin because his consciously directed efforts to the right hand are so recognisably inferior.

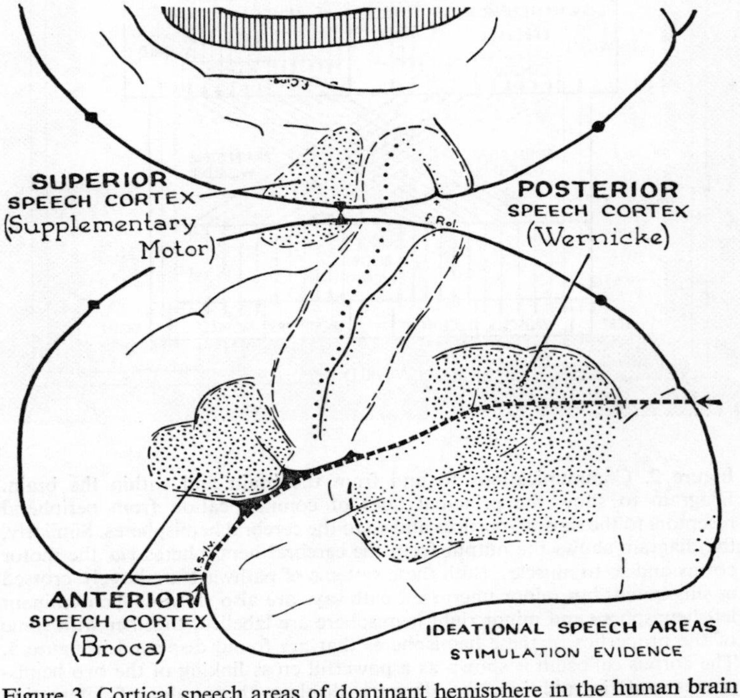

Figure 3. Cortical speech areas of dominant hemisphere in the human brain as determined by aphasic arrest by electrical interference (Penfield, 1966).

Speech and consciousness (Eccles, 1973)
We have seen already that with human subjects callosal transection reveals that the left hemisphere is the speech hemisphere for all subjects so far investigated (figure 3). In fact there is an identification of speech hemisphere with dominant hemisphere and association of this hemisphere with the conscious experiences of all the subjects, both as regards receiving from the world and acting upon it. There is thus strong evidence (*cf.* Popper, 1970, 1974) that we have to associate the dominant hemisphere, that is, the speech hemisphere, with the amazing property of being able to give rise to conscious experiences in perception, and also to receive from them in the carrying out of willed movements. Moreover, the most searching investigation discloses that the minor hemisphere does not have in the smallest degree this amazing property of being in liaison with the conscious mind of the subject in respect either of giving or receiving. One would predict with assurance that in subjects with the rarely occurring right hemispheric representation of speech, the right hemisphere would be dominant, as revealed after the callosal transection, and be alone associated with the conscious experiences of the subject.

The unique association of speech and consciousness with the dominant hemisphere gives rise to the question: is there some special anatomical structure in the dominant hemisphere that is not matched in the minor hemisphere? In general, the two hemispheres are regarded as being mirror images at a crude anatomical level, but recently it has been discovered (figure 4) that in about 80 per cent of human brains there are asymmetries with special developments of the cerebral cortex in the regions both of the anterior and posterior speech areas (Geschwind and Levitsky, 1968; Wada *et al.*, 1973; Geschwind, 1972). However, apart from such differences at a macro-level, one must assume that there are specially fine structural and functional properties as the basis for the linguistic performance of these speech areas. Undoubtedly most exciting work awaits the investigation of these areas by electron-microscopic techniques. It can be anticipated that eventually there will be electrophysiological analysis of the ongoing events in the speech areas of conscious subjects whose brains are exposed for some therapeutic purpose. In the evolution of man there must have been most remarkable developments in the neuronal structure of the cerebral cortex that have made possible the evolution of speech. One can imagine that progressively more subtle linguistic performance gave primitive man the opportunities for very effective survival, which may be regarded as a strong evolutionary pressure. As a consequence there were the marvellously rapid evolutionary changes transforming in two to three million years a primitive ape to the present human race.

In respect of the anatomically represented speech areas, and the associated linguistic ability and consciousness, the human brain is unique. Undoubtedly the experimental investigations on chimpanzees, with respect to their developing both a sign language (Gardner and Gardner, 1969) and a symbol language

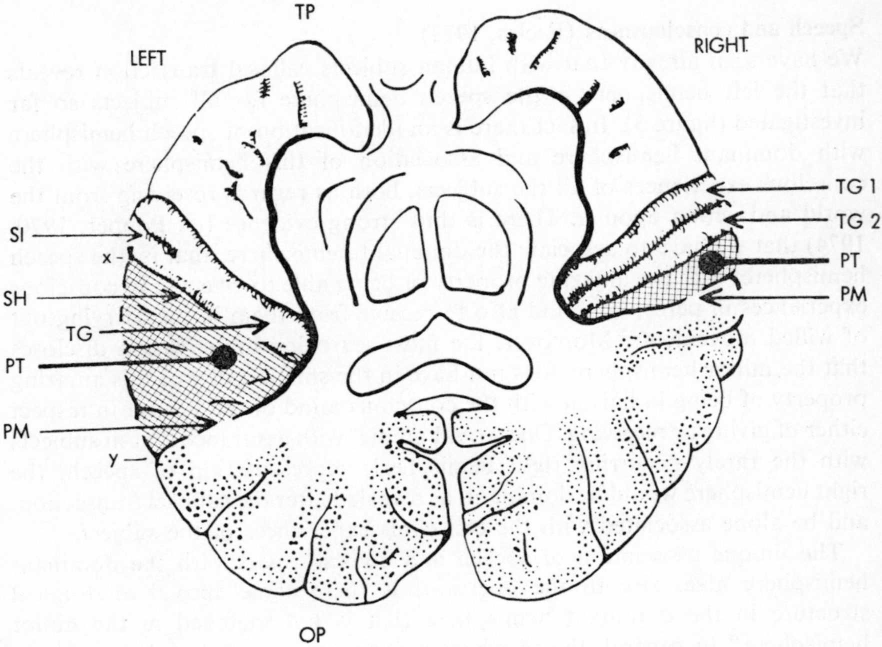

Figure 4. Upper surfaces of human temporal lobes exposed by a cut on each side as illustrated by the broken line in figure. 3. Typical left–right differences are shown. The posterior margin (PM) of the planum temporale (PT) slopes backward more sharply on the left than on the right, so that the end y of the left Sylvian fissure lies posterior to the corresponding point on the right. The anterior margin of the sulcus of Heschl (SH) slopes forward more sharply on the left. In this brain there is a single traverse gyrus of Heschl (TG) on the left and two on the right (TG_1, TG_2). TP temporal pole; OP, occipital pole; SI, sulcus intermedius of Beck. (Geschwind and Levitsky, 1968.)

(Premack, 1970), show that the chimpanzee brain exhibits considerable levels of intelligent and learned performance, but this chimpanzee communication is at a quite different level from human speech. Moreover, this linguistic performance is at a lower level than that of the minor hemisphere in Sperry's experiments.

Wada *et al.* (1973) have made the remarkable discovery that even the seven-month foetus has already developed hypertrophy of the eventual speech areas. Genetic instructions are thus building the speech areas long before they are to be used. The human brain at the stage of the infant is already giving evidence of incipient speech performance. A two-year-old child displays an extraordinary ability to develop the expression of sounds in relationship to meaning and intention. As I mentioned earlier, damage to the potential speech hemisphere at an early stage can result in the other undamaged hemisphere becoming the speech hemisphere. However, this transfer may not happen.

A hemisphere severely damaged in infancy may still develop as the speech hemisphere.

In a remarkable new conceptual development arising from study of split-brain subjects, Levy-Agresti and Sperry (1968) have proposed that the dominant and minor hemispheres have a division in their operational tasks. The various specific performances are listed in figure 5. It is suggested that this division of tasks enables each hemisphere to perform its particular general mode of processing information before there is synthesis and eventual appearance as conscious experience. Since neural events in the minor hemisphere do not directly give the subject conscious experiences, we have to postulate that the neuronal machinery concerned in these specific operational tasks works at an unconscious level, which would be in good accord with the psychiatric concept of the unconscious mind. For example, in listening to music it can be envisaged that initially immense and complex operational tasks such as decoding, synthesising and patterning are carried out in the temporal lobe of the minor hemisphere. Communication *via* the corpus callosum to the liaison areas of the dominant hemisphere with the consequent conscious experiences presumably is delayed until these most sophisticated neural operations have been carried out in the special musical centres. In their operational function these centres can be regarded as being analogous to the speech centres, but as yet they await a systematic and comprehensive neurological investigation.

DOMINANT HEMISPHERE	MINOR HEMISPHERE
Liaison to consciousness	No such Liaison
Verbal	Almost non - verbal Musical
Ideational	Pictorial and Pattern sense
Analytic	Synthetic
Sequential	Holistic
Arithmetical and computer like	Geometrical and Spatial

Figure 5. Various specific performances of the dominant and minor hemispheres as suggested by the new conceptual developments of Levy-Agresti and Sperry (1968). There are some additions to their original list.

Reductionism and free will

The most effective reductionist strategy in relation to the brain–mind problem is illustrated by one or other form of psychoneural parallelism. Examples already given are the hypotheses of psychoneural identity or of biperspectivism. It is postulated that conscious experiences run parallel with the brain events, and that for each brain event there is a unique conscious experience. The association is postulated to be automatic, requiring no mechanism of interaction. Parallelism gives a simple and attractive explanation of perceptual experiences. But it denies that consciousness can exert any effective action on neural happenings. Neural events and conscious experiences are postulated to exist in parallel without any trace of interaction in either direction. The experience that by willing it is possible to bring about an action—that is, free will—is thus regarded as an illusion. It is surprising that it is not generally recognised that the denial of free will is self-refuting. I will say 'I believe that my thoughts can lead to action, for example in the movements required for their expression in language—as I am now doing'. The parallelist rejoinder must be 'Thoughts cannot cause movements and so cannot be expressed in language; you are illuded'—and so is he! I have learned the futility of arguing about whether one has free will or not. The plain fact is that each of us experiences it and any attempt at denial can claim no more authority than a reflex built by operant conditioning. I do not argue with purely reflexing systems!

My position is that I have the indubitable experience that by thinking and willing I can control my actions if I so wish, *although in normal waking life this prerogative is exercised but seldom.* I am not able to give a scientific account of how thought can lead to action, but this failure serves to emphasise the fact that our present physics and physiology are too primitive for this most challenging task of resolving the antinomy between our experiences and the present primitive level of our understanding of brain function. When thought leads to action, I am constrained, as a neuroscientist, to postulate that in some way, completely beyond my understanding, my thinking changes the operative patterns of neuronal activities in my brain. Thinking thus comes to control the discharges of impulses from the pyramidal cells of my motor cortex and so eventually the contractions of my muscles and the behavioural patterns stemming therefrom.

It has been known for many years that electrical stimulation of the motor cortex of conscious subjects evokes actions which are disowned by the subject. As Penfield reports: 'When a subject observes such an action, he remarks, "that is due to something done to me and is not done by me".' Evidently a motor action emanating from the motor cortex in response to a voluntary command has some concomitants that are not present when a similar action is artificially evoked from the motor cortex.

A fundamental neurological problem is: how can willing of a muscular movement set in train neural events that lead to the discharge of pyramidal cells of the motor cortex and so to activation of the neural pathway that leads to the muscle contraction? This question has been investigated physiologically by searching for signs of neuronal activity in the cerebrum before the discharge down the pyramidal tract. The experimental requirement is to have a movement executed by the subject *entirely on his own volition*, and yet to have accurate timing in order to average the very small potentials. This has been solved by Deecke, Scheid and Kornhuber (1969), who use the onset of the movement to trigger a reverse computation of the potentials up to 2 seconds before the onset of the movement, which is standardised, the flexion of a finger for example. In this way it was possible to average the potentials evoked at various sites over the surface of the skull. The slowly rising negative potential, called the 'readiness potential', was observed over the whole cerebral surface except for positivities at the most anterior and basal regions. Usually it began about 850 ms before the onset of the movement and led on to sharper potentials, positive or negative, less than 100 ms before the movement. These were located more specifically over the area of the motor cortex concerned in the movement. We can assume that the 'readiness potential' is generated by complex patterned neuronal discharges that eventually project to the pyramidal cells of the motor cortex and synaptically excite them to discharge, so generating the waves just preceding the movement.

These experiments at least provide a partial answer to the question: what is happening in my brain at the time I am deciding on some motor act? They certainly show that complex neuronal operations are occurring in the brain prior to the neural discharge that brings about the required movement. The split-brain experiments demonstrate that the initial neural activities of a willed movement are generated in the dominant hemisphere even if occurring in the limbs programmed from the motor cortex of the minor hemisphere. They are transmitted to the minor hemisphere by the corpus callosum and so to the motor cortex of that hemisphere. It must be recognised that this transmission in the corpus callosum is not a simple one-way transmission. The fibres must carry a fantastic wealth of impulse traffic in both directions. We still have to consider how the willing of an action can act on the neuronal systems of the liaison brain and cause some correlated activity that builds up, as displayed by the 'readiness potential'.

The principal grounds for the theoretical belief that this control is an illusion are derived from the assumptions that both physics and neurophysiology give a deterministic explanation of all events in the brain and that we are entirely within this deterministic scheme. In this context reference may be made to the discussion by Popper (1950) in which he concludes that not only quantum physics but even 'classical mechanics is not deterministic, but must admit the existence of unpredictable events'. A similar argument has been developed by MacKay (1966). Moreover, the neurophysiology of a deterministic character is merely a primitive reflexology, and not related at all to the dynamic properties of the immense neuronal complexities of the brain. There are thus no sound scientific grounds for denying the freedom of the will which, ironically, must be assumed if we are to act as scientific investigators.

One can surmise from the extreme complexity and refinement of its organisation that there must be an unimagined richness of properties in the active cerebral cortex. In a situation where 'will' is operative, there will be a changed pattern of discharge down the pyramidal tract and this change must be brought about because there is a change in the spatiotemporal pattern of influences playing upon the pyramidal cells in the motor cortex. The 'readiness potential' bears witness to this cortical activity preceding the pyramidal tract discharge. If the 'will' really can modify our reactions in a given situation, we have somewhere in the complex patterned behaviour of the cortex to find that the spatiotemporal pattern which is evolving in that given situation is modified or deflected into some different pattern.

Evidently we have here a fundamental problem that transcends our present neurophysiological concepts. Some tentative suggestions have been made (Eccles, 1953, 1970). It is necessary to take into account the evidence that 'will' can act on the cortical neuronal network only when a considerable part of it is at a relatively high level of excitation—that is, we have to assume that, for 'will' to be operative, large populations of cortical neurones are subjected to strong synaptic bombardments and are stimulated thereby to discharge impulses which bombard other neurones. Under such dynamic conditions it may be conservatively estimated that, out of the hundred or more synaptic contacts made by any one neurone, at least four or five would be *critically effective* (when summed with synaptic bombardments by other neurones) in evoking the discharge of neurones next in series. The remainder would be ineffective because the recipient neurones would not be poised at this critical level of excitability, being either at a too low level of excitation, or at a too high level, so that the neuronal discharge occurs regardless of this additional synaptic bombardment. Thus at any instant the postulated action of the 'will' on any one neurone would be effectively detected by the 'critically poised neurones' on which it acts synaptically. When a region of the cortical neuronal network is at a high level of activity, the discharge of an impulse by any one neurone will have contributed directly and indirectly to the excitation of hundreds of thousands of other neurones within a few milliseconds.

As a restatement of the conclusion of the preceding section we may say that in the active cerebral cortex the patterns of discharge of large numbers of neurones would rapidly be modified as a result of an 'influence' that initially caused the discharge of merely one neurone. But further, if we assume that this 'influence' is exerted not only at one node of the active network, but also over the whole field of nodes in some sort of spatiotemporal patterning, then it will be evident that potentially the network is capable of integrating the whole aggregate of 'influences' to bring about some modification of its patterned activity, that otherwise would be determined by the pattern of afferent input and its own inherent structural and functional properties. Such integration would occur over hundreds of thousands of nodes in a few milliseconds, the effects exerted on any and every node being correlated in the resultant patterned activity of the surrounding hundreds of thousands of neurones. Thus, in general, the spatiotemporal pattern of activity would be determined not only by (i) the microstructure of the neural net and its functional properties as built up by genetic and conditioning factors, and (ii) the afferent input over the period of short term memory, but also (iii) the postulated 'field of mind influence'.

Thus, the neurophysiological hypothesis is that the 'will' modifies the spatiotemporal activity of the neuronal network by exerting spatiotemporal 'fields of influence' that become effective through this unique detector function of the active cerebral cortex. It will be noted that this hypothesis assumes that the 'will' or 'mind influence' has itself some spatiotemporal patterned character in order to allow it this operative effectiveness.

These concepts are closely related to those recently developed by Sperry (1969), who states:

> In the present scheme the author postulates that the conscious phenomena of subjective experience do interact on the brain processes exerting an active causal influence. In this view consciousness is conceived to have a directive role in determining the flow pattern of cerebral excitation.

This hypothesis of mind–brain liaison has the merits of relating the occasions when the mind can operate on the brain to the observed high level of neuronal activity during consciousness, and of showing how an effective action could be secured by a spatiotemporal pattern of minute 'influences'. If the neuronal activity of the cerebral cortex is at too low a level, then liaison between mind and brain ceases. The subject is unconscious as in sleep, anaesthesia, coma. Perception and willed action are no longer possible. Furthermore, if a large part of the cerebral cortex is in the state of the rigorous driven activity of a convulsive seizure, there is a similar failure of brain–mind liaison, the consciousness of the subject being likewise explicable by the deficiency of the sensitive detectors, the critically poised neurones.

Brain–mind interaction reconsidered

When I postulated many years ago (Eccles, 1953), following Sherrington (1940), that there was a special area of the brain in liaison with consciousness, I certainly did not imagine that any definitive experimental test could be applied in a few years. But now we have this distinction between the dominant hemisphere in liaison with the conscious self, and the minor hemisphere with no such liaison. It is this empirical discovery that I try to illustrate in figure 2. In it are shown the communication lines going both ways—out to the world *via* the motor pathways, and receiving from the world *via* receptors and afferent pathways.

The information flow diagram of figure 2 can form a background to the recent conceptual developments of Sperry in brain–mind interaction. In general terms he states (1970*a*):

Conscious phenomena in this scheme are conceived to interact with and to largely govern the physiochemical and physiological aspects of the brain process. It obviously works the other way round as well, and thus a mutual interaction is conceived between the physiological and the mental properties. Even so, the present interpretation would tend to restore mind to its old prestigious position over matter, in the sense that the mental phenomena are seen to transcend the phenomena of physiology and biochemistry.

Consciousness does do things and is highly functional as an important component of the causal sequence in higher level reactions. This is a view that puts consciousness to work. It gives the phenomena of consciousness a use and a reason for being and for having been evolved.

The split-brain investigations have, I think, falsified the psychoneural identity hypothesis. It is demonstrated that the minor cerebral hemisphere, with its ongoing activities that can be categorised as displaying memory, understanding even at a primitive verbal level, and concepts of spatial relations, does not give any conscious experiences to the subject, who remains in conscious liaison only with neural events in the dominant hemisphere. Evidently the concept of psychoneural identity has lost the feature of primitive simplicity wherein all neural activities of the brain are identified with conscious experiences derived therefrom. In particular, sophisticated, intelligent and learned activities of the minor hemisphere do not achieve liaison to the consciousness of the subject. Moreover, as Sperry has realised, the problems have to be approached at a new level of understanding, the holistic approach. And this occurs in special regions only of the cerebral cortex and in special states of these regions. Moreover, psychoneural parallelism has to be rejected, for on this view the mental states are ineffective, being merely spin-offs of neural activities that they cannot influence.

Brain evolution and consciousness

Evidently, as recognised by Dobzhansky (1967) and Popper (1974), immense and fundamental problems are involved in the evolution of the brain that occurred as man was gradually developing his means of communication in speech. One can imagine that speech and brain development went on together in the evolving process and that from these two emerged the cultural performance of man. Over hundreds of millenia there must have been a progressive development of language from its primitive form as expressive cries to a language that became gradually a more and more effective means of description and argument. In this way, by forging linguistic communication of ever increasing precision and subtlety, man must gradually have become a self-conscious being aware of his own identity or selfhood. As a consequence he also became aware of death, as witnessed so frequently and vividly in other members of the tribal group that he recognised as beings like himself. We do not know how early in the story of man this tragic and poignant realisation of death-awareness came to him, but it was at least a hundred thousand years ago, as evidenced by the ceremonial burial customs with the dead laid in graves with antlers, weapons, ornaments, etc.

The more recent developments of Palæolithic man are exemplified for us by the Lascaux cave paintings, for example. Such artistic representation of animals could only have been done by a kind of early art school operating by linguistic description and criticism. Language must already have been a highly developed form of communication between those cooperating in such artistic achievements in which the forms and movements of animals had to be visualised and remembered and graphically represented. Following Popper (1968*a*, 1968*b*, 1970) my theme is that in this long Palæolithic era of hundreds of millenia man was creating himself in all aspects of World 2 by creating his culture, which is World 3. These double processes of creation are indissolubly locked together as in a phenomenon of cross-catalysis. The progressive acceleration of cultural development can be attributed to a progressive skill and effectiveness in linguistic communication and the consequent development of technology that distinguishes the Neolithic age from the relatively slow development of the long Palæolithic era. In the maturing civilisations of Mesopotamia and Egypt and later Greece, the exigencies of survival were no longer dominant in the thoughts of men, and the creative imagination of man could instead be expressed in literature, in art, in architecture, and in the further developments in religion, in philosophy and in science that are associated with his attempts to understand the manner of being he was, his origin and his destiny.

Summary on reductionism and the brain–mind problem

As shown by the arrows in both directions in figure 2, there is an incessant interplay at a level of interaction between World 2 and the liaison brain that is beyond our present level of understanding. This is a tremendous challenge for future research. In this respect we can think of the whole range of psychiatry with such problems as those of the unconscious self, of sleep and dreams, of obsession. Despite our present ignorance of the precise neurological basis of all these problems of the psyche, we can have hope for some clearer understanding because it is now possible to define the liaison areas of the brain, and postulate that only in these areas and in certain dynamic states of the brain does this relationship occur. This insight, limited as it is, provides an immense challenge for the future.

The empirical support given to the concept of the liaison brain must discomfort the philosophers who like to be holistic about the brain–mind problem, as is for example proposed in the psychoneural identity hypothesis (*cf.* Feigl, 1967). It is there postulated that all neuronal activity in the cerebrum comes through to consciousness somehow or other and is all expressed there. An often used analogy is that neuronal activity and conscious states represent two different views of the same thing, one as seen by an external observer, the other as an inner experience by the 'owner' of the brain. This proposed identification, at least in its present form, is refuted by discovery that after commissurotomy none of the neuronal events in the minor hemisphere is recognised by the conscious subject.

I am in agreement with Popper (1974) that reductionism is a necessary strategy in research. Even at the level of the cerebral cortex of animals and man it is essential to investigate the responses of nerve cells in all their patterned complexity and in all their learnt responses on the basic postulate that it is all explicable in terms of biophysics, biochemistry, neurochemistry, molecular neurobiology, etc. Yet at the same time it must be recognised that reductionism fails when confronted by the brain–mind problem. Schrödinger (1958) and Wigner (1969) suggested that fundamental changes would be required in physics, which hitherto has ignored the world of conscious experience. Yet the philosophy of science is essentially concerned with observations (perceptions), explanatory ideas, arguments, criticisms and creative imagination, all of which are in the world of conscious experience.

References

Birch, C. (1974). Chance, necessity and purpose. In this volume.

Deecke, L., Scheid, P. and Kornhuber, H. H. (1969). Distribution of readiness potential, pre-motion positivity, and motor potential of the human cerebral cortex preceding voluntary finger movements. *Exp. Brain. Res.*, **7**, 158–68.

Dobzhansky, Th. (1967). *The Biology of Ultimate Concern*. New American Library, New York.

Eccles, J. C. (1953). *The Neurophysiological Basis of Mind: the principles of neurophysiology*. Clarendon Press, Oxford.

Eccles, J. C. (1965). *The brain and the unity of conscious experience* (Eddington Lecture). Cambridge University Press, London.

Eccles, J. C. (1970). *Facing Reality*. Springer, New York, Heidelberg, Berlin.

Eccles, J. C. (1973). Brain, speech and consciousness. *Naturwissenschaften*, **60**, 167–76.

Feigl, H. (1967). *The 'Mental' and the 'Physical'*. University of Minnesota Press, Minneapolis, Minnesota.

Gardner, R. A. and Gardner, B. T. (1969) Teaching sign language to a chimpanzee. *Science*, **165**, 664–72.

Geschwind, N. (1972). Language and the brain. *Scientific American*, **226**, 36–8.

Geschwind, N. and Levitsky, W. (1968). Human brain: left-right asymmetries in temporal speech region. *Science*, **161**, 186–7.

Globus, G. G. (1974). Biological foundations of the psychoneural identity hypothesis. *Philosophy of Science* (in the press).

Huxley, J. (1962). Higher and lower organization in evolution. *Journal of the Royal College of Surgeons of Edinburgh*, **7**, 163–79.

Lazlo, E. (1972). *Introduction to Systems Philosophy*. Gordon and Breach, New York and London.

Levy-Agresti, J. and Sperry, R. W. (1968). Differential perceptual capacity in major and minor hemispheres. *Proc. Nat. Acad. Sci. USA*, **61**, 1151.

MacKay, D. M. (1966). Cerebral organization and the conscious control of action. In *Brain and Conscious Experience* (ed. J. C. Eccles). Springer, New York.

Moruzzi, G. (1966). The functional significance of sleep with particular regard to the brain mechanisms underlying consciousness. In *Brain and Conscious Experience* (ed. J. C. Eccles). Springer, New York.

Penfield, W. (1966). Speech and perception—the uncommitted cortex. In *Brain and Conscious Experience* (ed. J. C. Eccles). Springer, New York.
Penfield, W. and Roberts, L. (1959). *Speech and Brain-Mechanisms*. Princeton University Press, Princeton, New Jersey.
Polten, E. P. (1973). *Critique of the Psycho-Physical Identity Theory*. Mouton, The Hague.
Popper, K. R. (1950). Indeterminism in quantum physics and in classical physics. *Brit. J. Phil. Sci.*, **1**, 117–33.
Popper, K. R. (1968*a*). Epistemology without a knowing subject. In *Logic, Methodology and Philosophy of Science*, III (ed. van Rootselaar and Staal). North-Holland, Amsterdam.
Popper, K. R. (1968*b*). On the theory of the objective mind. In *Akten des XIV. Internationalen Kongresses für Philosophie*, Vol. 1, Wien.
Popper, K. R. (1970). Personal communication.
Popper, K. R. (1974). Scientific reduction and the essential incompleteness of all science. In this volume.
Premack, D. (1970). The education of Sarah: a chimp learns the language. *Psychology Today*, **4**, 55–8.
Schrödinger, E. (1958). *Mind and Matter*. Cambridge University Press, London.
Sherrington, C. S. (1940). *Man on His Nature*. Cambridge University Press, London.
Skinner, B. F. (1971). *Beyond Freedom and Dignity*. Knopf, New York.
Sperry, R. W. (1968*a*). Hemisphere deconnection and unity of conscious awareness. *Am. Psychol.*, **23**, 723–33.
Sperry, R. W. (1968*b*). Mental unity following surgical disconnection of the cerebral hemispheres. In *The Harvey Lectures*, **62**, 293–323. Academic Press, New York.
Sperry, R. W. (1969). A modified concept of consciousness. *Psychological Review*, **76**, 532–6.
Sperry, R. W. (1970*a*). Perception in the absence of the neocortical commissures. In *Perception and its Disorders*. Res. Publ. A.R.N.M.D., vol. XLVIII. The Association for Research in Nervous and Mental Disease.
Sperry, R. W. (1970*b*). Cerebral dominance in perception. In *Early Experience in Visual Information Processing in Perceptual and Reading Disorders* (ed. F. A. Young and D. B. Lindsley). Nat. Acad. Sci., Washington, D.C.
Sperry, R. W., Gazzaniga, M. S. and Bogen, J. E. (1969). Interhemispheric relationships: the neocortical commissures; syndromes of hemisphere disconnection. In *Handbook of Clinical Neurology*, chap. 14, 273–90. Wiley, New York.
Teilhard de Chardin, P. (1959). *The Phenomenon of Man*. Harper, New York.
Wada, J. A., Clarke, R. J. and Hamm, A. E. (1973). Morphological assymetry of temporal and frontal speech zones in human cerebal hemispheres: observation on 100 adult and 100 infant brains. Tenth Int. Con. Neurol., Barcelona.
Wigner, E. P. (1969). Are we machines? *Proc. Am. Phil. Soc.*, **113**, 95–101.

Discussion

Rensch

Chimpanzees have not developed a *speech centre*, but they can use other parts of the forebrain to replace the lacking language which is based on the use of sounds. Gardner's and Premack's chimpanzees learned a symbolic language to some degree by using gestures or plastic symbols. Premack's chimpanzee Sarah even learnt to use a series of abstract symbols like 'red', 'yellow', 'size', 'shape', 'colour', and she could give a right answer when she was asked: does 'yellow' mean a size, a colour or a shape? Helen Keller, who could neither speak nor hear, nor see, also seems to have replaced her lacking motoric speech centre by other parts of the forebrain.

Eccles

No! I believe the speech centre, more properly the linguistic centre, is concerned in humans in all manner of linguistic communication. For example it is used in writing, which can be called a sign language. And also it would be used in other sign languages such as in typing or the hand signs used by deaf mutes, or the Braille signs used by the blind. Helen Keller is a very good example. In all these examples the linguistic areas of the left hemisphere are concerned. With respect to the Gardner and Premack chimpanzees, I doubt if these clever learned responses can be regarded as a language even remotely resembling human language. With respect to Helen Keller, I would state that she was not lacking a 'motoric speech centre', but was merely lacking the input for its development until her teacher gave her a sign language.

Rensch

Although it is a good custom that the chairman is the last one to enter in the discussion, I would like to contribute a brief remark in order to avoid misunderstandings of *terminology*. In his fascinating lecture Sir John made a distinction between 'outer sense', 'inner sense' and 'self-consciousness'. Corresponding to classical epistemology his outer sense could be called sensations, and his 'inner sense' mental images and thought processes. Both types must be regarded as conscious phenomena.

As the *split-brain patients* have sensations also in the right hemisphere and as they seem to have also some simple memory there, we must attribute consciousness also to the right hemisphere.

With regard to the concept of the own self I already emphasised yesterday that this concept becomes gradually developed in human children and to some extent also in higher animals. As *Homo sapiens* is mainly thinking in words, this concept is probably developed with the help of the speech centre in the left hemisphere. We are forced to use words and think with their mental images because they represent symbols of concepts which circumscribe abstractions and generalisations, and among them also symbols for causal and logical relations. All our voluntary thinking and planning involves a certain connection with the concept of the own self. Normally the left hemisphere is therefore predominant in thought processes, although these may correspond to physiological processes of large parts of the whole brain in an intact person.

Eccles

I agree with the first remarks of Professor Rensch, with the exception that my third category is the 'pure ego' or 'self', not 'self-consciousness'. It is of course my thesis that both outer sense and inner sense are components of conscious experience. I do not deny consciousness to the right hemisphere, though as with animals we have no proof of this. It may be objected that we have no proof of self-consciousness of other human beings, and so are condemned to solipsism. I avoid this fate by formulating a criterion for establishing the existence of self-consciousness (World 2) in another living being. If I can relate to such a being at a level of World 3 communication in any mode, then I maintain that that being has self-consciousness, and so has a World 2 existence; that is, World 3 and World 2 relate to each other. This is only possible with human beings. It is worthy of note that human beings with grossly subnormal mentality, such as Mongoloids, easily qualify as self-conscious beings on this test because they have a true linguistic ability, whereas this has not been demonstrated for non-human primates.

8. Reductionism in Biology

W. H. THORPE

Before commencing a discussion of this subject, it seems to me essential to consider the evidence for reductionism in the physical sciences. But first we must make clear what we mean by the term 'reductionism', which is after all a word that has only recently come into general use in this connection. Thus it will be found that a great many writers who in fact discuss this topic deeply do not use the word at all.

Clerk Maxwell in 1877 pointed out that the term 'Physical science' is often applied 'in a more or less restricted manner to those branches of science in which the phenomena considered are of the simplest and most abstract kind, excluding the consideration of the more complex phenomena such as those observed in living beings'. Basing his thesis on this, Carl Pantin has classified the sciences into two groups: first, the 'restricted', which are the physical sciences, in which, it might seem, the practitioners thereof need employ only the concepts and methods characteristic of that science and have no need to turn for explanations to other sciences such as geology and the biological sciences. Those, by contrast, who study the 'unrestricted sciences' must be prepared to follow their problems into any other science whatsoever—including, of course, the physical sciences—in their explanations and descriptions, and because excursions into physics so often provide us with knowledge and formulations of the greatest value, there is a very strong tendency for biologists to consider this process of seeking physical explanations the most important part of their work and therefore to be overwhelmingly reductionist in their methods and outlook.

From the philosophical point of view, Hume can, I suppose, be regarded as one of the founders of modern reductionist thought. As Paul Weiss has said, to the superficial observer of nature, perhaps we might say the naive observer of nature, the universe presents itself as an immense cohesive continuum. But the scientific method inevitably entails the focusing of attention on small parts of the continuum and thus beginning a process of analysis into discrete objects, fragments or tendencies. When physics arrived at a reasonably satisfactory concept of atoms it began to be widely felt that the atomic underpinning of things must have greater reality than things directly observed.

As Garstens (1971, p. 88) says, this has undoubtedly provided a strong reason for the attempt in the reductionist philosophy to explain everything in terms of atoms. If we come to find this process scientifically or philosophically unsatisfactory, then we have to show that complex macroscopic structures are as basic in some ways as the atomic substructures. And so we may reach a modern definition of the term 'reductionism' as '*the attributing of reality exclusively to the smallest constituents of the world, and the tendency to interpret higher levels of organisation in terms of lower levels*'. (Barbour, 1966, p. 52). But if we find, as indeed we do find, that this process is incomplete and unsatisfactory as a general methodology of science, then we come to realise that besides this process of analysis we must all the time maintain the opposite process of synthesis; for without these two together all our activities are likely to be partial and limited in their results. The fact of the matter is that the analytical thinking which underlies reductionism is itself an abstraction from a more elaborate and complex reality, and in this sense the term 'restricted sciences' is a good one. As Bertrand Russell once said, 'Physics is mathematical, not because we know so much about the physical world, but because we know so little: it is only its mathematical properties that we can discover'. There is, however, a very important reason, so commonplace that it is often overlooked, why analysis is secondary to synthesis. This is that our perceptual systems are primarily adapted or 'designed' for synthesis (Gibson, 1966). Therefore our inevitable first response is to synthesise; to look for 'wholes'. Not until we have gone some way along this path of synthesis, of 'whole-making', of recognising 'objects', must we, or indeed can we, engage in the analytic process. But not only is analysis misleading, or worse, without prior synthesis; it is also meaningless and sometimes extremely dangerous without continuing, or at least recurrent periods of, synthesis.

If, then, reductionism is inadequate in spite of the enormous advances that it has yielded and continues to yield in science, we have to look very carefully for precisely those areas where it seems to fail us. In this connection it appears to me that the idea of 'emergence' in its strictest form is important, because it is so very easy to assume uncritically that we have in fact achieved a complete reduction to a simpler stage of reality when often we have indeed achieved nothing of the kind. I myself have found the formulation of C. D. Broad, given in his book *The Mind and its Place in Nature* (1925), of especial value. Broad did not doubt that he had successfully refuted what he called 'reductive materialism' in relation to mental qualities and used the term emergence to describe features in the world which could not be satisfactorily dealt with by a reductive process. It is, I am sure, worth quoting Broad's definition of emergence in full; it runs as follows: 'Emergence is the theory that the characteristic behaviour of the whole *could* not, even in theory, be deduced from the most complete knowledge of the behaviour of its components, taken separately or in other combinations, and of their proportions and arrangements in this whole.' Broad goes on to point out that if we want to explain the behaviour of any whole in terms of its structure and components, 'we *always* need two independent kinds of information. (*a*) We need to know how the parts would behave separately, and (*b*) we need to know the law or laws according to which the behaviour of the separate parts is compounded when they are acting together in any proportion and arrangement.' As Broad points out, it is extremely important to notice that these two bits of information are quite independent of each other in every case.

And so we see that the requirements for the complete proof of a reductionist conclusion are very severe indeed. The physicist Pattee summarises the general position admirably as follows: if we ask 'What is the secret of a computing machine?' no physicist would consider it in any sense an answer to say, what he already knows perfectly well, that the computer obeys all the laws of mechanics and electricity. If there is any problem in the organisation of a computer, it is the unlikely constraints which, so to speak, harness these laws to perform highly specific and directive functions which have of course been built into the machine by the expertise of the designer. So the real problem of life is not that all the structures and molecules in the cell appear to comply with the known laws of physics and chemistry. The real mystery is the origin of the highly improbable constraints which harness these laws to fulfil particular functions. This is in fact the problem of hierarchical control. And any claim that life has been reduced to physics and chemistry must in these days, if it is to carry conviction, be accompanied by an account of the dynamics and statics and the operating reliability of enzymes ultimately in terms of the present-day ground work of physics, namely quantum mechanical concepts. So we have two questions, 'how does it work?' and 'how does it arise?'. The second question has, in fact, two facets; (*a*) how does it arise in the development of the individual organism during the process of growth from the moment of fertilisation of the egg? and (*b*) how does the egg itself come to get that way?—that is to say, how can we conceive of evolution as having designed the cell?

It is a necessary concept of the hierarchy in biology which pinpoints the problem. And the problem is one of hierarchical interfaces. In common language a hierarchy is an organisation of individuals with levels of authority—usually with one level subordinate to the next one above and ruling over the next one below. For an admirable account of this, see Koestler (1969). So any general theory of biology (which must include the concept of hierarchy) must thereby explain the origin and operation, the reliability and persistence of these constraints which harness matter to perform coherent functions according to an hierarchical plan. Pattee says

> it is the central problem of the origin of life, when aggregations of matter obeying only elementary physical laws first began to constrain individual molecules to a functional, collective behaviour. It is the central problem of development where collection of cells control the growth of genetic expression of individual cells. It is the central problem of biological evolution in which groups of cells form larger and larger organisations by generating hierarchical constraints on subgroups. It is the central problem of the brain

where there appears to be an unlimited possibility for new hierarchical levels of description. These are all problems of hierarchical organisation. Theoretical biology must face this problem as fundamental since hierarchical control is the essential and distinguishing characteristic of life.

He goes on to point out that a simpler set of descriptions at each level will not suffice. Biology must include a theory of the levels themselves.

Pattee expresses himself as satisfied neither with the claim that physics explains how life *works* nor the claim that physics cannot explain how life *arose*. In his view, (I) the concept of autonomous hierarchy involves collections of elements which are responsible for producing their own rules as contrasted with collections which are designed by an external authority to have hierarchical behaviour. He then (II) assumes, of course, that they are part of the physical world and that all the elements obey the laws of physics. He limits his definition of hierarchical control (III) *to those rules or constraints which arise within such a collection of elements but which affect individual elements of the collection*. Finally, and perhaps most important, he points out (IV) that collective restraints which affect individual elements always appear to produce some integrated function of the collection. In common language this is to say that such hierarchical constraints produce specific actions or are 'designed for' some purpose. It is in considering the third of the above four statements in relation to classical mechanics that the difficulties are seen to be at their greatest. Classical physics appears to provide no way in which an explanation can be reached because it requires a 'collection' of particles which constrains individuals particles in a manner not deducible from their individual behaviour. However, it has been pointed out that in quantum mechanics the concept of the particle is changed and the fundamental idea of the continuous wave description of motion produces the stationary state or local time-independent collection of atoms and molecules. So it seems to be conceivable that hierarchical structures could be reducible to quantum mechanics, although as we shall see later the whole scheme of quantum mechanics is now in such confusion that, to the outsider, it seems far from clear to what extent they will be able to help.

To the ordinary working scientist there is an obvious course of action; perhaps one should call it a temptation. Having first assumed that there is a basic set of fundamental laws, the temptation is to proceed from there to what seems an obvious corollary, that as everything obeys the same fundamental laws then the only scientists who are studying anything really fundamental are those who are working on these laws. A physicist colleague of mine, to whom I am much indebted (Anderson, 1972), has pointed out that if this were so then the only scientists who would certainly be regarded as carrying out 'fundamental' work would be some astrophysicists, some elementary particle physicists, some logicians and other mathematicians and a few more. This reductionist point of view which seeks knowledge by analysis almost inevitably leads its proponents to assume, quite unwarrantably, that all that is then required is to work out the consequences of these laws by the prosecution of what is called 'extensive science' whereupon all truth will be revealed! But there is a tremendous fallacy here. For the success of the reductionist hypothesis in certain areas does not by any means imply the practicability of a 'constructionist' one: to reduce everything to simple fundamental laws does *not* imply the ability to start from those laws and reconstruct the universe. In fact, 'the more the elementary particle physicists tell us about the nature of the fundamental laws, the less relevance they seem to have to the very real problems of the rest of science, much less of society'. The constructionist hypotheses breaks down in the twin difficulties of scale and complexity. The behaviour of large and complex aggregates of elementary particles, so it turns out, is not to be understood as a simple extrapolation of the properties of a few particles. Rather, at each level of complexity entirely new properties appear, and the understanding of these new pieces of behaviour requires research which is as fundamental as, or perhaps even more fundamental than, anything undertaken by the elementary particle physicists. The standard procedure is to start from the laws which govern the motion of individual atoms and electrons and attempt to understand how large collections of them—particularly macroscopic solid bodies—behave. It is a truism, in physics at least, that as we look at phenomena of a very different scale we often find that the basic laws which govern the motion change. For example, on the cosmic scale Newtonian mechanics has been replaced by the more accurate Einstein theory; on the atomic scale Newtonian mechanics has been replaced by quantum mechanics; and on the subnuclear scale the laws are still changing. But it is not this kind of change in the underlying mechanical laws which is characteristic of the accompanying change of scale. In fact it is a mistake to be too analytical in one's approach and to assume that all new and fundamental laws come from logical analysis. They do not. Take the arguments for the building of a thousand billion electron volt accelerator cited by Anderson. We often hear it argued that, in short, intensive research goes for the fundamental laws, extensive research for the explanation of phenomena in terms of known fundamental laws. It is often assumed to follow that once new fundamental laws are discovered a large and ever-increasing activity begins in order to apply the discoveries to hitherto unexplained phenomena; thus the frontiers of science extend all along a long line from the newest and most modern intensive research, over the extensive research recently spawned by the intensive research of yesterday, to the broad and well developed web of extensive research activities based on intensive research of past decades. Thus, on this view, ordinary physicists are applied particle physicists, chemists are applied physicists, biologists are applied chemists, psychologists applied biologists, social scientists applied psychologists, etc. Anderson states

I believe this is emphatically not true: I believe that at each level of organisation, or of scale, types of behaviour open up which are entirely new, and basically unpredictable from a concentration on the more and more detailed analysis of the entities which make up the objects of these higher level studies. True, to understand worms we need to understand cells and macromolecules, but not mesons and nucleons. And even the comprehension of cells and macromolecules can never tell us all the important things that need to be known about worms.

At each level in fact there are fundamental problems requiring intensive research which cannot be solved by further microscopic analysis but need, as Anderson says, '*some combination of inspiration, analysis and synthesis*'. Another point is that clearly fundamental questions always seem to cluster around just the area where the scale changes. Thus in biology exciting things seem to occur at the interfaces with chemistry and with psychology—both scale changes in the sense intended. To take another example, aerodynamics: turbulence represents a real problem in principle because it spreads in scale from atomic to microscopic; meteorology is enormously interesting and difficult because the scales of importance stretch from very local to world-wide; and yet in both cases the fundamental laws of motion of the air have been known for decades or centuries.

The argument against reductionism from the biological point of view is perhaps best set out by stating that one can never hope to observe the whole repertoire of an organism if it is kept in isolation and observed solely in an artifically simplified environment. Thus a worker honey-bee could never be fully understood—that is to say, its possible responses and reactions could never be elucidated and itemised—if it were studied in isolation in an experimental cage or chamber. This is because the range, elaboration and specialisation of conditions and stimuli which it encounters in its natural life could *never* be duplicated experimentally without having first observed them in nature, and probably not fully even then. Similarly a nerve cell from the brain of a higher vertebrate could never be put through all its paces isolated in a tissue culture or other experimentally simplified situation. The fact is that a highly complex organ or system has capacities and potentialities that are the properties of the system and not merely of the components of the system.

As we go up the scale we find ourselves talking about increasing complication: and so, ascending the hierarchy of the sciences, we find fundamental questions leaping out at us at each stage as we try to fit together the less complicated pieces into a more complicated system and as we try to understand the basically new types of behaviour which can result. Thus students of animal behaviour find themselves concerned about sensitivity, contractility, habituation, learning, perception, memory, orientation, perceptual synthesis and mind—all topics which can be illuminated but not fully resolved by reductionist techniques. Coming up to the human level, since human beings can build and programme electronic computers it is *incontestable* that a human being is intrinsically a more complicated machine than a computer—'for instance, he has the possibility of *choosing* to do this thing himself, which the computer does not'.

Theoretical physicists point out that the fundamental laws of physics are completely independent of the sign of time. An electron cannot tell whether it is heading from tomorrow into yesterday or *vice versa*—cannot tell in any way. On the other hand we can tell time subjectively and in fact any large enough piece of matter can tell the time objectively. The rain falls down, not up; the sun shines, does not absorb light, and so on. This is the arrow of time referred to by Bronowski. Anderson himself comes back to the old argument about free will *versus* determinism. And he injects into this argument a new twist which some will dispute and which will no doubt require much debate. *In principle* it might still (according to some theoreticians) be argued that if a Cartesian superbrain really knew the wave function of the universe *now* it could calculate its subsequent history for ever. But the point is that the complication of this universe is so great that the superbrain or supercomputer would (and this is an exact statement) have to be infinitely bigger than the universe it was computing, and would never even be capable of gathering the appropriate information. So our computer, infinitely bigger than the universe, would keep falling behind just because it would be unable to transmit information faster than light.

This argument seems to restore free will at least in the Kantian sense, that is, a determinacy which can mean nothing to us: whether or not everything is a mechanism, it is a mechanism the details of which are in principle hidden from us. Anderson ends his discussion with the remarks that when he was asked to write something about his personal scientific philosophy he discovered for the first time that he had one! Moreover he found that one of the very central tenets was entirely different from what he would have expected: namely that the whole can be greater than and very different from the sum of its parts. He believes the trouble with so many physicists and others who expatiate upon matters of scientific philosophy is that they have never got round to asking themselves what their own philosophy really is.

But of course a great many physicists of great distinction, and especially quantum physicists, have indeed got round to asking themselves what their philosophy is. And the result to the enquiring biologist at least is both surprising and disturbing. The basic problem which confronts them is that of measurement (which is an event in time) and the 'timeless' logic of the world of quantum theory. It has been said (Bastin, 1971) that in Dirac's theory these difficulties are settled in one stroke at the outset. The mathematical principle of superposition of states is ascribed directly to the necessity of regarding a single particle or a single photon as being in each of two states prior to observation and as being definitely in one or other state as the instantaneous result of the observation. Dirac postulates no collapse of the wave-function which in other accounts is invoked to give some physical meaning to the alleged effects of the observation process. Dirac's method is reminiscent of one theological tradition for dealing with a mystery and is probably the best that can be done for no-nonsense formalism. He leaves the mystery as stark as possible and says that the facts in the physical case (it is revelation in the theological) demand it.

Bastin concludes that one is therefore forced to contemplate a situation, which must be unique in the history of science, where the practitioners of a scientific theory which has reached the stage of being regarded as a finished product habitually work with a jumble of elements taken from a variety of different conceptual frameworks, none of which, singly, is adequate to present the facts that are known, and each of which is partly or even largely incompatible with the rest. What keeps these practitioners feeling that it is one discipline they are operating is quite a puzzle. Partly it is a faith in the existence of a growing body of knowledge which constitutes physics and which persists massively unchanged whatever we may add on to it in the way of revolutionary principles or discoveries.

Professor O. R. Frisch (1971) points out that the measurement is not completed till something irreversible has happened, and that means that *irreversibility is an essential part of a measurement*. Information is something which is obtained at some time and continues into the future. He goes on to say that quantum theory predicts that whenever one puts two polaroids in any arbitrary direction, at right angles to each other, if one photon were to go through one of them, then 'another' photon would be sure to go through the other one. This is like saying that when one photon arrives at its polaroid, if it goes through, it telegraphs back—with speed exceeding that of light—to tell the other photon 'go through your polaroid'; or 'please adopt the polarisation corresponding to the one which has just been measured for me'.

Should we then say that space is three-dimensional but that particles are in collusion; that they, as it were, communicate in some way or another over great distances? Frisch says that if you try to work out a signalling system with more than the speed of light on this basis it does not work. I have cited these varied examples to show that to the perhaps naive biological enquirer there are some physicists who seem to be clinging hopefully to reductionist determinism as at least a theoretically possible position, while others, probably the majority, conclude it to be in principle inconceivable. Doubtless others here, with far more competence than I, will go further into this issue.

As for me I wish now to look at this problem of reductionism in biology as posing three questions—(1) where can it reasonably be assumed? (2) where can it reasonably be doubted? and (3) where, if at all, is it impossible to conceive?

In discussing reductionism in biology I intend to omit almost completely the greatest of all biological questions; namely, that of the origin of life. I think it fair to say that all the facile speculations and discussions published during the last 10–15 years explaining the mode of origin of life have been shown to be far too simple-minded and to bear very little weight. The problem in fact seems as far from solution as ever it was. To take simply one difficulty to be overcome in the process of the build-up, from amino acids in aqueous solution, of complex proteins or enzymes: Dixon and Webb (1964) calculate that an amount of protein of M.W. 60 000 equal in weight to the earth would contain 6×10^{46} molecules. The number of possible proteins of this M.W. is 10^{650}. Therefore at any instant not more than one in 10^{603} possible proteins could exist at all. So the simultaneous formation of two or more molecules of any given enzyme purely by chance is fantastically improbable. Since we yet really know so little about the age, dimensions or complexity of the universe, we can really make no useful calculations about the probability of the occurrence of life's origin. For reasons such as these it is entirely reasonable to consider that the origin of life may have been an unique occurrence. And about a truly unique event science can say nothing.

The origin of even the simplest cell poses a problem hardly less difficult. The most elementary type of cell constitutes a 'mechanism' unimaginably more complex than any machine yet thought up, let alone constructed, by man. As Monod (1971) says, to theorise about the origin of the first 'cell' perhaps one thousand million generations back poses several Herculean problems—such as the synthesis of the total metabolic system of the cell, the origin of the genetic code and its translation mechanism (which in turn accounts for the organism being self-programming and self-reproducing) and the development of the selectively permeable membrane. There is no real clue as to the way in which any of these riddles were solved, so it is open to anyone to espouse any theory which he finds helpful. But for anyone to state as a matter of conviction that the reductionist approach is the proved key to the understanding of any of these biological riddles justly exposes him to being branded as a charlatan. For belief in such a theory, however firmly it may be held, amounts to nothing more than blind faith unsupported by valid scientific evidence.

These above-mentioned examples are then instances of our question 2 above.

Question 1 need scarcely detain us because it is overwhelmingly obvious that immense areas of modern biology provide clear evidence for at least the partial success of the reductionist approach. One need only instance the unravelling of the story of the role of hormones in physiology and behaviour or Hodgkin's analysis of the nature of the nervous impulse as a physico-chemical system. But granting all this there is a sense in which all discoveries of this type are partial in that each forward step reveals a series of further problems each calling for more sophisticated and elaborate techniques, a seemingly infinite regress. It is this feeling that has given rise to the bitter aphorism that 'a specialist is one who makes no errors in his steady approach to the grand fallacy'!

Let us now turn our attention to the other great frontier of biological knowledge and consider its problems in the light of the reductionist hypothesis.

This is, of course, the problem of the nervous system, particularly as it functions in the brain of man and of the higher animals and the correlated problem of the nature of 'mind', self-consciousness and 'knowing'.

Living organisms can be defined with much cogency as self-reproducing and self-programming systems which store and organise information. Molecular biology has thus thrown an immense flood of light on the first or second of these points. But the storage and organisation of information is perhaps the most basic feature of animals. This aspect can be expressed by saying that 'animals are perceiving organisms'. That is to say animals are characteristically capable, in vastly varying degrees, of 'perceptual synthesis'. The very storage of information implies, as the quantum physicists have emphasised, irreversible events in time. It follows that organisms, particularly those provided with organs of perception, show behaviour which is '*directive*' in the sense that they have been programmed during evolution to be adapted to their environment (see Thorpe, 1963, especially Chapter I). If this is accepted there remains the key question—how far are animals also purposive in the sense that human beings are aware of a purpose, however short-term it may be? This amounts to enquiring how far animals possess 'consciousness'? —how far are they 'experiencing selves'?

This topic is best approached then by considering the evidence for perceptual synthesis. To show how complex this may be, even in minute animals which have only exceedingly simple sense organs and very elementary nervous systems, *Microstomum*, an aquatic, free-living flatworm of the order Rhabdocoela is worth careful consideration. This minute creature represents an evolutionary stage in which the nervous system is hardly more complex than that of the sea anemone. Its 'brain' is little more than a thickening of each ventral cord or a bi-lobed mass springing from the dorsal surface of these cords. Yet, as Lashley says, 'here in the length of half a millimetre are encompassed all the major problems of dynamic psychology'. This worm is equipped with nematocysts or stinging cells like those of the hydroids, which it discharges in defence and in capture of prey. It does not grow its own weapons but captures them from the freshwater *Hydra*, which is eaten and digested until the undischarged stinging cells lie free in the *Microstomum*'s stomach. These pass through the walls into the mesoderm where they are again picked up by wandering tissue-cells and carried to the skin. When the cell carrying the nematocyst gets to the surface, it turns round so as to aim the poison tube outwards; it then grows a trigger and sets the apparatus to fire on appropriate stimulation. When *Microstomum* has no stinging cells, it captures and eats *Hydra*. When, by contrast, a sufficent concentration of the cells is reached the worm loses its appetite for *Hydra*, and will starve to death rather than eat any more of the polyps, which are apparently not a food, only a source of weapons. Here, then, is a specific appetite satisfied only by a very indirect series of activities, a recognition and selection of a specific object, recognition of the undischarged stinging cells by the wandering tissue-cells, and some sort of 'perception' of its form so that it may be aimed. The uniform distribution of the nematocysts over the surface suggests a *Gestalt*. So striking are these facts that Kepner was driven to postulate a group mind amongst the cells of the body to account for the internal behaviour of the *Microstomum*. Such a conclusion seems to us absurd: but it is to be remembered that behaviour such as this, while striking the ethologist with amazement, is a commonplace of embryology—though the embryologist has often no better theory for explaining it than has the ethologist.

The perceptual synthesis of many insects is quite as remarkable, though, in the higher Hymenoptera for instance, the sense organs (compound eyes and organs of chemical sense) are highly elaborated. *Ammophila* and similar hunting wasps dig burrows in which they place a series of paralysed caterpillars for their offspring to devour when they hatch from the egg. When leaving to fetch another caterpillar the wasp closes the entrance with great care. Yet on return it locates the exact spot without difficulty. It does this by observing the position of familiar objects like pine cones or stones, trees or small bushes in the neighbourhood. In fact the wasps learn the details of the 'landscape' in the region of their burrow with extraordinary exactness. If, while a wasp is away hunting, one removes one or two apparently obvious landmarks in the neighbourhood of the entrance, the animal will still nevertheless find the hole just by using other landmarks which the experimenter had not yet thought of. In fact the wasp's visual perception is organised in a very highly detailed manner so that in order to confuse it on its return you have to do quite a lot of 'landscape gardening'! This perceptual synthesis can be accomplished by many of the higher Hymenoptera but particularly the hunting wasps and the colony-forming bees. Every experienced beekeeper knows that if you move the hive from one site to another, then those bees which are out foraging at the time that the hive is moved of course return to the old place and are very unlikely to be able to find the hive in its new location unless it is very near. If, however, you move the hive at night or before opening it in the morning while all the bees are inside, one notices that the foragers instead of leaping out of the entrance and flying straight off as they normally do will now hesitate and then perhaps circle around the hive for a few seconds or a minute, clearly learning the new landmarks, by what one may call a 'survey flight', before they go off. So in the honey bee and many related forms it is clear that the perceptual and communicative abilities are very highly organised indeed. But one can find essentially the same thing much lower in the animal kingdom. It can be found in a very elementary form in limpets and many other molluscs and of course in many fish; in fact this phenomenon occurs in virtually every kind of animal which has a nest or home or shelter which is important for it to be able to return to.

One of the most remarkable examples of exploratory learning, as it is often convenient to term this perceptual synthesis, is provided by the Gobiid fish *Bathygobius soporator*, a species which inhabits tidal pools in the Bahamas and other tropical shores. Aronson (1951, 1971) finds that the fish are so well orientated that they are able to jump from pool to pool at low tide without running any significant risk of finding themselves on dry land. That is, at low tide it is obvious that they know the layout of the pools sufficiently well to enable the fish to leap correctly even though it is perfectly clear that they cannot see one pool while they are swimming in another. It seems hard to avoid the conclusion that these gobies swim over the rock depressions at high tide and thereby acquire an effective memory of the general features and topography of the limited area around the home pool—a memory which they are able to utilise at low tide when restricted to the pool. This makes abundantly clear the remarkable precision of perceptual knowledge which the fishes have acquired. Even more remarkable are the perceptual achievements during the migratory and homing performances of fish such as salmon, trout, white bass, etc. Most fish migrations consist of (*a*) a dispersal of eggs, or larvae, or young fishes, either by drifting passively with the current or by an active search for normal habitats; (*b*) return journey—an active movement usually against the current to the spawning grounds; and (*c*) a dispersal of the 'spent' fishes, a process which may again be either passive with the current, or active, in search of fresh feeding ground.

Little can be said as to the means of maintenance of ordinary periodic migration movements. In some cases it is impossible that learning of the topography can play any part in this, and there seems to be a general lack of evidence that schools of young fish follow, or are in any way led by, experienced adults. So it must appear that the regularity and consistency of the main migration routes of fish are a product of the innate organisation of the fish, together with a sensitivity to current, water temperature, food supply and other characteristics of the environment. As soon as fish-marking techniques were sufficiently well developed to yield results, evidence began to come in showing that some species of fish, particularly of the genera *Salmo* and *Oncorhynchos*, were returning to spawn, not merely in the geographical area in which they were hatched, but actually in the same stream, and even identical portions of the stream, which they had inhabited in their earliest youth. There is now clear evidence that a high proportion of some species of salmon and trout succeed in returning to the stream in which they had emerged as youngsters; and that very few individuals of these species enter streams a hundred or so miles away from the streams in which they were reared. Transplantation experiments carried out on Chinook salmon (*Oncorhynchus tschawytcha*) showed that all the fish from transplanted eggs which were recovered were found in the river in which they were brought up and not in the river in which they were spawned. Again it was found that over four years of experiments 97·9 per cent of steelhead trout (*Salmo gairdneri*) returned to the home stream and only 2·1 per cent were seen four miles distant. Similar figures for silver salmon (*Oncorhynchus milktschitsch*) in the same streams were obtained. It is now quite clear that the return is not to be explained on the assumption that the intervening period has been spent in the near neighbourhood of the native stream. In fact quite the contrary—as research shows for the Chinook salmon. It is now evident that one species (*O. nerka*) normally travels approximately 100 miles a day when the urge to return is upon it; and there is a record of a salmon (*S. salar*) which travelled at the rate of 60 miles a day for 12 days.

The next problem concerns the sensory stimuli which serve as guides or reference points during these astounding homing performances. There are certainly a large number of possible stimuli to be considered, of which water currents are amongst the most important. However, fish will very rapidly learn to keep a particular course and direction, employing orientation by the sun as a basis and using many different clues for keeping steady on course. But it seems clear that, in many cases, the accurate return cannot be the result of visual memory of the home or the route therefrom. The young salmon descending for the first time proceed slowly, 'playing about' along near the shore, probably drifting passively for large stretches of the route. The returning adult swims strenuously in deeper waters of the same river, and the two pathways must then often be separated by a distance greater than the visual range. So we have to fall back on the assumption that it is the ability of the salmon to perceive and remember the chemical characteristics of the water of the stream bed which enables the animal to achieve the apparent *tour de force* of returning to the stream of its nativity, anything from two to six years after having left it. It has been shown that the organs of chemical sense can appreciate small, but probably constant, differences in the characteristics of the water from different streams and currents—including of course differences in salinity. Moreover a learned olfactory preference for the waters of the home creek has been demonstrated experimentally. Then there are also differences in temperature, hydrogen ion concentration, proportion of dissolved gases, temperature and density stratification and the general turbulence. Added to this there might be recognition of the characteristic sounds made by waterfalls and rapids, the memory of the general nature of the river bottom (which of course might be partly visual), and perhaps the memory of the type of food to be obtained there. Finally, as regards the route in deeper waters of the sea, we must not forget the possibility that many fishes produce noises which may provide for echo-sounding against the sea bed. Or—in the absence of noise produced by the species itself—there is the possibility of the perception of the resonance effect of the surface wave-noise on the bottom. Taking into account all these sources of stimuli for orientation, let us consider what the homing performance of *Salmo* entails. Translated into laboratory terms, it is very like learning to run a gigantic and complex maze in reverse as a result of one experience of that maze 2 to 6 years before! It is not necessary to assume that the fish remembers every detail of the hundreds of miles which its journey may cover; but one must suppose it remembers the characteristics of the different sections and particularly the features of the various 'choice points' constituted by the junctions of the tributaries of the main rivers. It may of course be that many of these junctions do not offer as free a choice as may appear at first sight. Yet, however much we try in our imagination to simplify the problem posed by the returning fish, it remains a most astonishing performance.

I now wish to discuss a facet of bird behaviour which illustrates more dramatically the astounding powers of perceptual organisation which many of these creatures possess. Everyone of course knows that some birds migrate and that in the northern hemisphere they tend to go south-west in autumn and north-east in spring. They do this by maintaining a particular direction relative to the sun, allowing for the sun movement during the daylight hours. It goes without saying that in any species where the young birds migrate to winter quarters before the adults, the former must possess the ability to fly in a constant direction. It has now been well established that, as a result of experiments in 'orientation cages' with birds under the hormonally controlled urge to migrate, many birds will flutter in that compass direction in which they 'should' be flying in at that time of year. Under these conditions some birds such as starlings (*Sturnus vulgaris*) remain orientated, that is, flutter in the correct direction, only so long as they can see the sky and the sun is not obscured. If the apparent direction of the sun is changed by mirrors, the orientation of the birds changes accordingly. The direction of the sun of course changes with the time of day, but the migration direction does not. The bird must therefore be able to correct for the movement of the sun—that is, it must have some sort of internal clock mechanism. The existence of this clock or chronometer which depends on the light–dark cycle, and can be upset by providing an artificial cycle out of phase with the natural one, has been well established by experiments. The exact method by which the sun is used, however, is still somewhat doubtful. It seems, however, that some species such as starlings can orient by using the azimuthal direction alone while others may be able to extrapolate the sun's observed path to find the highest point, which in the northern hemisphere is always due south and so can serve as a fixed reference point. Nocturnal migrants of course face other problems. Various warblers are able, when migrating in autumn, to maintain correct direction even when they can only see the central part of the sky. However, they become disoriented when the stars are hidden by cloud. In these cases it seems fairly clear, though perhaps not certainly proved, that the birds are responding to the form or *Gestalt* of stimuli provided by the star pattern, especially those in the neighbourhood of the north star. Even the relatively simple task of maintaining a direction on a migratory flight involves formidable sensory and perceptual problems which are not yet fully understood. It has often been claimed that the birds have a magnetic sense serviceable as a compass, and at the time of writing some evidence is being produced that, under certain conditions, some species of birds may be able to detect magnetic forces sufficiently exactly to render possible their use in orientation (Alder, 1971; Keeton, 1971).

Birds can succeed in far more complex orientational tasks than this. It has long been known that birds forcibly removed from the nesting areas during the breeding season, and transported long distances before release, are often able to return to the nest with such speed and reliability that random search for familiar landmarks cannot possibly be the sole explanation. A few instances will show this. Experiments with oceanic birds provide the most telling examples. Thus the Manx Shearwater (*Procellaria puffinus*) was the first species used in critical experiments (Matthews, 1968). Shearwaters taken from their breeding burrows on the island of Skokholm, off Pembrokeshire, transported in blacked-out boxes to Cambridge, and there released, covered the return journey to Skokholm (approximately 290 miles) in some cases in no more than 6 hours. Since this species never normally flies over land, almost the whole of this route must have been completely unfamiliar. Even more remarkable was a bird of the same species which, taken by air to Boston Harbour in the United States and then released, returned to its Skokholm burrow in 13 days having covered the journey of 3050 miles at an average speed, assuming daylight flight only, of over 20 mph. In this case again the bird can never have been familiar with the East coast of North America and the western Atlantic since this area is beyond its normal geographical range. One other case, and a very striking one, among oceanic birds may be given, namely that of the Laysan Albatross (*Diomedia immutabilis*). In a number of experiments the longest homing flight was from the Philippines to Midway Island, a line of 4120 statute miles covered in 32 days. The fastest flight by this species was from Whidbey Island, Washington: 3200 statute miles in 10·1 days, which equals 370 miles per day—though the performance of the Leach's petrel which covered nearly 3000 miles at an average of 217 miles per day (Billings, 1968) is in some respects even more astonishing. Again with some of these birds releases were outside the normal range of the species. In a significant majority of these cases the birds on release set out immediately in approximately the correct direction for home. This implies that the bird released from an unknown point is already goal-oriented and that it can perform the equivalent of fixing its present position on a grid of at least two coordinates, calculating the course to steer to regain the coordinates characteristic of home, and steering it. This does not mean of course that the bird goes through the sequences of calculations necessary for the human navigator who uses say, a wireless position line, a sextant observation of the sun, nautical almanac, chart, a ruler and protractor and so on. The knowledge of how this is achieved is yet so incomplete that we are unable to say what measurements the bird takes to fix its position on release. It is, however, highly improbable that it is responding to forces resulting from the earth's magnetic field, and it is not likely to be responding to forces relative to the earth's rotation (Coriolis force). Observation of the sky seems to be important, for many species are disorientated when the sky is overcast and they home less well if confined where they cannot see the horizon from the point of release than they do if the whole sky is visible. Matthews (1968) suggested that diurnal birds obtain the necessary information from observation of the sun's arc, provided they could 'remember' the characteristics of the sun's arc at home. For this system to suffice, the following 'automatic' measurements and comparisons will be required:

(1) The observation of the sun's movements over a small part of this arc, and by extrapolation the determination of the highest part of the arc. This will give the geographical south and local noon.

(2) Comparison of the remembered noon altitude at home with the observed noon altitude. This will give the difference in latitude.

(3) Comparison with home position's azimuth at local noon. This will give the difference in longitude which alternatively might be appreciated as a direct time difference which, on present evidence, seems more probable.

Astonishing though these conclusions are, they seem to be inescapable on present information and in fact this theory is simpler than many other hypotheses that have been advanced. The honey bee (and a great many other invertebrates), as is well known, has the ability to detect the polarisation pattern of light reflected from the blue sky to aid it in its orientation flights; but birds have no such abilities.

The question of the actual path followed by birds on homing flights has been much illuminated by the development in recent years of radio transmitters sufficiently minute to be attached to migrating birds without interfering with flight abilities. When this is done, it often transpires that the tracks are much straighter than we previously had any reason to suspect. Still more extraordinary are some of the flight paths of nocturnal spring migrants carrying radio transmitters weighing only 3 grams. Thus a Swainson's thrush (*Hylocichla ustulata*) took off on its spring migration at 20.00 hours and was shadowed through the night until it landed 8 hours later 450 miles north-west, achieving a performance even more remarkable than seems at first sight. For the actual track flown measured only 453 miles. Even if the bird was flying a compass course and not homing to its breeding quarters, such accuracy, as Matthews remarked, is the envy of the human navigator. In another case, a grey-cheeked thrush (*Hylocichla minima*) was tracked for 400 miles. It took off at 19.55 hours under a clear sky and flew NNE at 20 mph helped by a tail wind of 22 mph. After 140 miles its course took it over Lake Michigan, while the plane which was following it, for safety, had to go round over land. However, contact with this particular bird was renewed in Wisconsin at 02.48 hours, but then the plane had to turn back because of a thunderstorm! It must be accepted of course that the bird had an initial innate ability to perceive and coordinate from various stimuli in the environment such as rate of movement and altitude of the sun and possibly also of the stars. In the case of the thrushes just discussed it is to be presumed that they have some visual knowledge of the environment and may have flown something like this course before, at least in the opposite direction, a few months previously. But with night fliers it is extremely doubtful whether visual aids would be much help; particularly in the case of the monumental flight in a thunderstorm over Lake Michigan! And with the homing experiments, quite clearly the bird can have nothing in the way of a plan or map in its head based on previous experience to guide it. With homing birds, we can say with conviction that they have compass orientation to start with and landmarks nearer home towards the end of the flight to guide them: between these all must, so it seems, be accomplished by true navigation. With night fliers, and indeed often with day fliers, celestial cues, whether from stars or sun, will often be lacking. And even when it would seem possible that nocturnal migrants might be orienting by recognisable and known landmarks, a bird may often in practice seem to disregard landmark features as navigational aids; for birds migrating over the Gulf of Mexico do sometimes appear to alter their direction on the basis of topographical cues though, when inland, large rivers do not seem to be employed as guides. There are many other problems of perceptual synthesis which might be discussed and are certainly relevant. But it can be said with little fear of contradiction that the homing achievements of birds and fishes are unsurpassed by the mammals, and no more remarkable examples of perceptual organisation are to be found in the animal kingdom until we come to higher ranges of that group, namely the Primates. Particularly disturbing to the neurophysiologist is the fact that bees and ants, with a brain made up of certainly not more than 10^5 cells (and very much fewer in the ants), display perceptual abilities so little inferior to those of the higher vertebrates.

My main reason for giving so much weight to problems of spatial position, of orientation and of direction-finding is that they lead us naturally to the problem of conscious self-awareness (the experiencing self, to quote Eccles). And this is the problem above all problems for the reductionist since many, if not most, philosophers of science (as well as many neurophysiologists) would agree that 'mind' as known to our own consciousness is something irreducibly different from the awareness that science has built up of the animal body as a material mechanism. In fact I find dualism based on this conviction to be very widespread indeed in the scientific fraternity—including most of those present at this conference. So the problem arises, since we can only 'know' consciousness in ourselves as individual persons, how far (if at all) down the animal scale are we scientifically justified in considering 'consciousness' to be present?

It has frequently been assumed that in the more primitive animals the beginning of consciousness was in some way connected with the first appearance of the ability to combine different sensory modalities to provide an elementary map of the environment so that for the first time in evolutionary history animals could 'find their way about' (see Thorpe, 1963; Pantin, 1968). In that it suggests that the development of proprioception marks an especially important stage, this view receives much support from recent work in the USSR summarised by Razran (1971) who comes to the conclusion that 'the entire reach of consciousness inheres in the proprioceptive sequelae, or kinestheses of sensory-orienting reactions: liminal consciousness, or sensation, in those of unintegrated or even individual reactions, and organised consciousness, or perception, in the integrate transform of the reactions'. He goes on to propound the view that the liminal consciousness of a sensation is anchored to a particular modality or quality, whereas perception is largely intermodal or supermodal in quality. He also quotes with approval Gibson's statement that 'proprioception or self-sensitivity is . . . an overall function, common to all systems, not a special sense, and that the activity of orienting, exploring and selection extracts the external information from the stimulus flux while registering the change as subjective feeling'.

This highly simplified and very brief account of the development of a conscious experience will serve to refute dramatically the view implied by the Identity Hypothesis that as soon as neural activity lights up in the cerebral cortex there is a conscious experience. In fact there is an intense ongoing activity in the cortex of the awake subject in the total absence of any specific sensory inputs; and as everyone knows there is even activity during sleep, an activity which is enhanced during periods of dreaming. So it is clear that time is required both for synaptic transmission and for the enormous development and elaboration of neural patterns which are apparently needed to establish conscious experience.

Then it must be recognised that intention comes into the picture. Only a very small fraction of the complex ongoing patterned response of the cerebral neurones is experienced. The remainder fades away unobserved; and of course it is essential that it should do so, for one can imagine the confusion that would result if we actually experienced the totality of the cerebral patterns in our cortex. Now a very obvious feature of our conscious experience is that with training and learning, from infancy upwards, we have the power of relegating many very elaborate actions and responses to the subconscious or unconscious levels where they are, so to speak, so automated that they can take charge of an enormous preponderance of our life activities, to our great relief and contentment. Everyone knows how car-driving becomes automatic, just as does piano-playing, with the result that we can drive safely while at the same time conversing and observing the scenery, with just enough unconscious or subconscious attention to the highway to arrive safely. Then, when an accident threatens, suddenly consciousness takes over, and many who have been in this unenviable situation report that time seems suddenly to have stretched until seconds apparently become minutes so that at least we have ample time to decide on the best strategy.

Sherrington dramatically imagines the behaviour of the cortex as like a loom weaving 'a dissolving pattern, always a meaningful pattern, though never an abiding one; a shifting harmony of sub-patterns'. This is his 'enchanted loom'. So we have all the unimaginable complexities of operation in a neuronal machinery of the brain generating a level of complexity transcending any human evaluation. Over against this there is the problem defined by Sherrington of the conscious experiences that are quite different in kind from any goings-on in the neuronal machinery; nevertheless, events in the neuronal machinery are, *as far as the physiologist is concerned*, assumed to be a necessary condition for the experience. Here I agree with Eccles that though they are a necessary condition they are not a sufficient condition; for even the most complex dynamic patterns played out in the neuronal machinery of the cerebral cortex are in the matter–energy world, whereas transcending this level and in emergent relationship from it is the world of conscious experience.

These remarks make clear the importance, indeed the absolute necessity, of attention and awareness in building up our own mental unity and individuality throughout our lives from babyhood onwards. We are in fact creating our minds step by step all our lives. Some of this creation, perhaps much of it, is certainly at the subconscious level, particularly in our earlier days. But although the subconscious or unconscious mind can do marvellous things and can cope with much learning of the elementary conditioning sort, yet it is unthinkable that anything like a mature, responsible, fully socialised human being could develop without the continual and overriding control established by our conscious will. Thus although looked at from the strictly neurophysiological point of view interactions may seem to be one-way—from neurological states to conscious experience—there must in fact be two ways, for if it were not so, no human being could ever develop. So whatever theory we may hold, no thinking person in his senses can deny the existence and the importance of consciousness. If I may quote a few remarks that I made in a discussion on this topic in 1966:

> Consciousness is a primary datum of existence and as such cannot be fully defined. The evidence suggests that at the lower levels of the evolutionary scale consciousness, if it exists, must be of a very generalised kind, so to say unstructured. And that with the development of purposive behaviour and a powerful faculty of attention, consciousness associated with expectation will become more and more vivid and precise.

As these remarks reveal, I am thinking all the time of the possibility of the existence of consciousness in animals other than ourselves. And indeed, as you will have gathered, I do indeed find it essential to assume something very similar to consciousness and conscious choice in many of the highest animals. So to me one of the important considerations is (even admitting that organisational changes and developments, including those of great complexity, can occur in both the human and animal brain below the level of consciousness) that conscious choice and awareness have an absolutely essential function in individual development, and therefore, I think we can confidently assert, exist in at any rate the later stages of the evolutionary process. So one of the key questions to be asked (a topic worth brooding on) is how far and at what levels can we regard the development of consciousness as a valuable factor conferring selective advantage in the evolution of animals and men? William James in his *Principles of Psychology* (1890) says 'Consciousness is what we might expect in an organ, added for the sake of steering a nervous system grown too complex to regulate itself.' This now sounds to us very naive, so much of the complex steering systems of the higher animals having been explained on reasonably self-regulating and automatic bases. And where there are adequate cybernetic principles at work consciousness may seem redundant. Indeed, if self-regulation were the chief or sole criterion as evidence for consciousness, then all animals and plants must be conscious—a conclusion which strikes the modern biologist as absurd. The evolutionary problem is, given a brain of a given degree of elaboration, is it likely to be a more effective mechanism—more effective in an evolutionary sense—if it has consciousness, so to speak attached, than if it has not?

I am of the opinion that the answer must be 'yes'. This answer implies the assumption that, for a given complexity and elaboration of neural machinery, greater efficiency will be attained if it is accompanied by some form of conscious choice and conscious will. It may be that the degree of conscious development is linked directly to brain size. At least it seems clear that in the human brain 'mentality' cannot be explained merely on the basis of neural convergence 'pursued to culmination in final supreme convergency on one ultimate pontifical nerve cell . . . as the climax of the whole system' (Sherrington, 1940). The brain region we may call 'mental' is not a concentration on to one cell but an enormous expansion into millions. 'Where it is a question of "mind" the N.S. does not integrate by centralisation upon one pontifical cell. Rather it elaborates a millionfold democracy.'

I come now to the thorny question of language and its relation to speech. The general colloquial use of the term 'language' is so vague as to be of little value. But there has been for many years a widespread tendency both among students of animal behaviour and a good many physiological psychologists (such as Teuber, 1967) to use the term language for many of the more elaborate examples of communication amongst animals, especially for the transfer of social information and particularly when the transfer is vocal or auditory. While there is much to be said for this, the usage in the past has perhaps been too naive (Thorpe, 1968). Hebb and Thompson (1954) proposed that the minimal criterion of language is twofold. First, language combines two or more representative gestures or noises purposefully for a single effect, and secondly it uses the same gestures in different combinations for different effects, changing readily with circumstances. It will be quite obvious from the current work of R. A. and B. T. Gardner or of D. Premack that not only the communication of the chimpanzee Washoe with her associates but also the communication of many birds and some other animals comes within this definition of language (see tables 1–3). There remains, however, the problem of intent or purpose. To express my own view, and I do not wish to in any way saddle the Drs Gardner with it, I would say that no one who has worked for a long period with a higher animal such as a chimpanzee, particularly in the circumstances of the Gardners' work, is justified in doubting the purposiveness (not merely the directiveness) of such communication. I believe such purposiveness is also clear to the experienced and open-minded observer with many of the Canidae, with some, probably many, other mammals, and with certain birds (Thorpe, 1966, 1969). Some of the philosophical depth and significance of this question of purposiveness, both for students of animal behaviour and for philosophers such as Price and Whitehead, will be found in Pantin (1968).

It would no doubt be easy to devise definitions of language such that no examples of animal communication could readily find inclusion therein. There have always been, and no doubt there will continue to be, those who resist with great vigour any conclusions which seem to break down what they regard as one of the most important lines of demarcation between animals and men. We must surely be justified in accepting such preconceived definitions only with the utmost caution. One of the tasks of the scientific student of animal behaviour is to attempt to establish whether there are such hard and fast dividing lines, and if so what and where they are. Of one thing we can be certain: it is that work such as that of the Gardners and of Premack is only the beginning of the application of an important and powerful new technique from which we stand to learn much in years to come. I believe that no one should have anything to fear from its cautious and objective application. Personally I believe it is safe to conclude that if chimpanzees had the necessary equipment in the larynx and pharynx they could learn to talk at least as well as can children of three years of age, and perhaps older.

Finally, 'Could they count?' The answer is, I believe, 'Yes, up to the number seven'. For the now classic work of Otto Koehler and others on the 'unnamed number' concept in birds and mammals has demonstrated beyond doubt the existence in animals of an abstract 'concept of number'. But man can manipulate abstract symbols to an extent far in excess of any animal, and that is the difference between bird 'counting' and our mathematics. I think we can sum up this matter by saying that although no animal appears to have a language which is propositional, fully syntactic and at the same time clearly expressive of intention, yet all these features can be found separately (to at least some degree) in the animal kingdom. Consequently, bearing in mind the work on chimps discussed above, we can say that the distinction between man and animals, on the ground that only the former possess 'true language', seems far less defensible than heretofore.

There is another topic of great social and practical concern in the field of psychiatry with which I wish to conclude. There seems little doubt that the

Table 1. A comparison of the communication systems of animals and men based on the design features of Hockett (from Thorpe, 1972)

Design Features (all of which are found in verbal human language)	1 Human paralinguistics	2 Crickets, grasshoppers	3 Honey bee dancing	4 Doves
1. Vocal–auditory channel	Yes (in part)	Auditory but non-vocal	No	Yes
2. Broadcast transmission and directional reception	Yes	Yes	Yes	Yes
3. Rapid fading	Yes	Yes	?	Yes
4. Interchangeability (adults can be both transmitters and receivers)	Largely Yes	Partial	Partial	Yes
5. Complete feedback ('speaker' able to perceive everything relevant to his signal production)	Partial	Yes	No?	Yes
6. Specialisation (energy unimportant, trigger effect important)	Yes?	Yes?	?	Yes
7. Semanticity (association ties between signals and features in the world)	Yes?	No?	Yes	Yes (in part)
8. Arbitrariness (symbols abstract)	In part	?	No	Yes
9. Discreteness (repertoire discrete not continuous)	Largely No	Yes	No	Yes
10. Displacement (can refer to things remote in time and space)	In part	—	Yes	No
11. Opennesss (new messages easily coined)	Yes	No	Yes	No
12. Tradition (conventions passed on by teaching and learning)	Yes	Yes?	No?	No
13. Duality of patterning (signal elements meaningless, pattern combinations meaningful)	No	?	No	No
14. Prevarication (ability to lie or talk nonsense)	Yes	No	No	No
15. Reflectiveness (ability to communicate about the system itself)	No	No	No	No
16. Learnability (speaker of one language learns another)	Yes	No (?)	No (?)	No

5 Buntings, finches, thrushes, crows, etc.	6 Mynah	7 Colony nesting sea birds	8 Primates (vocal)	9 Canidae non-vocal communication	10 Primates–chimps, e.g. Washoe
Yes	Yes	Yes	Yes	No	No
Yes	Yes	Yes	Yes	Partly Yes	Partly Yes
Yes	Yes	Yes	Yes	No	Yes
Partial (Yes if same sex)	Yes	Partial	Yes	Yes	Yes
Yes	Yes	Yes	Yes	No	Yes
Yes	Yes	Yes	Yes	Yes	Yes
Yes	Yes	Yes	Yes	Yes	Yes
Yes	Yes	Yes	Yes	No	Yes
Yes	Yes	Yes	Partial	Partial	Partial
Time No Space Yes	Time No Space Yes	No	Yes	No	Yes
Yes	Yes	No?	Partial	No?	Yes?
Yes	Yes	In part?	No?	?	Yes
Yes	Yes	No?	Yes	Yes	Yes
No	No (?)	No	No	Yes	Yes
No	No	No	No	No	No
Yes (in part)	Yes	No	No?	No	Yes

Table 2. Washoe's signs grouped into six overlapping categories (from Gardner and Gardner, 1971)

Appeal		*Action*								*Agent*	*Attribute*
gimme		in	come	bed	look	baby	cheese	hurt	spoon	me	black
hurry		out	go	brush	oil	banana	climb	key	string	you	green
more		up	help	catch	tooth-brush	berry	clothes	leaf	sweet	Washoe	red
please	there	down	hug	clean	ride	bib	cow	light	tree	Dr G.	white
	Location		kiss	comb		bird	dog	meat	window	Mrs G.	enough
			open	cover		book	flower	pants		Greg	funny
			peekaboo	dirty		bug	fruit	pencil		Naomi	good
			spin	drink		car	grass	shoes		Roger	quiet
			tickle	food		cat	hammer	smell		Susan	sorry
				listen		chair	hat	smoke		Wende	mine
							Object				

Table 3. Parallel descriptive schemes for the earliest combinations of children and Washoe (from Gardner and Gardner, 1971)

Brown's (1970) scheme for children		The scheme for Washoe	
Types	Examples	Types	Examples
Attributive: Ad+N	big train, red book	*Object–Attributable* *Agent–Attribute*	drink red, comb black Washoe sorry, Naomi good
Possessive: N+N	Adam checker, mommy lunch	*Agent–Object* *Object–Attribute*	clothes Mrs G., you hat baby mine, clothes yours
N+V	walk street, go store	*Action–Location* *Action–Object* *Object–Location*	go in, look out go flower, pants tickle baby down, in hat
Locative: N+N	sweater chair, book table		
Agent–Action: N+V	Adam put, Eve read	*Agent–Action*	Roger tickle, you drink
Action–Object: V+N	put book, hit bell	*Action–Object*	tickle Washoe, open blanket
Agent–Object: N+N	mommy sock, mommy lunch	*Appeal–Action* *Appeal–Object*	please tickle, hug hurry gimme flower, more fruit

belief in reductionism, subconscious though it may be and perhaps all the more serious for that, has an effect on the mental health and indeed the sanity of modern man which is difficult to exaggerate. Dr Viktor Frankl, Professor of Psychiatry in the University of Vienna and widely known for his therapeutic work, finds that one of the major threats to health and sanity is what he calls the existential vacuum (Frankl, 1969). Dr Frankl describes how more and more patients in all parts of the world crowd into the clinics and consulting rooms disrupted by a feeling of inner emptiness, a sense of the total and absolute meaninglessness of their lives. And he believes this is the direct and disastrous result of the denial of value which is regarded as characteristic of modern scientifically oriented society. It is the result, so he argues, of a widespread assumption that, since science in its technique is largely reductionist, reductionism is the only philosophy in which one can believe.

Dr Frankl gives some very telling examples: he says that L. J. Hatterer, a Manhattan psychoanalyst, pointed out in a paper that many an artist has left a psychiatrist's office enraged by interpretations which suggest that he writes because he is 'an injustice collector', or a sado-masochist; acts because he is an exhibitionist; dances because he wants to seduce the audience sexually; etc. Frankl says that the unmasking of motives is indeed perfectly justified, but it *must stop* where the man who does the unmasking is finally confronted with what is genuine and authentic within a man's psyche. If he does not stop here, what this man is really unmasking is his own cynical attitude to his own nihilistic tendency to devaluate and depreciate that which is human in man. Frankl argues that there is an inherent tendency in man to reach out for meanings to fulfil and for values to actualise. Alas, man is offered (to quote two outstanding American scholars in the field of value psychology) the following definitions: 'Values and meanings are nothing but defence mechanisms and reaction formations'. Frankl adds that for himself he is not willing to live for the sake of his reaction formations and even less to die for the sake of his own defence mechanisms. (It is perhaps worth mentioning that Dr Frankl spent some time in a concentration camp where he lost his wife and family.) He goes on to say that reductionism today is a mask for nihilism. Contemporary nihilism, he says, no longer brandishes the word 'nothingness'; today nihilism is camouflaged as 'nothing but-ness'. Human phenomena are thus turned into mere epiphenomena. He argues that, contrary to a widely held opinion, even existentialism is not nihilism; the true nihilism of today is reductionism. The true message of existentialism is *not* nothingness, *but* the no-thingness of man—that is to say, *a human being is no thing*, a person is not one thing amongst other things.

Frankl gives another example of his own experience as a junior high school student when his science teacher used to walk up and down the class explaining to the pupils that life in its final analysis is nothing but combustion, an oxidation process. In this case reductionism actually took the form of oxidationism! The boy Frankl at once jumped to his feet and said, 'Dr Fritz, if this is true, what meaning then does life have?' At that time he was twelve. Now imagine what it means that thousands and thousands of young students are exposed to indoctrination along such lines, are taught a reductionist concept of a man and a reductionist view of life.

Frankl states that about twenty per cent of neuroses today are noogenic (originating in thought) by nature and origin. He defines the existential vacuum as the frustration of what we may consider to be the most basic motivational force in man, and what we may call (by a deliberate over-simplification) '*the will to meaning*'—in contrast to the Adlerians' will to power and the Freudians' will to pleasure. There is also some evidence in the same direction produced by the results of brain surgery. Dr Hyman, a California brain surgeon, states that he is again and again confronted with patients whom he has completely relieved from intractable pain by stereotactic brain surgery, and who then say to him: 'Doctor, I am free from pain—but now more than ever I ask myself what the meaning of my life is, because I know that life is transitory, particularly in my situation.' *The conclusion seems to be that people do not care so much for pleasure and avoidance of pain, but they do care profoundly for meaning.*

Conclusion

If then we grant that, as epistemology seems to demand, mind, or conscious awareness, is in principle irreducible to other modes of existence, then further conclusions of far-ranging importance are inevitable. Not only is dualism established but many characteristics and results of man's mental life must be seen in a new perspective. Among these are those which appear to be shared by some animals, such as aesthetics and perhaps ethics in some slight degree. Others still appear peculiarly human—death awareness (and hence in large measure religion, which seems to have existed since at least the late Pleistocene 0·75 million years ago), morals, the recognition of a difference between what is and what ought to be, and a host of duties and moral imperatives which stem from this. In view of the wide spectrum of interests and expertise represented by this conference I happily leave these in the expectation that others far better qualified than I will take them up.

References

Adler, H. E., (1971). Orientation: sensory basis. *Ann. NY Acad. Sci.*, **188**.

Anderson, P. W. (1972). More is different: broken symmetry and the nature of the hierarchical structure of science. *Science*, **177**, 393–6.

Aronson, L. R. (1951). Orientation and jumping behavior in the gobiid fish *Bathygobius soporator*. *Amer. Museum Novitates*, **1486**, 1–22.

Aronson, L. R. (1971). Further studies on orientation and jumping behavior in the gobiid fish *Bathygobius soporator*. *Ann. NY Acad. Sci.*, **188**, 378–92.

Barbour, I. G. (1966). *Issues in Science and Religion*. S.C.M. Press, London.

Bastin, T., ed. (1971). *Quantum Theory and Beyond*. Cambridge University Press, London.

Billings, S. M. (1968). Homing in Leach's Petrel. *Auk*, **85**, 36–43.

Broad, C. D. (1937). *The Mind and its Place in Nature*. Kegan Paul, Trench and Trubner, London.
Bronowski, J. (1969). *Nature and Knowledge: the philosophy of contemporary science*. Condon Lectures, State System of Higher Education. Eugene, Oregon.
Dixon, M. and Webb, E. C. (1964). *Enzymes* (2nd edition). Longman, London (see p. 668).
Eccles, J. C., ed. (1966). *Brain and Conscious Experience*. Springer-Verlag, New York.
Frankl, V. E. (1969). Reductionism and nihilism. In Koestler and Smythies (1969), 396–408.
Frisch, O. R. (1971). The conceptual problem of quantum theory from the experimentalist's point of view. In Bastin (1971), 13–21.
Gardner, R. A. and Gardner, B. T. (1971). Teaching sign language to a chimpanzee. In Schrier and Stollnitz (1971).
Garstens, M. A. (1971). Measurement theory and complex systems. In Bastin (1971), 85–90.
Gibson, J. J. (1966). *The Senses considered as Perceptual Systems*. George Allen and Unwin, London.
Hebb, D. O. and Thompson, W. R. (1954). The social significance of animal studies. In Lindzey (1954), Chapter 15.
Hinde, R. A., ed. (1972). *Non-Verbal Communication*. Cambridge University Press, London.
James, W. (1890). *The Principles of Psychology*. Holt, New York.
Keeton, W. T. (1971). Panel discussion. *Ann. NY Acad. Sci.* **188,** 401.
Kepner, W. A., Gregory, W. J., and Porter, R. J. (1938). The manipulation of the nematocysts of Chlorohydra by Microstomum. *Zool. Anz.*, **121**, 114–24.
Koestler, A. and Smythies, J. R., eds. (1969). *Beyond Reductionism*. Hutchinson, London.
Lashley, K. S. (1949). Persistent problems in the evolution of mind. *Quart. Rev. Biol.*, **24,** 28–42.
Lindzey, G., ed. (1954). *Handbook of Social Psychology*. Addison-Wesley, New York.
Matthews, G. V. T. (1968). *Bird Navigation* (2nd edition). Cambridge University Press, London.
Monod, J. (1970). *Le Hasard et la Nécessité*. Editions du Seuil, Paris. Translated as *Chance and Necessity*. Knopf, New York (1967), Collins, London (1972).
Pantin, C. F. A. (1968). *The Relations between the Sciences* (ed. A. M. Pantin and W. H. Thorpe). Cambridge University Press, London.
Pattee, H. H. (1970). The problem of biological hierarchy. In Waddington (1970), 117–36.
Premack, D. (1970). The education of Sarah. *Psychology Today*, **4**, 55–8.
Razran, G. (1971). *Mind in Evolution*. Houghton Mifflin, Boston, Mass.
Roslansky, J. D. ed., (1969). *Thc Uniqueness of Man*. North Holland Publishing Company, Amsterdam & London.
Schrier, A. and Stollnitz, F., eds. (1971). *Behaviour of Non-Human Primates*, vol. 3. Academic Press, New York.
Sherrington, C. S. (1940). *Man on his Nature*. Cambridge University Press, London.
Thorpe, W. H. (1963). *Learning and Instinct in Animals*. Methuen, London.
Thorpe, W. H. (1966). Ethology and consciousness. In Eccles (1966), 470–505.
Thorpe, W. H. (1968). Perceptual bases for group organisation in social vertebrates, especially birds. *Nature* (London), **220**, 124–8.
Thorpe, W. H. (1969). Vitalism and organicism. In J. D. Roslansky (Ed. *q.v.* pp. 71–99).
Thorpe, W. H. (1972). Comparison of vocal communications in animals and man. In Hinde (1972), 27–47.
Waddington, C. H., ed. (1970). *Towards a Theoretical Biology: three drafts*. Edinburgh University Press, Edinburgh.
Weiss, P. A. (1969). The living system: determinism stratified. In Koestler and Smythies (1969), 3–55.

Discussion

Rensch

Like Sir John Eccles, you believe that conscious phenomena can *steer* physiological brain processes. However, then the law of the conservation of energy is infringed, because the gapless flow of causal, that is to say material, events in the brain would get additional impulses, and the kinetic or potential energy would increase. It seems to me as if you regard consciousness as a kind of supplementary physiological process. This would mean that you explain consciousness in a materialistic manner. I find it difficult to combine this conception with your dualistic opinion. If one assumes that psychic and physiological processes are identical, such difficulties do not arise.

Thorpe

The point you raise is one which has been in the minds of philosophically minded biologists, and those who hold any form of theory implying interaction between two entirely different 'substances', for a very long time. However, I feel that we need no longer be apprehensive about this; for the laws of thermodynamics now bear a very different aspect to those which they presented fifty or more years ago. Thus, as Professor Popper reminded me, Schrödinger argued that the First Law is statistical only. In any case it is now clear that laws such as these break down at the frontiers of physical science—for instance they certainly cannot be relied upon to apply under conditions which obtain in collapsed stars subject to hitherto unimagined pressures. Monod and Popper have suggested in their different ways that the mind–brain frontier is one which must surely mark a 'discontinuity' at least as great as any other barrier facing scientific thought. This view receives support from R. W. Sperry (1970) who says 'the present interpretation would tend to restore mind to its old prestigious position over matter, in the sense that the mental phenomena are seen to transcend the phenomena of physiology and biochemistry'. Eccles, in his paper presented at this conference, expresses the view that the split-brain investigations have falsified the psychoneural identity hypothesis. Moreover, as Eccles again points out, both Schrödinger and Wigner have suggested that the brain–mind problem is likely to require fundamental changes in the hypotheses of physics. Indeed C. D. Broad (*Scientific Thought*, Routledge and Kegan Paul, London, 1923) found it necessary to postulate a new spatial dimension to accommodate the facts of the existence and nature of mind. Of course we still know remarkably little about the mind–brain frontier. Nevertheless I am not at all disturbed if knowledge reveals a situation in which the 'First Law' must be presumed to be broken.

Rensch

A *concept of the own self* is not a primarily 'given' phenomenon. Such a concept, or at least non-verbal prestages, can very probably also be attributed to higher animals. In human children, as in higher animals, this central concept becomes gradually developed during the first years of life. A young child learns to discriminate between two kinds of phenomena: those deriving from its own body and those from the environment. A concept of the own body becomes developed by proprioceptive sensations, reciprocal sensations when touching its own body, hearing its own voice, and by intense feelings, mainly pain, hunger and thirst. This primary concept of the own self becomes enhanced by gradually storing memories about experiences concerning his own body, memories of how far the child can grasp and jump and memories concerning many experienced actions and reactions with regard to the living and nonliving environment. All this can arise in a corresponding manner in higher animals. Gardner's chimpanzee Washoe even learned to use correctly a gesture to express its own self, and Premack's chimpanzee Sarah learnt to use a plastic symbol for herself.

Later on, human children develop a special human kind of concept of the 'own self' because this becomes connected with a special word or the mental image of this word. All volitional thinking and planning is then connected with this central concept. And this is probably also the case when a chimpanzee plans a more complicated action.

Thorpe
I agree that the concept of 'own self' is not a given phenomenon, in the sense that it springs full-blown into existence at birth, or at any other clearly definable point in the foetal or post-foetal life history. But the fact that it is acquired gradually, and is to some degree dependent on environmental circumstances, in no way contradicts the view that it is an essential, basic feature, a *sine qua non* of all human accounts of the world and all human pretensions to knowledge. I am much attracted by the 'three-level' view of conscious experience propounded by Polten (1972) and discussed by Eccles in his paper at this conference. Moreover, Popper has argued in several of his papers that mankind has been creating himself, through countless millenia, in all aspects of 'World 2', by creating his culture, which is in 'World 3'.

9. Unjustified Variation and Selective Retention in Scientific Discovery

DONALD T. CAMPBELL

Are you awed by the exquisite fit between organism and environment, and find in this fit a puzzle needing explanation?

Does the power of visual perception to reveal the physical world seem so great as nearly to defy explanation?

Do you marvel at the achievements of modern science, at the fit between scientific theories and the aspects of the world they purport to describe? Is this a puzzling achievement? Do you feel the need for an explanation as to how it could have come about?

The epistemological setting

Many philosophers and no doubt some biologists do not have these problems. The impressive fit that others claim to see they deny. For how can one claim that perception maps with accuracy a world of independent objects when we can know of these objects only through perception? Are we not creating a needless shadow world from the evidence of perception, and then claiming an undemonstrable fit between the two? Again, how can we claim a fit between the theories of science and the real world, when we know that real world only through the theories of science? Especially how can we, now that we recognise (as I too do) that the facts against which theories are checked are themselves theory-laden, even down to the simplest visual perceptions? With regard to the fit of organism and environment, are not our descriptions of the ecological niches of animals in fact *post hoc* explanations designed just to explain the same specific characteristics of the animals which we now claim as evidence of fit (Scriven, 1959; Campbell, 1960, 396–7)?

This work was supported in part by National Science Foundation Grant GS30273X. In order to create space for the note on downward causation in hierarchically organised biological systems, this version differs from the privately circulated one in the elimination of sections entitled 'Is there a logic of discovery?' and 'Thomas Kuhn's natural-selectionist epistemology'.

I myself have no quarrel with such philosophers. I must concede that I do not have compelling evidence of fit in any logically entailing sense, neither by deductive nor by inductive logic (were such to exist). In seeing the fit, and the puzzle, I do so only on the basis of presumptions which go beyond my capacity to verify or compellingly to demonstrate to another person.

In so far as this essay is epistemological, it is an exemplar of what can be called *descriptive* epistemology, in contrast with *analytic* epistemology. (I use the term analytic epistemology to include the central problem core of classical epistemology as well as the modern interests going under that name.) Quine calls it 'epistemology naturalised' (1969) and Shimony (1971*a*, 1971*b*) calls it 'Copernican epistemology', meaning that he starts the problem viewing man as a small part of a large world rather than as the creator and centre of a phenomenal world. The later epistemological efforts of Bertrand Russell were of this type (1948; 1944, 700–2). Many modern philosophers are participating in such endeavours (for example, Toulmin, 1961, 1972; Maxwell, 1972, 1973; Wartofsky, 1968; Stemmer, 1971; Ackermann, 1970; and, without the specific focus on evolution, Pasch, 1958; Hirst, 1959; Wallraff, 1961; Mandelbaum, 1964), although it is still a minority heresy. More typical is Dretske's (1971) reply to Shimony, a reply with which I do not necessarily disagree, since I see descriptive epistemology as indeed undertaking a different task from that of traditional analytic epistemology.

More than Quine, Russell and Shimony, I see descriptive epistemology as consistent with and building upon the magnificent intellectual achievements (albeit negative ones) of classical analytic epistemology. Descartes, Locke, Berkeley, Hume and Kant attempted to answer the problem of knowledge without the circularity of assuming knowledge in the process. To the question 'how is knowledge possible?' they answered 'there is no knowledge', or 'knowledge is impossible', in the sense of certain, incorrigible, true belief about the nature of the world. The intellectual achievements going under such names as solipsism, the argument from illusion, the problem of other minds, the distinction between noumena and phenomena, the semblance screen, the scandal of induction, etc., have given us a proper modesty about the 'knowledge' we have of the world through either vision or science, and have loosened us from the naive realism or the direct realism towards which our perceptual and cognitive systems bias us. The fashionable emphasis on the radical underjustification of scientific truth (for example, Kuhn, 1962; Feyerabend, 1970; Quine, 1969; Toulmin, 1961; Polanyi, 1958; Petrie, 1969) is but our modern phrasing of Hume's scepticism. Scientific induction is scandalous in that it involves presumptions that are unverifiable. In spite of our longing for certain knowledge, such is not to be our lot. In particular, the achievements of analytic scepticism help reconcile us to the profoundly indirect, equivocal and presumptive processes of knowing which descriptive epistemologies of the evolutionary variety present.

In contrast to the traditional epistemological approach of holding all knowledge in abeyance until the possibility of knowledge is first established, descriptive epistemology addresses the problem of knowledge assuming that the biologists' description of man-the-knower, and the physicists' description of the world-to-be-known, are approximately correct, although corrigible.

This *does* beg the traditional epistemological question. It does undertake the problem of knowledge within the framework of contingent knowledge, and by assuming such knowledge. Descriptive epistemology is a part of science (broadly conceived) rather than philosophy, at least for those who see the two as quite separate enterprises. Such a descriptive epistemology, however, is a major concern for a number of modern philosophers (in this case as defined by university department, professional association and journals), and is present as a part of most of the classic epistemological writings.

Of course, the enterprise involves a choice of analytic epistemological positions, just as any science implicitly or explicitly does. The joint acceptance of the puzzling fact of fit and the emphasis on corrigibility involves a commitment to an epistemological dualism (Lovejoy, 1930; Köhler, 1938), and to a critical realism (Feigl, 1950; Mandelbaum, 1964; Maxwell, 1972).

By 'epistemological dualism', I refer to the conceptualisation of beliefs ('knowledge') as distinct from the referents of the beliefs, so that fit, non-fit, degrees of fit, error and correction are conceivable. This is related to the biologist's duality of organism and environment. The 'critical realism' is a commitment to the reality of an external world, even though the beliefs about ('knowledge' of) that world are conceded to be imperfect. (I include in this view 'metaphysical realism' (Popper, 1972), 'hypothetical realism' (Lorenz, 1941, 1959, 1969; Campbell, 1959), 'structural realism' (Maxwell, 1973, 1972) and 'logical realism' (Northrop, 1949, 1964).) These views are no doubt characteristic of most scientists, although only a minority of philosophers adhere to them, even of philosophers of science. (More problematic is a critical realism of ideal forms. The term 'realism' has had such seemingly incompatible usages in the history of philosophy. In recognition of one aspect of Popper's 'Third World' realism, I will tentatively accept this too.) Probably descriptive epistemology must also be an 'epistemology of the other one' (Campbell, 1959, 1969), giving up the effort to solve the problem of knowledge for oneself, working instead on the problem of how people in general, or other organisms, come to know.

In descriptive epistemology (as Russell (1944), Quine (1969) and Shimony (1971) make clear) an ontology must go along with an epistemology. The nature of the world and the possibility of knowing it are intimately interdependent. It is in this area that the recurrent efforts to justify induction (Reichenbach, 1938; Salmon, 1961; Williams, 1947; Harrod, 1956) (which, joining the analytic epistemologists, I regard as misguided) can be salvaged as ontological hypotheses about the nature of the world, and perhaps about the nature of any inductively knowable world. They thus can be interpreted as contributing to the basic (ontological) presuppositions of induction (Russell, 1948) underlying our knowing activities.

One other point of contact with traditional epistemology needs to be noted. It should be recognised that there is not just one analytic epistemology, but instead many consistent analytic epistemologies. Of these many, we are interested in that one that is most descriptively appropriate. But the descriptive requirement is not in place of the analytic one, but is rather an additional requirement. The natural selection epistemology here offered has one special analytic feature: if one is expanding knowledge beyond what one knows, one has no choice but to explore without benefit of wisdom (gropingly, blindly, stupidly, haphazardly). This is an analytic truth central to all descriptive epistemologies of the natural selection variety (for example, Popper, 1959, 1963, 1972; Toulmin, 1972). We will return to this theme.

The all-purpose explanation of fit

For the three problems of fit that began this inquiry, and indeed for all problems of fit, there is available today only one explanatory paradigm: blind variation and selective retention. In application, it describes weak, tedious and improbable (albeit not impossible) mechanisms, quite out of proportion to the magnificent achievements to be explained. We believe in it because it describes a possible route, and because there are, at present at least, no rival explanatory theories.

Not only is it the sole explanatory conception for the achievement of fit between systems, it is also an essential explanatory component for all instances of purposiveness, of teleological achievement or teleonomy. It is also an essential part of processes producing and maintaining form and order, as in crystal formation and in atomic and molecular cohesion. Hierarchical structures require hierarchical replications of the process, with a node of selection for every emergent level of organisation.

We who believe this way are rightly called physicalists, materialists and reductionists. But we must distinguish ourselves from a common class of reductionists who deny the existence of fit, design, purpose, emergent higher levels of organisation, etc., just because they are difficult to explain. Thus, by and large, we accept the vitalists' facts, we disagree with their explanations. Thus we join with Bertalanffy (1967, 1969), Whyte (1965), Polanyi (1966, 1969), Thorpe (1969), Koestler (1969) and others in our awe for hierarchical organic form and human intellect. But we find these authors insufficiently awed to feel the need to explain these wonders. But while we must join them in admitting that natural selection and its analogues produce only puny, tedious and improbable explanations, we judge these to be the only ones available. If this makes these achievements seem unlikely, rare and precious, this is quite compatible with our awe.

It is well to reconsider Darwin's relation to the natural theologians such as Paley (1802) and Ray (1691) who were widely read in his day. They supported their theology with the 'argument from design', evidenced by instance after instance of the wonderful fit of specialised organic form to environmental problem and opportunity. Darwin did not in the least doubt or undermine this evidence of design and fit. Rather, he added to it exquisite detail, as in the fit between orchids and the insects that pollinate them. Equally awed by the facts of intricate adaptation, he provided an explanation that was compatible with the physical science model of causation. That explanation, 'natural selection', is by all odds the paradigmatic illustration of the variation and selective retention model. As Zirkle (1941) has shown, the concept was many times hit upon, but it was Darwin and Wallace who made it a part of our own continuous central intellectual current.

The application of the model to creative thought and scientific discovery is the central focus of my paper today. As a continuous part of our intellectual tradition, it is even older than natural selection. Alexander Bain, using the phrasing 'trial and error', was making this application as early as 1855. But in spite of the many eminent independent discoveries and advocates, it has remained a minority position, and is probably the one application of the model least readily accepted. One of the purposes of this paper is to try to clarify the reasons for this resistance.

The model has seen application at many other levels, from morphogenesis and wound healing to learning and science. My own contribution has been primarily as a reviewer of these varied literatures (1956*a*, 1956*b*, 1959, 1960, 1965, 1974). In so far as I have made a novel contribution, it has primarily been in a stubborn effort to apply the model to visual perception (1956*b*, 1966), although even here I have been abetted by prior applications of a broadcast variation and selective reflection model to radar and echolocation devices, such as the lateral-line organ in fish (Pumphrey, 1950). Related is my emphasis on a hierarchy of *vicarious* variation and selective retention processes.

Selection theory

There are three major components to the model:

(1) Variations, a heterogeneity of alterations on existing form.

(2) Systematic selection from among the variations. Systematic elimination.

(3) Retention, preservation (and, in many systems, multiplicative duplication) of selected variations.

Lacking any one of these, no increase in fit or order will occur. The concurrence of all three is a rare event—fit and order are rare events. Variation and retention are at odds in most exemplifications of the model. Maximising either one jeopardises the other. Some compromise of each is required. (This conflict is somewhat relaxed in the vicarious blind-variation-and-selective-retention processes to be discussed below.)

In epitomising the process, there has often been a disproportionate emphasis upon one of the features at the expense of the rest. The critics (for example, Bertalanffy, 1967, 1969; Blachowicz, 1971; Whyte, 1965) overemphasise the variation component. It is well to remember that Darwin focused on selection, artificial and natural. James Mark Baldwin, due to be rediscovered as the first brilliant generaliser of the model (his obscure *Darwin and the Humanities* (1909) is the most efficient introduction) used the phrase 'selection theory'.

We must remember the prior understandings that Darwin and his generation were contending with: the pervasive assumption was that evolution had proceeded by wise changes, by deliberate, planned, appropriate variations. These foresighted changes were either the work of God or, in the case of Lamarck, based upon the animal's own wise survey of his needs. (It is this feature of Lamarckianism that James (1880) and Baldwin most opposed.) In these models, the source of improved fit was in the design of the variations. Darwin changed all this by making the improved fit a function of selection after the fact. Rather than to foresighted variations, design is due to the hindsight of a selective system.

It is when the role of selective systems (especially hierarchical ones) is emphasised that the natural selection model is most compatible with the factual claims of the vitalists. Discussion of a sample of Bertalanffy's protests will help make this clear (1967, pp. 82–7):

> ... selection, *i.e.*, favored survival of 'better' precursors of life already presupposes self maintaining, complex, open systems which may compete; therefore selection cannot account for the origin of such systems.
>
> But the alternative, either 'scientific' explanation by random events directed by the environment, or else vitalistic (teleological, purposive, perfectionist, etc.) agents, is patently false as I have said for more than thirty years. Nobody presumes that an atom, crystal or chemical compound, is the handiwork of a vitalistic demon; but neither is it the outcome of accident. Structure and formation of physical entities at any level—atoms, molecules, high-molecular compounds, crystals, nucleic acids, etc.—follow laws which are progressively revealed by the respective branches of science. Beyond this level, we are asked to believe, there are no 'laws of nature' any more, but only chance events in the way of 'errors' appearing in the genetic code, and 'opportunism' of evolution, 'outer-directed' by environment. This is not objectively founded science, but preconceived metaphysics.
>
> ... There are good reasons to believe (which can be put forth in detail) that the code does have organisational and regulative properties, not well known at present, but indicating that not all mutations are equiprobable.
>
> ... these aspects deserve emphasis and investigation equal to those given to undirected mutation and selection. Evolution then appears essentially co-determined by 'internal factors' . . . , or 'inner-directed'.

May it first be noted that many of us see in crystal formation a chance variation and selective retention process. For example, in a saturated salt solution of intermediate temperature, most of the adjacencies between one salt molecule and another are as easily moved out of as entered into, and the thermal noise, Brownian movement or whatever produces a continuing change of adjacencies. But while this is true of most adjacencies, there are a few which result in a particular fit, in which the force fields of the two molecules summate to produce an adjacency exceptionally hard to dislodge. These particular adjacencies require less energy to enter than to disrupt, and thus, though they are rare, they are selectively retained and accumulated, forming the orderly crystal pattern. In this process the three essentials to the model are present: a system producing variations, a systematic selection of certain variations after they have happened to occur, and a preservation of the variations. Crystal formation is limited to the rare combination of these three requirements. Extreme heat, such as to liquify or vaporise salt, will increase the variations component, but destroy the retention system by continually producing energy inputs that exceed the disruption threshold. Extreme cold will remove the variation component. It is only when these two are in a compromised balance that the selective-retention negentropic process of crystal formation can take place. This model is also applicable to the other levels of order and structure that Bertalanffy stresses, from atoms on up. It is the selective system that explains the repeated condensation of the common elements from the pre-atomic plasmas of the stars. Iron atoms and molecules in different parts of the universe have been repeatedly independently invented by a convergent evolution, to be explained by the shared selective system.

In this case, the selective system is internal, in Whyte's (1965) terminology. It puzzles me that Whyte, Bertalanffy (1967, 1969), Thorpe (1969), Blachowicz (1971) and others see the undoubted importance of internal selectors as a contradiction to the theory of natural selection in either its original or modern form. Weiss (1969) is understandably protesting against the dominance of a molecular biology that is reductionist to the extreme of denying all order above the molecular level, denying the vitalists' fact. The protest of the others, however, includes rejection of those who use natural selection to explain the emergence of increased complexity and higher order levels of organisation.

It might be objected that the mystery of this has been solved only by creating a new mystery of selective systems. To this I would plead guilty. The selective systems we posit need in their turn explanation, just as do the physical processes generating variations. (In all scientific explanation, we explain by explanators which will themselves need explanation.) But I do join those who feel a great conceptual gain in moving from teleological causation to natural selection explanations of teleological achievements.

In dealing with life and thought, we will need to consider at very least three general classes of selectors, *structural*, *external* and *vicarious*. It seems useful to divide Whyte's internal selectors in two subtypes, here called *structural* and *vicarious*. In the salt crystal formation, the internal selectors mentioned were *structural*. Had we done a similar analysis of the smooth curvature of the side of the crystal that rested against the test tube, this would have involved a selective blocking of molecular movements representing an external, environmental selector. The dominant patterns in snowflakes are a result of structural selectors. But if in falling through the atmosphere the looser lattices and string formations are torn apart, then at least part of the compactness of the snowflakes we examine is due to an external selector removing the more fragile ones. In diatoms, the shaping by structural selectors is still very visible, although editing by external selectors has had a much larger hand in their shape than in the case of the snowflake. When we get to forms like the giraffe and camel, the influence of external selectors on form is much more obvious, but editing by structural selectors is still pervasive.

Consider a gene, a DNA molecule, being bombarded by cosmic rays that rearrange its prior structure. A most important selective requirement is molecular stability. The great bulk of the disruptions produce rearrangements that are unstable as molecules, being fragmented or imbalanced. Only a few move the atomic material from one stable molecular form of DNA to one of the other stable DNA molecules. This stability requirement is a structural selective process. Before there can be a mutant gene, the requirements of being a gene must be met. Once the mutant gene has survived the structural selective process, if it then produces survival-relevant variation in the somatotype of the organisms in which it appears, it then may be subject to *external* selection through the average of the lives and deaths and effective fertilities of the organisms that carry it. It is to this class of selectors that 'fit', 'adaptation', 'correspondence', 'objectivity' or 'descriptive truth' is due. If Darwin and the neo-Darwinists have neglected to emphasise the structural selectors it is because adaptation to an external environment was the problem they were trying to solve, not because they regarded it as the only source of restraint and selection.

Vicarious selectors are a class of internal selectors which are related to adaptation in that they 'represent' external selectors vicariously. Take, for example, a salamander with legs capable of regeneration. The regrowth of a lost leg seems to continue 'until it reaches the ground', until it is the adaptive length. In actuality, leg growth is not curbed by the external selector of hitting the ground. Rather, the leg length is curbed by internal selectors, internal inhibition processes. Another example: the pleasures and pains, tastes and fears, rewards and punishments that mediate learning provide another class of vicarious selectors. Thus an animal's learning and eating is guided by the purely signal values of sweetness and bitterness, rather than the external realities of nourishment which sweetness and bitterness imperfectly represent. The governance-settings of the vicarious selectors are themselves in turn subject to mutation, and the variations are subject to external selection. While internal-vicarious selection is profoundly indirect, and potentially biased as a result of approximate structure (oversimplification) and changes in external environment, it results in much more precise selection than do the vagaries of life-and-death that mediate the external selectors. Those vicarious selectors that are represented in conscious experience are likewise compellingly 'immediate' at the phenomenal level, their profound indirectness being totally disguised.

Vicarious selection needs much more elaboration than it has been given here, or elsewhere. Waddington's chreods are predominantly of this nature, although containing structural selection features. Both in their morphological development and in their knowledge processes, organisms represent nested hierarchies of vicarious selectors, with a node of selection preserving every level of order against entropic tendencies, including the quasientropic effects of genetic and somatic mutation, etc. This removes the descriptive or factual disagreement with Bertalanffy, Thorpe, Whyte *et al.* Does it also concede their argument by making the very modifications in the selective retention model which they call for? I think not. Two features of the vicarious selectors should be noted. First, each of them is itself a discovery produced by natural selection, and containing partial or general knowledge about the nature of the world. Second, each of them embodies in its own operation a variation-and-selective-retention process. Edelman's paper in this volume provides an excellent detailed analysis of a vicarious variation and vicarious selection process.

'——' variation and selective retention

Perhaps the slogan 'variation and selective retention' should suffice. However, most advocates have felt the need to add some adjective to the word 'variation'. Their goal in so doing, as I empathise it, has been to accent the point that the resulting adaptation and order does not come from the variation simply. But the adjectives chosen have been the basis for much misunderstanding.

It is perhaps easiest to specify the adjectives that are explicitly to be denied: wise, designed, prescient, informed, foresighted, clairvoyant, intelligent, preadapted. Also rejected is a trial-and-correction relationship among successive variations: later adaptations are no more wise or preadapted than earlier ones (except in so far as selection has operated).

Even for this set of rejected adjectives, the form of the rejection must be specified: they are rejected as useful in explaining new, improved adaptations. It is not denied that if we observe variability (in problem-solving behaviour, for example) we may find the variations intelligent, preadapted, etc. Rather, if we find them thus, this is taken as evidence of already achieved adaptation, rather than a relevant explanation of further increases in adaptation.

To put the point negatively, the variations could be by chance and the process would still work. The source of the variations is irrelevant. To put it more positively: increasing knowledge or adaptation of necessity involves exploring the unknown, going beyond existing knowledge and adaptive recipes. This of necessity involves unknowing, non-preadapted fumbling in the dark.

Among the adjectives that have been acceptable are the following: *chance*, *random*, *contingent* (in an older usage), *aleatory*, *fortuitous*, *spontaneous*, *haphazard*, *happenstance* and *blind*. Of these, *random* has the greatest usage by both advocate and opponent alike. But it has many undesirable or exaggerated features which have confused opponents or have been opportunistically misinterpreted by them.

Equiprobability of all variations is an aspect of the modern statistical concept of randomness which is unnecessary, and in fact unrepresented in any of the actual implementations of the model. At the level of the genes, certain mutations, including lethal ones, are much more frequent than others. Even if the cosmic rays were as likely to hit one part of a DNA molecule as another, the novel recombinations of the ingredients would be more frequent for proximal substitutions than for remote ones. Structural selection adds to such statistical bias. A natural selection for mutation sequences is possible. But while such departures from randomness may reflect already achieved adaptations, may restrict new adaptations whether due to retaining prior adaptations or through structural limitations, these biases do not explain further adaptation. A cat in Thorndike's puzzle box is far from random in his response emission, primarily because of innate and acquired preferences for certain responses over others, a partial wisdom appropriate to the ecology of past traps in evolution and ontogeny, but also because of structural biases against generating certain kinds of novel variations. If these predilections are strong enough, the cat will not solve the puzzle box, because Thorndike has deliberately designed it to be puzzling, to have a counterintuitive solution. In particular, the cat's strong expectation that locomotor permeability accompanies transparency, his tendency to try to get out by going through the walls that light is coming through, have been rendered counterproductive by Thorndike; otherwise it would not have been a puzzle box at all. To solve Thorndike's puzzle, the cat has to generate some very low probability responses that it cannot generate 'wisely'. After frustrating itself with stubborn repetitions of 'intelligent' responses, it may 'by chance' and 'inadvertently' (Guthrie, 1954; Guthrie and Horton, 1946) generate a number of low-probability responses, among which may be one that releases the trick door. Equiprobability is both descriptively wrong and analytically nonessential. But variability reaching into responses beyond the already adaptive is essential.

Unrestrained variability is a connotation of *randomness* held by many critics. Unfortunately Monod, in an overly dramatic advocacy, has furthered this misunderstanding by speaking of 'pure chance, absolutely free' (1971, 112). Unspecified under equiprobability was for what range of variations this equiprobability held, and analogous problems of range exist even without equiprobability. Some critics write as though advocates of random mutation believed that the mutations from an octopus could include a giraffe, that the whole range of animal forms and more was available in the population from which each randomisation drew. More specifically, they argue as though each restriction on the range of variations was proof of the falsity of neo-Darwinian theory (Bertalanffy, 1969; Thorpe, 1969; Whyte, 1965; Blachowicz, 1971; etc.). This is of course, absurd. Darwin dealt with the observable range of variations within a breeding group, and no neo-Darwinist modifications have been made on this point with the incorporation of Mendelian unit characteristics. Furthermore, these critics argue as though the restricted range of variations somehow explained subsequent adaptations.

There are, in the neo-Darwinist theory, several major sources of restriction on the variations that may be generated. First, variation and retention are, as we have noted, inherently at odds, and the loyal retention of selected past mutations puts great limits on variation. Second, there are structural limits on variation, limits both of physical impossibility and of viability. Moving above the gene level to species change, selection (both internal and external) continues to counter variability by weeding out mutant genes and unorthodox recombinations. It is in keeping with, rather than counter to, neo-Darwinian theory that no evolutionary progression take place for a well-adapted form in a stable ecological niche. But where evolutionary movement, or increased adaptation, do take place, it is variability rather than prior restraint on variability that makes this possible, by providing new raw material of novelty upon which selection can operate.

As I read them, the neo-Darwinists have always believed that the source of variations when discovered would be deterministic and causal in a physical science sense, just as the faces that show up when a pair of dice are rolled conform to the laws of macrophysics, without recourse to quantum indeterminancy. It takes very special machinery, such as dice, dice cups and shaking, to approximate randomness in a statistical sense. As we understand it today, such machinery is lacking at the level of mutation. It is also lacking in genetic recombination for genes that are on the same chromosome. But for genes on separate chromosomes a mechanical randomisation process may be very nearly approximated. The chunking of the genetic material into numerous discrete chromosomes is almost certainly due to the selective advantage of providing more variable variations. Edelman (in this volume) describes a mechanical system for generating antibody types. Speculation on quantum indeterminancy effects in generating variability in human thought processes is interesting, but macrophysical mechanisms are probably adequate.

Statistical independence among successive variations is an aspect of randomness that is very desirable, but not essential. Sweep scanning, as by a radar, is a blind variation and selective retention process vicarious for locomotor effects, which I feel belongs in the general paradigm. So also broadcast and selective echo-location processes, and trial and error learning in which the exploration is systematic, responses being tried out in some regular order, as long as the regularity is independent of knowledge of the correct response. In all of these, the process would work even if the emission of variations were random; the systematicness is not at all the secret of success, but rather a mechanical convenience.

These exceptional cases are all ones in which a vicarious selection, short of life and death, was involved. If genetic mutations were to occur in a fixed order for all members of a species, with most mutations being lethal, species extinction without adaptation would be the most likely outcome.

Independence of the environmental conditions to which adaptation is taking place is an important and appropriate connotation of randomness. Thus random does not mean uncaused or independent of determinate cause, but rather a relative independence, relative to the eventual fit or structured order that is to be explained. Biologists frequently use the word *random* with this qualification. Thus Dobzhansky (1963, 211) says 'Mutations do arise, apparently in all organisms, and they arise at random with respect to their usefulness to their carriers'. Thoday (1967, 29) states the orthodox view as 'random with respect to need'. Waddington (1969, 370) states it as follows:

> A gene mutation consists in some sort of alteration in the sequence of nucleotides in the DNA. From a chemical point of view these alterations in nucleotides are presumably not wholly at random. There may well be quite considerable regularities in the processes by which the alterations come about; however, we know very little about them as yet. But even if we fully understood the physical and chemical processes involved, this would not be very relevant to evolution. From that point of view, the important thing is that the occurrence of a mutation and the nature of the mutational change is not directly connected with the environmental circumstances which will exert natural selection on the result. As far as natural selection and evolution is concerned, mutations can therefore be considered as effectively at random.

An important implication of this is that the order of occurence of the variations be independent of their adaptiveness, that a useful variation be no more likely to occur early than late, etc. (A negative correlation between a variation and its own reappearance, an inhibition of repetition, might be useful, and is an advantage built into sweep scanning processes.)

But even were (and where) some degree of adaptive correlation to be found between a new environmental setting and the mutations which are concomitant with it, or, more likely, between a new puzzle situation for an animal and the responses it emits, this neither violates the model nor provides an explanation of an eventual improvement of fit. For this adaptive bias in variations is itself an evidence of fit needing explaining. And the only available explanation (other than preordained harmony) is through some past variation and selective retention process. Furthermore, if the animal's partially intelligent floundering is replaced by a still more efficient or errorless response pattern, this gain in fit is not at all explained by the prior useful non-randomness. Rather, it can only be due to a selection from among the limited range of intelligent but imperfect variations those that happen to be still more adaptive (see Dobzhansky, 1963).

Another useful connotation of random is that prior runs do not affect subsequent ones, and in particular the rejection of the notion that the wisdom of later variations is improved by the knowledge of the failure of the earlier ones. Where descriptively this does happen, it is due to additional knowledge. If there is known to be a solution and a finite number of alternatives, then the elimination of wrong alternatives improves the chances of successive guesses. Where an animal is visually monitoring its own responses, the vicarious visual search process may achieve the trial and correction that Hilgard (1948, 335–8) speaks of. But these are not exceptions to the basic rule.

It seems to me that the needed adjective should be an excluding one rather than a designator. If we are to use the word 'random' it should be in some such phrase as 'might-as-well-be-random' variation and selective retention, or 'even if random'. 'Nonprescient' variation has been suggested (Campbell, 1956) but, against my own advice, I have fallen into using the term 'blind' variation (for example, Campbell, 1960), in part because of an ambivalent preoccupation with vision, and to counter the predilection to direct realism which the phenomenal experience of visual perception induces. While the term *blind* avoids many of the difficulties the technical meanings and misunderstandings of random induce, I have to confess that it has not made acceptance of the concept any easier, and indeed one of my former students, Blachowicz, specifically employs 'blind variation' to exemplify what he rejects in the natural-selection model of scientific development (1971, 180):

> The role of random processes in biology is perhaps not as difficult to defend as their role in theory-construction, where human purpose and planned activity have always been thought to be centrally operative. This evolutionary model now demands that 'randomness' be taken seriously in this latter sphere as well. Donald T. Campbell, for example, who has advocated the universal application of this evolutionary model to all adaptive processes, speaks of 'blind' variation on the one hand, but then affirms that 'insofar as one's guesses are correct other than by happenstance, one is making use of already accumulated knowledge, however approximate. Theory in science reduces the tremendously costly blind exploration to the minimum, but in its valid aspects cannot do more than spell out the implications of what is already known' [1959, 164–5]. Of course, it may be legitimate to include within our understanding of scientific discovery the process of articulating the 'implications of what is already known', but this surely does not exhaust its meaning. And further, as already indicated, it seems equally unacceptable to ascribe what is discovered by other means to pure 'guesswork'. The problem, in short, is that we seemed forced to reduce the process of scientific development to deductive or *analytic* exfoliations of what is already known, and/or to the absoutely *synthetic* contingencies of purely 'blind variation'. In this light, scientific thought becomes something very much like an amalgam of *mathematics* and *prejudice*.

Of course, part of the problem is that we do have a counterintuitive, countercultural message. We do want to deliver a message that offends. There is a genuine disagreement. Though we may feel in our opponents a stubborn tendency to mistake teleological assertion for explanation, it is often only a preference for confessing the teleological facts to be inexplicable rather than accepting an explanation so implausible, so roundabout, so full of unverifiable presumptions, as ours. Thus though I feel that getting a better adjective before variation will help, I deliberately seek out one that will make a point offensive to some, an adjective that will be striking because of its unexpectedness. 'Unjustified' variation is my current candidate.

I choose 'unjustified' because my missionary field for this paper is philosophy, and because among modern philosophers a popular definition of *knowledge* is 'well justified true belief'. This definition is useful because it recognises that beliefs can be true for irrelevant reasons, by accident. And an unjustified belief is not one for which truth is precluded, merely one for which it is not yet established. Thus unjustified may be better than blind, if blind indicates a definitely established inappropriateness that one expects to persist. Focusing, as does Popper, on expansions or improvements to knowledge, I want to join him in emphasising that the justification of a new true belief comes after its generation and testing. Conceptual advance in science is through a trial and error process involving numerous not-yet-justified (and for the most part never-to-be-justified) beliefs, some perhaps selected in the competitive testing, this criticism–testing–selection process constituting their initial justification (an always inconclusive justification, at least as far as descriptive, contingent or synthetic beliefs are concerned).

Unjustified variation and selective retention of scientific theories

Applying the '——' variation and selective retention model to creative thought is always the hardest to do convincingly, even though its application here anticipated Darwin by three years (Bain, 1855; see Campbell, 1960). Do I really, with a straight face, want to advocate that the discovery of new scientific theories is through an 'unjustified' or 'blind' variation of theories and a selective retention process? Yes, I do, implausible as it may seem.

Let me confess that I concede most of an opponent's facts from the very beginning. With my nested hierarchy of vicarious variation and selective retention processes, there is no need for the observable products at any stage to be blind. Thought consists of a vicarious blind exploratory locomotion in a vicarious mnemonic map of the environment. As Poincaré emphasised in his brilliant presentation of the model for mathematical creativity (1908, 1913), this process may well be entirely unconscious, the selected 'solution' appearing in consciousness phenomenally as from nowhere, a gift of a supernatural muse. Of course there are wise restraints in the mental map and the surrogate exploration of it. Of course this involves variations on an already cumulated corpus of knowledge, rather than a discard of all that has gone before.

In addition, I recognise the practical value of heuristic principles: but I regard these as partial, general knowledge of a domain, already achieved through a trial and error of heuristics, and of highly corrigible truth value.

The unit of variation needs also to be explicit. The unjustified variation may at some stages be a variation of relatively complete theories, variations on a substantial portion of the corpus, and at other times variation of minor detail, instrumentation, etc. Structural selectors are of course present: if a competition of theories is involved, then it is a blind variation of theories, all of which meet the structural requirements of being theories, and from among which one may turn out to provide a superior fit than another—

but if it is a case of expansion of knowledge, not because it was wiser in its generation than the other.

The role of wild speculations in generating scientific hypotheses

The variations are, to be sure, bound to be restricted. But the wider the range of variations, the more likely a novel solution. The recommendation to speculate wildly thus belongs in the guide book to the *strategy* of discovery, if not in the logic. A surprising number of advocates of the model, ancient and modern, have testified to this point.

I will start with a modern, Paul Feyerabend, with particular reference to his paper 'Against method: outline of an anarchistic theory of knowledge' (1970). It is with some ambivalence that I use his testimony, for Feyerabend's delight in anarchistic advocacy leads him to neglect asserting his awe of the achievements of modern physical and biological science. I believe that he has such awe and that he uses the term 'knowledge' in the title to refer to an impressive achievement in need of explanation. His belief that we cannot make general rules about the selection criteria that future science will in fact use seems to me compatible with this, and I agree with his criticism of the *ad hoc* aspects of Lakatos' (1970) otherwise admirable position. Similarly, I believe he is a natural-selection epistemologist, although the only testimony that comes close to asserting this is his off-hand comments on Boltzmann (Feyerabend, 1970, 112–13):

> Later in the nineteenth century proliferation was defended by *evolutionary arguments*: Just as animal species improve by producing variations and weeding out the less competitive variants, science was thought to improve by proliferation and criticism. Conversely, 'well-established' results of science and even the 'laws of thought' were regarded as temporary results of adaptation; they were not given absolute validity. According to Boltzmann [1905, 398, 318, 258–9], the latter 'error finds its complete explanation in Darwin's theory. Only what was adequate was also inherited. . . . In this way the laws of thought obtained an impression of infallibility that was strong enough to regard them as supreme judges, even of experience. . . . One believed them to be irrefutable and perfect. In the same way our eyes and ears were once assumed to be perfect, too, for they are indeed most remarkable. Today we know that we were mistaken—our senses are not perfect.' Considering the hypothetical status of the laws of thought, we must 'oppose the tendency to apply them indiscriminately, and in all domains' [p. 40]. This means, of course, that there are circumstances, not factually circumscribed or *determined in any other way*, in which we must introduce ideas that contradict them. We must be prepared to introduce ideas inconsistent with the most fundamental assumptions of our science even *before* these assumptions have exhibited any weakness. Even 'the facts' are incapable of restricting proliferation, for 'there is not a single statement that is pure experience' [pp. 286, 222]. Proliferation is important not only in science but in other domains too: 'We often regard as ridiculous the activity of the conservatives, of those pedantic, constipated, and stiff judges of morality and good taste who anxiously insist on the observance of every and any ancient custom and rule of behavior; but this activity is beneficent and it must be carried out in order to prevent us from falling back into barbarism. Yet petrification does not set in, for there are also those who are emancipated, relaxed, the *hommes sans gêne*. Both classes of people fight each other and together they achieve a well-balanced society' [p. 322].

It would be an important part of such an epistemology to advocate as he does *counterinduction* and proliferation (1970, 26):

Taking the opposite view, I suggest introducing, elaborating, and propagating hypotheses which are inconsistent either with well-established *theories* or with well-established *facts*. Or, as I shall express myself: *I suggest proceeding counterinductively in addition to proceeding inductively*. . . . evidence that is relevant for the test of a theory *T* can often be unearthed only with the help of an incompatible alternative theory *T′*. Thus, the advice to postpone alternatives until the first refutation has occurred means putting the cart before the horse. In this connection, I also advised increasing empirical contents with the help of *a principle of proliferation*: invent and elaborate theories which are inconsistent with the accepted point of view, even if the latter should happen to be highly confirmed and generally accepted. . . . such a principle would seem to be an essential part of any critical empiricism.

But he is so in love with *variation* as to totally neglect *selection* and to see *retention* only as variation's enemy. He is thus a very incomplete and one-sided evolutionary epistemologist at best.

The other testimony to wild speculation can be taken in chronological order. I am tempted to such full citation just because, Feyerabend not withstanding, such evidence is extremely rare today and was not common even in its heyday when James, Mach and Poincaré were affirming it.

We can go back to Alexander Bain. Presumably the theme was in his first edition of 1855. It has not been convenient for me to examine this, so I quote from a third edition (1874):

Possessing thus the material of the construction and a clear sense of the fitness or unfitness of each new tenative, the operator proceeds to ply the third requisite of constructiveness—trial and error—. . . to attain the desired result. . . . The number of trials necessary to arrive at a new construction is commonly so great that without something of an affection or fascination for the subject one grows weary of the task. This is the *emotional* condition of originality of mind in any department (p. 593).

In the process of Deduction . . . the same constructive process has often to be introduced. The mind being prepared beforehand with the principles most likely for the purpose . . . incubates in patient thought over the problem, trying and rejecting, until at last the proper elements come together in the view, and fall into their places in a fitting combination (p. 594).

Faraday comes next, with his emphasis on the essential role of imagination, and by what he regards imagination to be. I borrow from Williams (1968, 236):

'I should be sorry, however,' he wrote, 'if what I have said were understood as meaning that education for the improvement and strengthening of the judgment is to be altogether repressive of the imagination, or confine the exercise of the mind to processes of a mathematical or mechanical character. I believe that, in the pursuit of physical science, the imagination should be taught to present the subject investigated in all possible, and even in impossible views; to search for analogies of likeness and (if I may say so) of opposition—inverse or contrasted analogies; to present the fundamental idea in every form, proportion, and condition; to clothe it with suppositions and probabilities, that all cases may pass in review, and be touched, if needful, by the Ithuriel spear of experiment. But all this must be *under government*, and the result must not be given to society until the judgment, educated by the process itself, has been exercised upon it' [Faraday, 1859, 480].

The point was put more forcefully in his Diary. 'Let the imagination go, guiding it by judgment and principle, but holding it in and directing it by *experiment*' [Faraday, 1936, 337].

The discipline, government and judgment which he repeatedly insists must be superimposed on imagination are not to repress or stifle it (quite the contrary) but rather to edit it, a selection process operating before retention or propogation. Williams' elitist interpretation of Faraday's view of scientific discovery seems to me unnecessary and irrelevant, although shared by William James (1880) in the essay cited below. Too many potential creators are inhibited by a belief that gifted others solve problems directly, as Polya (1945, 1954) has noted in the course of recommending uninhibited trial and error. Faraday's recommendation to teach one's imagination to generate impossible views and oppositions and to try out all conceivable combinations is democratising in that it removes an inhibiting version of a wide-spread elitist doctrine of special gift. That some persons can generate hypotheses less inhibitedly or more rapidly than others, or have a better model of the world against which to edit them, are minor matters of degree in talents that all men share.

Stanley Jevons (1874) is still more explicit on the wildness. (My quotations are extracted from a very modern criticism of Mill and Bacon. Jevons is due for a selective revival, as Medawar (1967) has noted.)

I hold that in all cases of inductive inference we must invent hypotheses, until we fall upon some hypothesis which yields deductive results in accordance with experience. Such accordance renders the chosen hypothesis more or less probable, and we may then deduce, with some degree of likelihood, the nature of our future experience, on the assumption that no arbitrary change takes place in the conditions of nature (1877, 228).

It would be an error to suppose that the great discoverer seizes at once upon the truth, or has any unerring method of divining it. In all probability the errors of the great mind exceed in number those of the less vigorous one. Fertility of imagination and abundance of guesses at truth are among the first requisites of discovery; but the erroneous guesses must be many times as numerous as those which prove well founded. The weakest analogies, the most whimsical notions, the most apparently absurd theories, may pass through the teeming brain, and no record remain of more than the hundredth part. There is nothing really absurd except that which proves contrary to logic and experience. The truest theories involve suppositions which are inconceivable, and no limit can really be placed to the freedom of hypothesis (1887, 577).

William James (1880) is our next example. He is the first who explicitly uses an evolutionary analogy. This is militantly natural-selectionist and throughout the article he rails against Spencer for his Lamarckianism, that is, for his belief that the environment *directly* impresses its shape upon the organism's body and mind, with mere repetition. For James the process is instead profoundly indirect, and limited to within the range of mutations and fancies emitted by the adapting, knowing organism. While brilliant in this regard, James is surprisingly unobservant as to the details of the selection process, treating it as though all the selecting were being done by encounters with the external world, neglecting the vicarious editorial processes, conscious as well as unconscious.

I now pass to the last division of my subject, the function of the environment in *mental* evolution. After what has already been said, I may be quite concise. Here, if anywhere, it would seem at first sight as if that school must be right which makes the mind passively plastic, and the outer relations actively productive of the form and order of its conceptions; which, in a word, thinks that all mental progress must result from a series of *adaptive* changes, in the sense already defined. . . . It might, accordingly, seem as if there was no room for any other agency than this; as if the distinction we have hitherto found so useful between the agency of 'spontaneous variation', as the producer of changed forms, and the environment, as their preserver and destroyer, did not hold in the case of mental progress;

as if, in a word, the parallel with Darwinism might no longer obtain, and Spencer might be quite right with his fundamental law of intelligence, which says, 'The cohesion between psychical states is proportionate to the frequency with which the relation between the answering external phenomena has been repeated in experience'.

But, in spite of all these facts, I have no hesitation whatever in holding firm to the Darwinian distinction even here. I maintain that the facts in question are all drawn from the lower strata of the mind, so to speak—from the sphere of its least evolved functions, from the region of intelligence which man possesses in common with the brutes. And I can easily show that throughout the whole extent of those mental departments which are highest, which are most characteristically human, Spencer's law is violated at every step; and that, as a matter of fact, the new conceptions, emotions, and active tendencies which evolve are originally *produced* in the shape of random images, fancies, accidental outbirths of spontaneous variation in the functional activity of the excessively unstable human brain, which the outer environment simply confirms or refutes, adopts or rejects, preserves or destroys—*selects*, in short, just as it selects morphological and social variations due to molecular accidents of an analogous sort.

It is one of the tritest of truisms that human intelligences of a simple order are very literal. . . . But turn to the highest order of minds, and what a change! Instead of thoughts of concrete things patiently following one another in a beaten track of habitual suggestion, we have the most abrupt cross-cuts and transitions from one idea to another, the most rarefied abstractions and discriminations, the most unheard-of combinations of elements, the subtlest associations of analogy; in a word, we seem suddenly introduced into a seething cauldron of ideas, where everything is fizzling and bobbing about in a state of bewildering activity, where partnerships can be joined or loosened in an instant, treadmill routine is unknown, and the unexpected seems the only law. According to the idiosyncrasy of the individual, the scintillations will have one character or another. They will be sallies of wit and humour; they will be flashes of poetry and eloquence; they will be constructions of dramatic fiction or of mechanical device, logical or philosophic abstractions, business projects, or scientific hypotheses, with trains of experimental consequences based thereon; they will be musical sounds, or images of plastic beauty or picturesqueness, or visions of moral harmony. But, whatever their differences may be, they will all agree in this—that their genesis is sudden and, as it were, spontaneous. That is to say, the same premises would not, in the mind of another individual, have engendered just that conclusion; although, when the conclusion is offered to the other individual, he may thoroughly accept and enjoy it, and envy the brilliancy of him to whom it first occurred. . . . The conception of a law is a spontaneous variation in the strictest sense of the term. It flashes out of one brain, and no other, because the instability of that brain is such as to tip and upset itself in just that particular direction. But the important thing to notice is that the good flashes and the bad flashes, the triumphant hypotheses and the absurd conceits, are on an exact equality in respect of their origin. . . . The forces that produce the one produce the other. When walking along the street, thinking of the blue sky or the fine spring weather, I may either smile at some preposterously grotesque whim which occurs to me, or I may suddenly catch an intuition of the solution of a long-unsolved problem, which at that moment was far from my thoughts. Both notions are shaken out of the same reservoir,—the reservoir of a brain in which the reproduction of images in the reactions of their outward persistence or frequency has long ceased to be the dominant law. But to the thought, when it is once engendered, the consecration of agreement with outward persistence and importance may come. The grotesque conceit perishes in a moment, and is forgotten. The scientific hypothesis arouses in me a fever of desire for verification. I read, write, experiment, consult experts. Everything corroborates my notion, which being then published in a book spreads from review to review and from mouth to mouth, till at last there is no doubt I am enshrined in the Pantheon of the great diviners of nature's ways. The environment *preserves* the conception which it was unable to *produce* in any brain less idiosyncratic than my own (James, 1880, 455–7).

In the following year, Souriau (1881) presented a rather complete variation-*au-hasard*-and-selective-retention model of creative thought. While he emphasised chance, and recommended oblique thinking rather than direct (*penser à côté*), he did not particularly encourage wild speculation, although he did note: 'For every single idea of a judicious and reasonable nature which offers itself to us, what hosts of frivolous, bizarre, and absurd ideas cross our minds' (1881, 43).

In 1895, Ernst Mach returned from the provincial universities to Vienna, and changed careers from physics to the Professorship of the History and Theory of Inductive Science. In his inaugural address, he chose the theme 'On the part played by accident in invention and discovery', showing, I believe, the importance he gave to this message, atypical of his total philosophy though it might be. Much of the article (as with Bain and Souriau) is about chance encounters in the physical manipulations of the laboratory. But thought trials are also involved:

> Newton, when questioned about his methods of work, could give no other answer but that he was wont to ponder again and again on a subject; and similar utterances are accredited to D'Alembert and Helmholtz. Scientists and artists both recommend persistent labour. After the repeated survey of a field has afforded opportunity for the interposition of advantageous accidents, has rendered all the traits that suit with the mood or the dominant thought more vivid, and has gradually relegated to the background all things that are inappropriate, making their future appearance impossible; then from the teeming, swelling host of fancies which a free and high-flown imagination calls forth, suddenly that particular form arises to the light which harmonises perfectly with the ruling idea, mood or design. Then it is that that which has resulted slowly as the result of a gradual selection, appears as if it were the outcome of a deliberate act of creation. Thus are to be explained the statements of Newton, Mozart, Richard Wagner, and others, when they say that thoughts, melodies, and harmonies had poured in upon them, and that they had simply retained the right ones. Undoubtedly, the man of genius, too, consciously or instinctively, pursues systematic methods, wherever it is possible; but in his delicate presentiment he will omit many a task or abandon it after a hasty trial on which a less endowed man would squander in vain his energies (1896, 174).

Poincaré's essay on mathematical invention (1908) is the most developed of the classic presentations of the model, and is frequently enough reprinted to not need extensive quotation here. More than the others, he emphasises unconscious processes. 'Ideas rose in clouds. I felt them collide until pairs interlocked, so to speak, making a stable combination' (1913, 387). 'Among the great numbers of combinations blindly formed by the subliminal self, almost all are without interest' (1913, 392). 'They flash in every direction through the space . . . as would a swarm of gnats, or, if you prefer a more learned comparison, like the molecules of gas in the kinematic theory of gases. Their mutual impacts may produce new combinations' (1913, 273). 'In the subliminal self, on the contrary, reigns what I should call liberty, if we might give that name to the simple absence of discipline, and to the disorder born of chance. Only, this disorder itself permits unexpected combinations' (1913, 394).

For the many others who have advocated this model, and for detailed reactions to its critics, see Campbell (1960). After all this emphasis upon wild variation it is well to remind ourselves again that variation is only one of the several essential components. Systematic selection and loyal retention and duplication of selected variants are also required. Variation and retention are inherently in opposition and neither can completely dominate the other if the process is to work.

There is a pervasive belief in or longing for an intelligent discovery process through which the gifted scientist would move directly to the discovery without wasted effort. While the ubiquitous occurrences of blind alleys and waste explorations are recognised, they are treated as unfortunate wastes that could have been avoided. As has been argued, a purely analytic examination of the problem of going beyond prior knowledge should rule out such a model of direct intelligent discovery. But there are strong resistances to this insight. Perhaps these recurrent testimonies to the inevitably wasteful variation process in scientific discovery will help reinforce the point.

Summary

To achieve a fit between scientific theory and the world described, just as for the fit of organism to environment, a wasteful nonprescient variation and selective retention process is required. As an aspect of descriptive epistemology, this is phrased as 'unjustified' variation. Testimony to such a process in scientific discovery comes from Bain, Faraday, Jevons, James, Souriau, Mach, Boltzmann, Poincaré and Feyerabend, among others.

References

Ackermann, R. (1970). *The Philosophy of Science*. Pegasus, New York.

Bain, A. (1874). *The Senses and the Intellect*, third edition. Appleton, New York. (First edition, 1855.)

Baldwin, J. M. (1909). *Darwin and the Humanities*. Review Publishing Company, Baltimore. (Allen and Unwin, London, 1910.)

Bertalanffy, L. (1967). *Robots, Men, and Minds*. Braziller, New York.

Bertalanffy, L. (1969). Chance or law. In *Beyond Reductionism* (ed. A. Koestler and J. R. Smythies). Beacon Press, Boston, 56–84.

Blachowicz, J. A. (1971). Systems theory and evolutionary models of the development of science. *Philosophy of Science*, **38**, 178–99.

Boltzmann, L. (1905). *Populaere Schriften*. Barth, Leipzig.

Campbell, D. T. (1956*a*). Adaptive behavior from random response. *Behavioral Science*, **1**, 105–10.

Campbell, D. T. (1956*b*). Perception as substitute trial and error. *Psychological Review*, **63**, 330–42.

Campbell, D. T. (1959). Methodological suggestions from a comparative psychology of knowledge processes. *Inquiry*, **2**, 152–82.

Campbell, D. T. (1960). Blind variation and selective retention in creative thought as in other knowledges processes. *Psychological Review*, **67**, 380–400.

Campbell, D. T. (1965). Variation and selective retention in socio-cultural evolution. In *Social change in developing areas: a reinterpretation of evolutionary theory* (ed. H. R. Barringer, G. I. Blanksten and R. W. Mack). Schenkman, Cambridge, Mass., 19–49.

Campbell, D. T. (1966). Pattern matching as an essential in distal knowing. In *The Psychology of Egon Brunswik* (ed. K. R. Hammond). Holt, Rinehart and Winston, New York, 81–106.

Campbell, D. T. (1969). A phenomenology of the other one: corrigible, hypothetical and critical. In *Human Action: conceptual and empirical issues* (ed. T. Mischel). Academic Press, New York, 41–69.

Campbell, D. T. (1974). Evolutionary epistemology. In *The Philosophy of Karl R. Popper* (ed. P. A. Schilpp). The Open Court Publishing Company (The Library of Living Philosophers), La Salle, Illinois.

Caws, P. (1969). The structure of discovery. *Science*, **166**, 1375–80.
Dobzhansky, Th. (1963). Scientific explanation—chance and antichance in organic evolution. In *Philosophy of Science: The Delaware Seminar*, vol. 1 (1961–2) (ed. B. Baumrin). Wiley, New York, 209–22.
Dretske, F. I. (1971). Perception from an epistemological point of view. *Journal of Philosophy*, **66**, 584–91.
Faraday, M. (1859). *Experimental Researches in Chemistry and Physics*. Taylor and Francis, London.
Faraday, M. (1936). *Faraday's Diary*. G. Bell and Sons, London, vol. 7.
Feyerabend, P. K. (1970). Against method: outline of an anarchistic theory of knowledge. In *Analyses of Theories and Methods of Physics and Psychology* (vol. 4 of *Minnesota Studies in the Philosophy of Science*) (ed. M. Radner and S. Winokur). University of Minnesota Press, Minneapolis, 17–130.
Feigl, H. (1950). Existential hypotheses: realistic *versus* phenomenalistic interpretations. *Philosophy of Science*, **17**, 35–62.
Greene, J. C. (1971). *Darwin and the Modern World View*. Louisiana State University Press, Baton Rouge. (Mentor Books, 1963).
Guthrie, E. R. (1952). *The Psychology of Learning* (revised edition). Harper, New York.
Guthrie, E. R. and Horton, G. P. (1946). *Cats in a Puzzle Box*. Holt, New York.
Harrod, R. (1956). *Foundations of Inductive Logic*. Macmillan, London.
Hilgard, E. L. (1948). *Theories of Learning*. Appleton-Century-Crofts, New York.
Hirst, R. J. (1959). *The Problems of Perception*. Allen and Unwin, London.
James, W. (1880). Great men, great thoughts, and the environment. *The Atlantic Monthly*, **46**, 441–59.
Jevons, W. S. (1892). *The Principles of Science* (first edition, 1874; second edition, 1877; reprinted with corrections, 1892). Macmillan, London.
Koestler, A. and Smythies, J. R. (1969). *Beyond Reductionism*. Beacon Press, Boston.
Köhler, W. (1938). *The Place of Value in a World of Facts*. Liveright, New York.
Kuhn, T. S. (1962). *The Structure of Scientific Revolutions*. University of Chicago Press, Chicago.
Lakatos, I. (1970). Falsification and the methodology of scientific research programmes. In *Criticism and the Growth of Knowledge* (ed. I. Lakatos and A. Musgrave). Cambridge University Press, Cambridge, 51–8.
Lorenz, K. (1941). Kants Lehre vom apriorischen im Lichte gegenwärtiger Biologie. *Blätter für Deutsche Philosophie*, **15**, 94–125. (Translated as Kant's doctrines of the *a priori* in the light of modern biology, *General Systems*, vol. VII.)
Lorenz, K. (1959). Gestaltwarnehmung als Quelle Wissenschaftliche Erkenntnis. *Zeitschrift für experimentelle und angewandte Psychologie*, **6**, 118–65. (Translated as Gestalt perception as fundamental to scientific knowledge, *General Systems*, vol. VII, 1962).
Lorenz, K. (1969). Innate bases of learning. In *On the Biology of Learning* (ed. K. H. Pribram). Harcourt, Brace and World, New York.
Lovejoy, A. O. (1930). *The Revolt against Dualism*. Open Court, Chicago.
Mach, E. (1896). On the part played by accident in invention and discovery. *Monist*, **6**, 161–75.
Mandelbaum, M. (1964). *Philosophy, Science and Sense Perception: historical and critical studies*. The Johns Hopkins Press, Baltimore.
Maxwell, G. (1972). Russell on perception: a study in philosophical method. In *Russell: a collection of critical essays* (ed. D. Pears). Doubleday Anchor Books, New York.
Medawar, P. B. (1967). Hypothesis and induction. In Medawar, *The Art of the Soluble*. Methuen, London, 128–55.
Monod, J. (1971). *Chance and Necessity*. Knopf, New York.
Northrop, F. S. C. (1949). Einstein's concept of science. In *Albert Einstein: Philosopher of Science* (ed. P. A. Schilpp). (The Library of Living Philosophers). Northwestern University Press, Evanston, Illinois, 683–4.
Northrop, F. S. C. (1964). The epistemological possibles. In *Cross-cultural Understanding: Epistemology in Anthropology* (ed. F. S. C. Northrop and H. H. Livingston). Harper and Row, New York, 206–15.
Paley, W. (1867). *Natural Theology, or evidences of the existence and attributes of the Deity collected from the appearances of nature*. Gould and Lincoln, Boston. (First published 1802.)

Pasch, A. (1958). *Experience and the Analytic*. The University of Chicago Press, Chicago.
Petrie, H. G. (1969). *The logical effects of theory on observational categories and methodology*. Monograph final report, Office of Education Contract 0–8–080023–3669(010).
Poincaré, H. (1908). L'invention mathématique. *Bull. Inst. Gen. Psychol.*, **8**, 175–87.
Poincaré, H. (1913). Mathematical creation. In Poincaré, *The Foundations of Science*. Science Press, New York.
Polanyi, M. (1958). *Personal Knowledge*. Routledge and Kegan Paul, London.
Polanyi, M. (1966). *The Tacit Dimension*. Doubleday, New York.
Polanyi, M. (1969). Life's irreducible structure. In Polanyi, *Knowing and Being*. Routledge and Kegan Paul, London, 225–39.
Polya, G. (1945). *How to Solve it*. Princeton University Press, Princeton, N.J.
Polya, G. (1954). *Mathematics and Plausible Reasoning*: vol. I, *Induction and analogy in mathematics*, vol. II, *Patterns of plausible inference*. Princeton University Press, Princeton, N.J.
Popper, K. R. (1959). *The Logic of Scientific Discovery*. Basic Books, New York.
Popper, K. R. (1963). *Conjectures and Refutations*. Basic Books, New York.
Popper, K. R. (1972). *Objective Knowledge: an evolutionary approach*. Clarendon Press, Oxford.
Pumphrey, R. J. (1950). Hearing. In *Symposia of the Society for Experimental Biology*: IV, *Physiological mechanisms in animal behavior*. Academic Press, New York.
Quine, W. V. (1969). *Ontological Relativity*. Columbia University Press, New York.
Ray, J. (1727). *The Wisdom of God manifested in the Works of the Creation*, ninth edition. Innys, London. (Original edition, 1691).
Reichenbach, H. (1938). *Experience and Prediction*. University of Chicago Press, Chicago.
Russell, B. (1944). Reply to criticisms. In *The Philosophy of Bertrand Russell* (ed. P. A. Schilpp) (The Library of Living Philosophers.) Northwestern University, Evanston, Illinois, 681–741.
Russell, B. (1948). *Human Knowledge: its scope and limits*. Simon and Shuster, New York.
Salmon, W. (1961). Vindication of induction. In *Current Issues in the Philosophy of Science* (ed. H. Feigl and G. Maxwell). Holt, Rinehart and Winston, New York, 245–56.
Scriven, M. (1959). Explanation and prediction in evolutionary theory. *Science*, **130**, 477–82.
Shimony, A. (1970*a*). Scientific inference. In *Pittsburgh Studies in the Philosophy of Science*, vol. IV (ed. R. Colodny). University of Pittsburgh Press, Pittsburgh, 79–172.
Shimony, A. (1971*b*). Perception from an evolutionary point of view. *Journal of Philosophy*, **68**, 571–83.
Souriau, P. (1881). *Théories de l'Invention*. Hachette, Paris. (Trans. E. L. Clark, privately distributed, duplicated, 1972.)
Stemmer, N. (1971). Three problems in induction. *Synthese*, **23**, 287–308.
Thoday, J. (1967). Chance and purpose. *Theoria to Theory*, **2**, 29–38.
Thorpe, W. H. (1969). Retrospect. In *Beyond Reductionism* (ed. A. Koestler and J. R. Smythies). Beacon Press, Boston, 428–34.
Toulmin, S. E. (1961). *Foresight and Understanding: an inquiry into the aims of science*. Indiana University Press, Bloomington.
Toulmin, S. E. (1972). *Human Understanding*: vol. I, *The evolution of collective understanding* Princeton University Press, Princeton, N.J.
Waddington, C. H. (1969). The theory of evolution today. In *Beyond Reductionism* (ed. A. Koestler and J. R. Smythies). Beacon Press, Boston, 357–95.
Wallraff, C. F. (1961). *Philosophical Theory and Psychological Fact*. The University of Arizona Press, Tucson.
Wartofsky, M. (1968). Metaphysics as heuristic for science. In *Boston Studies in the Philosophy of Science*, Vol. III (ed. R. S. Cohen and M. W. Wartofsky). D. Reidel, Dordrecht, Holland, 164–70.
Weiss, P. A. (1969). The living system: Determinism stratified. In *Beyond Reductionism* (ed. A. Koestler and J. R. Smythies). Beacon Press, Boston, 3–55.
Whyte, L. L. (1965). *Internal Factors in Evolution*. Braziller, New York.
Williams, D. C. (1947). *The Ground of Induction*. Harvard University Press, Cambridge, Mass.

Williams, L. P. (1968). Epistemology and experiment: the case of Michael Faraday. In *Problems in the Philosophy of Science* (ed. I. Lakatos and A. Musgrave). North-Holland, Amsterdam, 231–9.

Zirkle, C. (1941). Natural selection before *The Origin*. *Proceedings of the American Philosophical Society*, **84**, 71–123.

Discussion

Medawar

Although I am generally sympathetic to Dr Campbell's approach and think it original and stimulating, yet I have some misgivings about its applicability to the world of ideation. Genuine blind variation tends to produce a far higher proportion of disordering or nonsensical variation than seems to be the case in the world of ideas. Thus very many genetic mutations are disordering in character, and a good many of Dr Edelman's antibodies are probably nonsense antibodies, that is, antibodies to nothing. However, in scientific hypothesis formation and in comparable procedures—such as a mechanic's trying to figure out why a car won't run—very few of the conjectures will be of the idiotic quality which you would expect if the ideation process were essentially a random one. It can be argued, of course, that there is a selective process but that it is an unconscious one, but I rather think that the process of selection Campbell had in mind was an overt one, as it is in natural selection of genetic mutations and in antibody formation. This is not in any way to deny the notion that new ideas arise by what may be an essentially random conjunction of ideas in the mind. I do however think that the process, whatever it may be, is really rather different from that which we have become familiar with through genetic mutation and the generation of diversity in immunological systems. I should very much like to hear Dr Campbell's comments on these remarks.

Campbell

Very definitely, I do hold the blind variation and selective retention model for thinking, although I recognise that this is the application most hard for others to accept. The absence of any conceivable alternative is an important determinant of my faith. Joining Poincaré, I accept the likelihood that most of the blind permuting takes place unconsciously. This admission has the cost of making the theory evade phenomenological test. (Note, however, the recurrent experiential witness to such a process on the part of Poincaré, Mach, James and the others.) But general acceptance cannot come until we have a detailed neurological model for the processes James refers to in such phrases as 'because the instability of the brain is such as to tip and upset itself', etc., a model with detail such as Edelman has provided for immunological discovery and recognition processes.

10. Reduction, Hierarchies and Organicism

MORTON BECKNER

Discussions of the philosophical issues of reduction, holism, emergence, inter-level explanation, organicism, neovitalism—in other words, of the relations between levels of organisation (both in hierarchically arranged objects such as organisms and in hierarchically arranged sciences)—are always conducted in a context of insufficiently clarified ideas. And not only are they insufficiently clarified; they are ideas whose connections involve the most fundamental controversies of metaphysics, the theory of knowledge, and the philosophy of science. This means that when we look at the details of some reasonably definite question, such as 'What is micro-reduction?' or 'Is the structure of a living organism reducible to the principles of physics and chemistry?', we find that the ideas we need are located at the nodes of a ravelled background web. We are almost tempted to see something in the old idealistic view that the solution or clarification of one question involves the solution or clarification of all.

If this is so, an investigation of our topic, the problem of reduction in biology, will attempt to bring part of the web into clearer focus, with the recognition that what lies outside the field of vision, or at its dimly seen edges, might conceivably make all the difference in our ability to map the centre of the field. So I apologise at the outset for ignoring some potentially important ideas. I shall simply take it for granted that they are clear enough for our purposes. Among them are the ideas of explanation, cause, probability, description, event-identity and prediction.

In this paper, I am mainly concerned with philosophical problems connected with the hierarchical organisation of biological systems. Some of these problems stem from the simple point that, in hierarchies, objects and events at the lower levels of organisation *comprise* the objects and events at higher levels; but at the same time we employ different and in some degree independent languages in the description of the different levels. In Section I, I collect a number of points that need to be explicitly remembered in discussions of reduction, and point out some common confusions that result when they are momentarily ignored. Section II employs these points in an analysis of the organicist position; I introduce a concept of event-reduction which helps me drive the proponent of high-level autonomy unwillingly into the arms of vitalism. Section III is a somewhat cryptic suggestion that questions about reduction are special cases of a more general problem about the conditions of applicability of one science to another.

I. The description of hierarchies

Take as an example of hierarchically arranged sciences Medawar's sequence: physics, chemistry, organismic biology, ecology/sociology. Call any science in this list 'higher' than all the sciences (if any) to its left, so that in a normal case of reduction a higher science would be reduced to a lower (Medawar, 1974). And it will probably be better if we speak of the relations between *theories* which are parts of these sciences, for example, of the relations between Mendelian genetics (part of organismic biology) and the molecular theory of genetic coding and transcription (part of chemistry). For short, I will call any such pair of theories, related as higher to lower, T_h and T_l respectively.

In order to fix our ideas, let me now collect some of the familiar major distinctions and theses that have figured in discussions of reduction and novelty.

(1) There is a difference between a set of hierarchically arranged theories and a set of hierarchically organised systems (in the sense of genuine working parts of the world, such as organisms, machines, populations, etc.). The latter concept has been thoroughly investigated. We have a 'perfect hierarchy' if we can define a part–whole relation and a set of parts and wholes that satisfy the following conditions: we specify a series of 'levels' L_0 to L_n, and a set of 'parts' such that

(*a*) every part P_i is assigned to exactly one level L_i;

(*b*) every part P_i (except those of level L_n) is a part of exactly one part at each level above i;

(*c*) every part P_i (except those of level L_0) is exhaustively composed of parts at each level below i.

The part–whole relation is a special but important case. But other relations yield other kinds of perfect hierarchies; for example, if the taxonomic categories define the levels, and each P_i is a taxon of level L_i, then a standard taxonomic system is a perfect hierarchy with respect to the relation of class inclusion. But I shall attend only to the part–whole case.

It seems clear that the systems of biological interest are not perfect hierarchies. For example, not all the tissues of an organism are exhaustively composed of cells. But we can think of organisms, and other systems, as fitting the perfect case more or less approximately; and to save words I shall pretend we do study perfect hierarchies.

The hierarchy model has historically provided a large part of the framework of discussion in the philosophy of biology. It is involved in a wide range of connected ideas: levels of organisation, sequences of boundaries (Polanyi, 1968), autonomy at one level with respect to lower ones, a temporal order in the arrival of the higher levels on the cosmic scene, the emergence of higher-level entities, etc.

The existence of hierarchical systems is certainly connected with the hierarchical arrangement of theories. But I do think we lack a detailed philosophical account of the connections, in part because the relations between higher and lower-level theories are not too well understood. I will offer one tentative suggestion.

The orthodox view of the relation between a T_h and T_l has, I think, two components: (*a*) that T_l is of wider scope than T_h, and (*b*) that the principles or laws of T_h are special cases of the principles of T_l. 'Wider scope' in this context simply means that the set of systems which exhibit the phenomena described by T_l is larger than, but includes, the set described by T_h. And to say that L_h (a law or principle of T_h) is a 'special case' of L_l is to say that L_l entails L_h under a restricted range of boundary conditions; for example, Galileo's law of free fall is a special case of Newton's law of gravitation since the latter entails the former (with the help of a bit more theory) under the condition of a uniform gravitational field across the path of free fall.

Unfortunately, we cannot adopt either clause of this account of hierarchically arranged theories without begging all the controversial questions about reduction and emergence. Consider clause (*a*). A defender of a strong doctrine of emergence might hold that the laws of physics which describe, say, the behaviour of particles which are not parts of biological systems, no longer describe them when they are parts of biological systems. This is no doubt false, but it is not false by virtue of our concepts of physical and biological systems. If it were true, then clause (*a*) would not catch what is involved in a hierarchy of theories.

And if our strong emergentist is correct, clause (*b*) must be rejected also, since the phenomena described by L_h would not be described by L_l. But there is a more plausible reason to reject clause (*b*). Suppose it is the case that the laws of physics merely put constraints on the range of logically possible behaviour of a biological system, and leave some possibilities open. Then if the system exhibits additional laws, they will not be special cases of the laws of physics. I will consider this in more detail in Section II; but the point is that we cannot decide against emergentism just in our philosophical account of hierarchical structure.

My suggestion is this: our view of what constitutes a hierarchy of theories is, to be sure, influenced by the notions of scope and generality differences; but there is another important factor. Even if we use nothing more than the common knowledge available to the man in the street or jungle, we cannot help noticing that there are very many hierarchically organised systems. A hut is made of plant parts; this is a simple two-level hierarchy. Bee colonies comprise bees, and bees comprise bee organs; this is one of three levels. And on more sophisticated investigation, it turns out that these obvious hierarchies have still more levels. This is simply an empirical fact about the world. Such hierarchies provide us with a way of arranging sciences on a scale: an *i*-level science is one that is largely concerned with *i*-level systems. If we let the correspondence serve as a criterion for distinguishing higher and lower theories, we discover that, in addition, we have many clear cases of the scope and generality differences mentioned in clauses (*a*) and (*b*).

(2) We may distinguish between emergence and reducibility. I shall not enter into the debate over the nature of theory reduction, beyond endorsing the views of Nagel (1961). According to him, the central idea involved in the reducibility of T_h to T_l is the explicability of all the laws of T_h in the theory T_l. Explicability presupposes (*a*) that all the concepts or terms C_h of T_h can be systematically connected with some of the terms C_l of T_l, either by synthetic identity statements, explicit definitions or some other semantic relation (the 'condition of connectability'); and (*b*) that each law L_h is deducible from T_l, together with the statements that connect the C_hs with the C_ls, and a description of relevant boundary conditions in the vocabulary of T_l (the 'condition of derivability'). (See also Broad (1951), Hempel (1966), Kemeny and Oppenheim (1956), Schaffner (1967) and Woodger (1952).)

A common philosophical strategy is to define 'emergence' in terms of 'reducibility'. In the special case of hierarchically organised systems, an orthodox definition, neglecting refinements, would be something like this: *i*-level phenomena are 'emergent' with respect to lower-level theories when, and only when, the *i*-level theories are not reducible to the theories of the lower levels.

This definition, however, even though it is cautiously relativised to a set of theories, does not catch the idea of emergence which critics of reductionism want. To show clearly why this is so, I need to digress at some length about the description of events. The digression will also provide a vocabulary that I will use in my remarks about event-reduction and organicism, so it should not be skipped.

An event such as the battle of Austerlitz can obviously be described in an indefinite number of ways, such as 'an important battle in Napoleon's career', 'a victory for France', 'the battle of Austerlitz', etc. However, there are a number of points that need explicit notice if we are to avoid certain common confusions.

(*a*) A description of an event may or may not be sufficient for the unique identification of the event. The third description above does, and the other two do not, uniquely identify the battle of Austerlitz. Whether or not a description is uniquely identifying, and what it identifies (for a particular speaker of the language employed in the description), naturally depends on several factors, including his general background knowledge and the context in which the description is offered. For example, if someone does not know that historians conventionally use the phrase 'the battle of Austerlitz' to refer to a particular battle which Napoleon won, but does know of several battles fought at Austerlitz, the given description is not uniquely identifying; and if he knows of only one such battle, one fought in the sixteenth century, the phrase will identify a battle other than the one an historian would pick.

(*b*) Suppose that Mr Jones executes his will by certain bodily movements (moving a pen in a special way). On this occasion, we would correctly say that the event under the description D_1 'Mr Jones' executing his will' is the same as the event under the description D_2 'Mr Jones' moving a pen in a special way'. This does not mean, however, that *every* event answering to the description D_1 (or D_2) is the same as some event answering to D_2 (or D_1). Mr Jones could execute his will by quite different movements, such as making an X; and any number of bodily movements answering to D_2 could be identical with something other than executing a will, for example, signing a letter. In general: D_1 and D_2 each, unless they have the form of definite descriptions, determine a *class* of events; and an event in either class may or may not be the same as an event in the other.

(*c*) A description of an event will of course leave some aspects of the event undescribed. Thus, it might happen that D_1 and D_2, which in fact describe the same event, fail to give us sufficient information to establish the identity.

(*d*) Any description employs certain concepts; and their employment presupposes a 'conceptual scheme' (Strawson, 1959). Thus it is possible for the same event to be described with the help of two or more distinct conceptual schemes. Thus it seems fair to say that the descriptions D_1 and D_2 of Jones' signing his will represent the employment of very different schemes: the former is (roughly) physical, and the latter is legal. The differences between physical and legal languages are not mere differences in vocabulary, but are deep enough to count as differences in conceptual schemes. Notice that although I have been speaking of the description of events, the same remarks will hold about entities of other kinds—individuals, properties, relations, processes, etc.

None of these points, bordering as they do on truisms, should come as a surprise; indeed, the recent mainstream of discussion in the philosophies of mind and action is based in part on explicit awareness of them. However, they are easy to overlook in considering hierarchical organisation. I want to call attention to four confusions that can result. Suppose we have a set of systems W (wholes) each of which is a perfect hierarchy of n levels with respect to the part–whole relation (W is an n-level 'part'); that we have an i-level theory for each i; and that the differences between any pair of higher (T_h) and lower (T_l) theories are deep enough to qualify T_h and T_l as embodiments of distinct conceptual schemes. Let C_i, E_i, D_i and P_i be respectively shorthand for i-level descriptive or theoretical concepts, events, descriptions and parts; and let K_i be a general term for the *kind* of event, part, etc. of level i. This shorthand provides a general way of speaking of differences exemplified by the differences between, say, physical parts and events on the one hand, and biological parts and events on the other. Examples of the *K*s might be 'physical', 'chemical', 'biological', 'psychical', etc. With these suppositions, and in this vocabulary, the confusions I have in mind, in order of decreasing naivety, are as follows.

(1) If zero-level theory T_0 provides us with some cash value for the remark that every P_0 is of kind K_0, we are strongly tempted to say that *W* contains only K_0 parts. A hard-boiled reductionist might say, for example, that since physical theory provides a description ('elementary particle') of each P_0 in a person, persons are composed of nothing but K_0 ('physical') parts. This is an outstanding example of a 'nothing but' inference, and it is trivially incorrect. A person—a human being—may well be exhaustively composed of elementary particles, but these are not the only sorts of parts that compose him. He is also, if we regard him as a perfect hierarchy, exhaustively composed of atoms, molecules, cells, etc. None of these are elementary particles, so it follows that a person is not composed only of elementary particles. (If a person is an *n*-level *W* of which elementary particles are the zero-level parts, then a person can be decomposed, without residue, into elementary particles; this may be the point that reductionists are trying to make.) In short, there is a distinction between being *exhaustively* composed of parts K_i, and being composed *only* of parts K_i. Observance of this distinction would block a lot of loose talk in the philosophy of biology. The tendency to miss the distinction, obvious though it is, is due to a tendency to think that an exhaustive description of a system at the *i*-level must also be an exhaustive description at all levels above *i*.

(2) One might think that since every P_i is exhaustively composed of parts P_{i-1}, any theory T_i must be reducible to a theory T_{i-1}. This also is a fallacy, one that has been consistently opposed by the organicist and vitalist traditions. They have opposed it, however, because of a belief either in the emergence of irreducibly novel phenomena at higher levels of organisation, or because of a conviction that mind-like entities, not themselves parts of the hierarchy, serve a control function at the higher levels. These views are controversial and, indeed, unnecessarily sophisticated. It is enough to observe that alternative descriptions D_1 and D_2 of the same thing, whether or not they belong to theories of different levels, need have no logical connections. In particular, neither D_1 nor D_2 necessarily entails the other, even with the help of additional premises drawn from available theory. If such relations of entailment do not hold from the formulations in T_l to those in T_h, then *a fortiori* T_h is not reducible to T_l.

It might be supposed that no one would endorse such a bad argument. No one would, after he had seen it explicitly laid out; but vague versions can be extremely persuasive. I have often asked biologist friends exactly why they are convinced that biology is reducible to chemistry. The common answer is always something like this: 'Well, after all, biology and the chemistry of biological systems are concerned with the same things—complex chemical processes.' Variations on this theme all exemplify the fallacy.

(3) The next confusion has to be assigned mainly to the organicist camp. It is the fallacy of concluding, from the premise that since a j-level phenomenon is of kind K_j, it is not also of kind K_i, where i is less than j. This is an instance of 'nothing buttery'; reductionists can take a little comfort in the fact that their opponents are sometimes guilty of it.

Examples: 'Mouse-hunting is a biological, not a chemical phenomenon'; 'executing a will is a legal, not a physical action'; 'a battle is a social, not a biological phenomenon'. Consider only the first example. The position might be reinforced by observing that textbooks of chemistry do not mention mouse-hunting, and that the theoretical vocabulary of chemistry does not include the terms 'cat', 'mouse' and 'hunt'. These observations are true, but it does not follow that the process of mouse-hunting is not a chemical process. There are obviously lots of chemical processes, chemical substances, and particulars composed of chemical substances, that have not been assigned special names by chemists.

The point is that the fact of hierarchical organisation is sufficient ground for holding that events at the j-level, in addition to their status as kind K_j, are *also* events of each kind K_i where i is less than j. To deny this is to commit oneself to the absurd position that, for instance, signing a document is never the same thing as executing a will, endorsing a cheque, moving an arm, etc.

However, consider a concept C_j that applies to events E_j of kind K_j. Even though each E_j is also of kind K_i, it does not follow that C_j is also of kind K_i. For example, such things as meiotic division, gastrulation, excretion, predation, etc., are biological processes, but they are also chemical and physical processes as well; but the *concepts* of meiosis, gastrulation, etc., are biological but not chemical or physical. Whether or not a concept belongs to K_i is a matter of whether the concept is part of an i-level theory. A paradigm of this confusion, then, would be an argument like this: 'The concept of meiosis is not part of the theoretical machinery of any physical science; therefore meiosis is not a physical process.'

(4) I will indicate the last confusion only briefly and examine it in more detail in Section II. It, too, is a faulty inference, namely, concluding that if a j-level *theory* is autonomous with respect to an i-level theory (i less than j), then there are i-level processes (events, effects, phenomena) that are autonomous with respect to all i-level processes. An example would be concluding that since Mendelian genetical theory is autonomous with respect to quantum theory, then there are processes of inheritance that are autonomous with respect to the behaviour of elementary particles.

There are thinkers who find this line of reasoning, or something very like it, almost irresistible. It underlies, it seems to me, all of the following important positions: Kant's view that the will of a rational agent is determined both by natural causality and by respect for law; the doctrine that the evolution of the higher taxonomic categories is somehow independent of speciation; Polanyi's (1968), that biological phenomena are subject to the control of dual (or *n*-tuple) systems; Driesch's (1914), that there are two different forms of causality, 'individualising' and 'singular'; in short, all theories (in Rensch's (1974) phrase) of polynomistic determination.

The idea of autonomy needs clarification, and of course the conclusion—that there are higher level autonomous processes—might be true even though it does not follow from the existence of autonomous theories. At this point I will say just enough to indicate where the fallacy lies, and reserve comment on the other issues.

I have distinguished between what we may call 'theory autonomy' and 'process autonomy'. Organicists and vitalists have written a lot in praise of theory autonomy, but it is not always easy to see exactly what they have in mind. At the very least, they seem to mean that, for example, biological theory is not simply a branch of physics or chemistry, but has concepts and principles of its own; and that the biologist can develop higher-level theories without paying all that much attention to what is happening at the lower levels. All this may be true, but it is not very pointed. I think organicists have something stronger in mind, and accordingly adopt the definition: T_j is autonomous with respect to T_i if and only if T_j is not reducible to T_i. This definition makes theory autonomy a straightforward matter of partial logical independence, that is, a relation between parts of scientific languages.

Autonomy between processes, on the other hand, has nothing to do with the languages we choose to describe them, but rather with some sort of causal independence. The issue is complicated by the fact of hierarchical organisation. The case where we are interested in causal connections between two systems that have no common parts is quite different from the case of hierarchies, in which autonomous processes (if there are any) are taking place in numerically the same or overlapping systems. Thus, in a hierarchy, we cannot define the relevant sense of 'independence' in terms, for example, of the absense of energy or information transfer. Since, as far as I know, no one has proposed an analysis except in terms of theory-reducibility (off the mark, as I have argued), we are free to adopt the following definition: in a hierarchy, *j*-level process P_j is autonomous with respect to *i*-level processes if and only if the laws of P_j are not fully determined by the laws of *i*-level processes (not of kind K_j).

If these characterisations of theory and process autonomy do, as far as they go, capture the relevant notions that have appeared in the philosophy of biology, it is easy to see that process autonomy does not follow from theory autonomy: for the irreducibility of T_j to T_i may be due to differences in their conceptual structure, and not to the lack of *determination* of processes P_j by processes P_i.

II. Organicism and event-reduction

This section, unlike the first, has a main thesis, namely that there are two main strands in the organicist tradition, which I call 'strong' and 'weak', and that strong organicism is toothless unless it is supplemented by vitalism.

Weak organicism may be more or less weak. It may hold no more than that the biologist, or anyone interested in higher-level phenomena, has a perfect right to construct theories about higher-level objects without attempting reduction to lower levels; or that, as a matter of fact, much of biology has developed historically without much concern for possible reductions; or, finally, that undue preoccupation on the part of biologists with lower-level phenomena is likely to stand in the way of discovery and understanding of the higher levels. The first two points are, it seems to me, correct. The third—on the heuristic value of a certain policy of method—is an extremely complex issue of historical and philosophical assessment. My own tentative sympathies are with the organicists, but I often think that the hortatory strain one finds in these discussions is misplaced: organicists such as Mayr and Simpson are capable of looking after their own interests without enlisting the aid of Crick and Monod.

Strong organicism, as I see it, is the view that there are higher-level processes that are autonomous with respect to lower levels. I have pointed out that theory autonomy provides insufficient warrant for strong organicism; I want now to consider whether there may be other reasons for adopting it. It will be helpful to clarify the notion of autonomous processes. In doing so, I shall introduce the technical concept of 'event-reduction'. This concept might prove useful in discussions of any of the major issues about reductionism, but in this paper I apply it towards a more limited goal.

Imagine that someone has constructed a large mechanical bug, internally powered, able to move about, and equipped with some sort of sensory device and control circuitry. We may say that the bug is a hierarchy with distinguishable levels; to highlight my argument we can consider only the 'physical' and 'robotic' levels of organisation. At the physical level we have theories that describe the behaviour of the sensory and control devices, the steering mechanism, the power supply, etc. All of this is part of theory T_l.

When we observe the behaviour of the 'whole bug', we notice that whenever it reaches a point where there is a cliff to its left it turns to the right; this is a law (L_h) at the robotic level—the only one we will be concerned with. It is part of T_h. In addition, we describe right turns under these circumstances as 'self-preservative'; call this description 'functional'.

Now there is a minimal but noteworthy sense in which a right turn E_h is 'determined' by events at the physical level, namely, the sense in which E_h is identical with a set of events describable in available T_l: the rotating of wheels, the pivoting of a servomechanism, etc. If we can explain each of these events E_l in T_l, we shall have explained E_h. However:

(1) It is possible that even though E_h is 'determined' by l-level events in this sense, T_l provides no explanation of E_h. Some event E_l might be inexplicable in T_l, for example because of quantum effects.

(2) Even though T_l were to provide an explanation of each E_h, T_l might still provide no explanation for our *law* L_h. Why should it always be a right turn when there is a cliff to the left? T_l is silent about this. (I am presenting conceptual possibilities, not asking the reader to imagine concrete mechanisms.) I can put this in a way that might seem startling: each higher-level event is determined by lower-level events, but the laws of the higher-level events are not explicable by the laws of lower-level theory.

(3) The explanation of an event involves (among other things) the derivation of a description of that event from other descriptions and laws. Recall the points that I laboured above, that descriptions of the same event can be drawn from the resources of distinct conceptual schemes, and that there may be no important logical relationships between descriptions D_h and D_l of the same event. Accordingly, we can envisage these contingencies:

(*a*) E_h is (or is not) explicable in T_l under every description of E_h in the vocabulary of T_h;

(*b*) E_h is (or is not) explicable in E_l under every description of E_h in the vocabulary of T_l;

(*c*) given any description of E_h in the vocabulary of T_h, it can be shown by some no-holds-barred method that E_h under D_h is identical with an event which is (or is not) explicable in T_l.

(I regret the scholastic atmosphere of all these distinctions, but I think they can be important.)

I can now define 'complete' and 'restricted' event-reduction.

(1) T_h is subject to complete event-reduction to T_l if and only if every event which exhibits the phenomena described by T_h is explicable in T_l, under every description of the event in the vocabulary of T_h.

(2) T_h is subject to restricted event-reduction to T_l if and only if every event which exhibits the phenomena described by T_h is explicable in T_l, under some description of the event in the vocabulary of T_h.

These definitions offer a short way of providing explications of some traditional points of view. I think the spirit of the tough-minded reductionist is captured in this: 'For every *h*-level difference we care to distinguish in *h*-level vocabulary and ascribe to an event, then the event is explicable under that description by a suitably enriched physicochemical theory; or, in other words, the development of theories T_h cannot in principle outstrip event-reduction to T_l—subject only to the unavailability of event explanations at the physicochemical level'. This is a very strong thesis. It entails the theory-reducibility of every T_h, not to some established T_l, but to an in-principle-possible *big* T_l. But theory-reduction implies no more than explanation of the laws in T_h, whereas the descriptive power of the conceptual apparatus of our theories in practice far outstrips our ability to assign the events so described their lawful place. Accordingly, this version of reductionism entails more than theory-reducibility. It says that given any property we have reason to assign to E_h (in the vocabulary of T_h) there is a physicochemical explanation telling why it has that property, that is, a derivation under the description which assigns that property to the event. In the bug case, it implies that L_h is explicable, and also that E_h, the right turn, is explicable under the description 'self-protective movement'. In general, we can say that if the ascription of functional significance is allowed by the vocabulary of T_h, and T_h is event-reducible to T_l, then in a clear sense (having nothing to do with the translation of functional into nonfunctional statements) functional ascriptions are reducible to T_l.

A contemporary materialist might hanker after this reductionist thesis, but he need not go so far. He need claim only a kind of restricted event-reducibility, namely that there is some *h*-level description D_h of every event E_h describable in T_h, such that E_h is explicable in an enriched T_l under D_h, if E_h is explicable at all, that is, under some description in the vocabulary of T_l.

The bug case again: it might be that we have a T_l in which the right turn E_h is explicable under some description in T_h. But, assuming that neither the term 'cliff' nor 'self-protection' is part of T_l but belongs to T_h, L_h might be irreducible to T_l; and although we can explain in T_h an event E_h which is in fact a self-protective movement, its self-protective function is not exhibited in the explanation. In sum: our languages may be through-and-through non-materialistic, and yet speak only of systems—working parts of the world—that are through-and-through material.

If T_h is either completely or restrictedly event-reducible to T_l, then the processes described by T_h are not in any interesting sense autonomous with respect to T_l. Of course T_h may contain laws that describe *h*-level processes, and it may be that everything the weak organicist says about the discovery of these laws is true. But if every higher-level event that figures in these laws is explicable in T_l, under some description or other, then the higher-level processes are not autonomous with respect to the lower level—even if T_l, for some reason, fails to explain the lawful sequences at the higher level. Thus, strong organicism must deny both complete and restricted event-reducibility.

Let us now look at the implications of strong organicism within the terms of the bug example. We are asked to imagine that L_h—'Such bugs always turn right when there is a cliff to the left'—qualifies as a high-level law at the robotic level of organisation; that L_h cannot be given a physical explanation; and that at least some of the events that constitute the right turn E_h are not susceptible of physical explanation. Under what circumstances might all these clauses be true? I think there are only two possibilities of philosophical interest. We shall consider only the ostensible case most unfavourable to materialism, the inexplicability in T_l of E_h. (If E_h is inexplicable, so is L_h.)

(1) Some of the laws of physics that apply to physical systems when they

are not low-level parts of a bug no longer apply when they are. For example, Driesch (1914), in a passage that is really not characteristic of his position, suggests that conservation of momentum fails for particles that are parts of organisms. If the relevant physical theory fails when we apply it to bugs (but keep it anyway because of its power elsewhere), then we would know why we cannot explain E_h. But this is too preposterous to detain us any longer.

(2) There may be low-level events that are inexplicable because of indeterminacy at low levels. Suppose for example, that the bug contains an indeterministic randomising device such as a pair of alpha particle detectors A_1 and A_2, and an alpha particle source so arranged that we can assign a definite probability to their activation in a given time interval. And suppose that when and only when A_1 (or A_2) is activated, the bug executes a left (or right) turn. Under these circumstances, E_h is inexplicable in T_l.

Now L_h is true if and only if A_2 is activated whenever there is a cliff to the left of the bug. The strong organicist can obviously take no comfort if this happens by chance alone, for his law L_h would not in any way be part of a general higher-level theory. The only joy for him would lie here: for some reason, when the randomising device is a low-level part of the bug, it always happens, but not just by chance, that A_2 is activated when there is a cliff to the left. This entails that under robotic organisation, alpha particles are guided into A_2 whenever there is a cliff to the left. We have entered the realm of the paranormal. We must suppose that there is something that exerts control over alpha particles (realistically, both their emission and their paths), and does it in the light of information about the relative position of the bug and the cliff. A nonmaterial controlling entity, 'mind-like' in nature, which operates 'on given, preformed, material conditions' by 'suspending possible change and relaxing suspension': the quoted phrases are Driesch's description of an entelechy. The strong organicist is driven to vitalism.

III. Reduction and application

The preoccupation of this paper has been the consistent pursuit of a distinction between (*a*) the logical relations that hold between theories, descriptions, conceptual schemes and other instances of language; and (*b*) what we may as well call real relations (causal, identity, spatial, temporal, part–whole, etc.) between the events and other phenomena that our languages describe. Although the distinction is part of the intellectual equipment of everyone, it can slip away in the course of even the most careful philosophical analysis or scientific thinking. It is especially slippery when we consider hierarchical organisation. And hierarchies are not all that special a problem for the philosophy of biology; any theory large enough to warrant one man's full attention will have to deal with them.

Suppose—what is not yet the case—that we have the distinction straight, and know all about the possible reducibility of higher to lower-level theories, and the reducibility of higher-level events. What methodological or programmatical injunctions would follow in train? What effects should such knowledge have on scientific practice? I think few. If the strongest formulable reductionist thesis were true, would this mean that all young students of biology ought to become molecular biologists? The question is absurd. At the very least, everyone will admit that the employment of the sophisticated and varied resources of molecular biology benefits by directions, hints and problems filtering down from higher-level theories. The successes of molecular biology are inconceivable apart from their context in Mendelian genetics, post-Fisher evolution theory, immunology, neurophysiology, etc.

If, on the other hand, the weakest (but still interesting) formulable reductionist thesis were false, would this mean that we should abandon molecular biology? Although some molecular biologists endorse a version of reductionism, it is still a simple point of logic that each of them could consistently close his papers with the footnote 'Nothing I have said here implies reductionism'. His findings might lead to the explanation of further higher-level laws or events, and that would be a gain. The antireductionist simply holds that there will always be an irreducible residue; he does not tell us where the boundaries are.

It might be objected to the above paragraphs that the truth about theory and event-reduction ought reasonably lead to a shift, if only a small one, in research priorities. I am dubious about even this. Research priorities in a scientific community will in fact be influenced by a variety of scientific, ideological and political factors; but surely the major emphasis, for maximum health, ought to be determined by our sense of where the action is.

In spite of all this, philosophical investigation of the concept of reduction, and historical investigations of examples of reduction, have been of great benefit in increasing our understanding of scientific methods. The important thing is not so much the truth about various reductionist theses, but insight into the conditions and strategies of the *application* of one science to another.

These questions about the applicability of a lower to a higher-level theory are very broad; reduction is a special case. But for historical reasons, the special case has commanded an undue share of explicit attention. However, attention to the special case has shed light on the wider problems. When Professors Dobzhansky and Ayala suggest the topic 'reductionist strategies in modern biology', I suspect their interest really extends over the whole range of application strategies. And cases of application other than reduction do deserve deliberate investigation. I will conclude with a few sketchy suggestions.

It is clear that T_1 is applicable to T_2 if they are respectively theories of systems that are hierarchically organised so that T_2 stands to T_1 as higher to lower-level theory. It is not clear, however, that the 'if' can be strengthened to an 'if and only if'. A lower-level theory cannot be reduced to a higher; but is it inconceivable that a higher-level theory is applicable to a lower, or that theories not standing in a hierarchical relation are applicable to each other? If so, the applicability has other grounds than the fact that physics is applicable to biology because biological systems are also physical.

The self-conscious seeking of theory reduction is a rarity in the history of science. When someone applies, say, organic chemistry to a biological problem such as allergic reactions, he is interested in something rather more like event-reduction: a chemical explanation of events described at the biological level; it does not really matter to him if immunologists have discovered laws at the biological level about allergic reactions. If this is so, it suggests what seems to me in fact a major application strategy: the revision of higher-level theory in a manner that facilitates event-reduction; that is, the introduction of higher-level descriptions with an eye towards the lower-level explanation of events under those descriptions.

References

Broad, C. D. (1951). *The Mind and Its Place in Nature*. Routledge and Kegan Paul, London.
Driesch, H. (1914). *The History and Theory of Vitalism*. Macmillan, London.
Hempel, C. G. (1966). *Philosophy of Natural Science*. Prentice-Hall, Englewood Cliffs, N.J.
Kemeny, J. G. and Oppenheim, P. (1956). On reduction. *Philosophical Studies*, **7**, 6–19.
Medawar, P. (1974). A geometric model of reduction and emergence. In this volume.
Nagel, E. (1961). *The Structure of Science*. Harcourt, Brace and World, New York.
Polanyi, M. (1968). Life's irreducible structure. *Science*, 1308–12.
Rensch, B. (1974). Polynomistic determination of biological processes. In this volume.
Schaffner, K. H. (1967). Approaches to reduction. *Philosophy of Science*, **34**, 137–47.
Strawson, P. F. (1959). *Individuals*. Methuen, London.
Woodger, J. H. (1952). *Biology and Language*. Cambridge University Press, Cambridge.

Discussion

Rensch

I would like to ask a question in order to understand better your somewhat uncommon expression '*downward determination*'. Let us take a simple example. When we combine the light metal sodium with the poisonous gas chlorine we get salt. In this case we can regard both kinds of atoms as a lower-level and the compound salt as a higher-level object. On this higher level some characteristics of the atoms are suppressed or altered. Salt has no metallic structure and it is not poisonous. This apparently comes about by the new systemic relations in this compound. Independently of whether we can or cannot understand this coming about of new characteristics, the problem is open to future analytic investigations in the sense of normal causal reductionism.

In a similar way all new chemical and physical complexes, and all new structures and functions in an organism, show such new systemic relations which are often unpredictable. If you would understand this altering of the characteristics of the lower level by the originating of the higher level as 'downward determination', then I would fully agree. New systemic relations and new characteristics are nothing unintellegible or mystical, although it is sometimes difficult to find out how they come about.

Beckner

'Downward determination' is a term that cropped up in discussion, but not in my paper. I did use it there, however, and I do think there are baffling aspects of the idea it represents. Professor Rensch is quite right in pointing out, for example, that whereas gaseous chlorine is poisonous, chlorine can be found in organised wholes that are not themselves poisonous; and that if this is what is meant by 'downward determination', then downward determination certainly exists, even if the non-poisonousness of the whole were not explicable in chemical theory. However, this point is simply a case of the general point that an object may have properties in one context that it does not have in another. A familiar sort of case would be: I am the tallest man in the room until Wilt Chamberlain enters; then I am no longer the tallest.

I think the question about downward determination is of another sort. Given that an event is at a high level of organisation, and that it is identical with a set of events at a lower level, can there be laws of the higher-level events which determine the events at the lower levels? I argue that the answer is no, unless we adopt *both* a doctrine of indeterminacy at the lower levels, *and* a doctrine of vitalistic control of indeterministic events at the lower levels. This is the upshot of my example of the mechanical bug. Since I would reject the latter doctrine on several grounds, I conclude that downward determination is not possible.

Rensch

We spoke about the necessity of defining *truth* several times, and also Sir Karl Popper just mentioned it. Being a reductionist I would propose the following definition: truth is the maximum possible approach to the understanding of extramental reality and mental relations and operations. It can only be an approach to trans-subjective reality, because we can only judge in the realm of our human capabilities according to the principle of immanency of Kant. Psychic phenomena are always a primary reality, but they must not necessarily be 'true', as for instance optical illusions and hallucinations show.

11. 'Downward Causation' in Hierarchically Organised Biological Systems

DONALD T. CAMPBELL

It is convenient to append as a comment on Beckner's stimulating paper some remarks made in part in reaction to other presentations, including those of Medawar, Edelman, and Monod, and in a final section to the presentations of Dobzhansky, Thorpe, Eccles, Birch, Rensch and Skolimowski. Beckner, Medawar, Edelman and Monod are among my fellow 'reductionists', in so far as there are any 'reductionists' present at this conference.

Certainly the Darwinian theory of natural selection has historically been and is still 'reductionist' *par excellence*, as in the views of such anti-reductionists as Bertalanffy, Polanyi, Koestler and Whyte. To that degree at least we are reductionists, even if we are not all 'microparticulate-derivationists' who believe that all the phenomena of the world are revealed in the interactions studied by the subatomic physicists. Instead, many of us even qualify as emergentists, albeit antivitalistic ones, in the sense of Montalenti's introductory statement of a reasonable consensus for our conference.

What concerns me is the possibility that we reductionists of today may repeat the mistake of the reductionists of the past by denying true facts to which the vitalists point. Remember, Darwin did not deny 'design'—he repudiated instead 'the argument from design'. Like Mayr and Monod, we must admit and admire 'teleonomy' while rejecting teleological explanation (Ayala, 1970; Wimsatt, 1972).

All at this conference have gone part of this way, by admitting the factuality of hierarchical organisation in biological systems. The organisational levels of *molecule*, *cell*, *tissue*, *organ*, *organism*, *breeding population*, *species*, in some instances *social system*, and perhaps even *ecosystem* (though most would be better designated eco-aggregates) are accepted as factual realities rather than as arbitrary conveniences of classification, with each of the higher orders organising the real units of the lower level. In accepting the fact of hierarchical biological systems, all of us reductionists emphasise with Beckner and Medawar two reductionist principles:

(1) All processes at the higher levels are restrained by and act in conformity to the laws of lower levels, including the levels of subatomic physics.

(2) The teleonomic achievements at higher levels require for their implementation specific lower-level mechanisms and processes. Explanation is not complete until these micromechanisms have been specified.

These assertions, with which I firmly agree, are enough to provoke real disagreement with many antireductionists, though again, perhaps not any at this conference.

What I would like to see all of the reductionists present affirm (or affirm more clearly and emphatically) is assent to two more of the vitalist's facts that go along with the vitalist's emphasis on the hierarchical organisation of biological systems, and to thereby agree that the two reductionist points above, while true, are not enough. I will try to state these two additional principles in a reductionist tone of voice. These two points are alleged to hold specifically for biological systems in which natural selection is involved (blind variation, selective retention *and reproduction*, following Dobzhansky). The achievements represented in points 3 and 4 below are to be explained by natural selection operating in complete compatibility with the physical model of causation:

(3) (The emergentist principle) Biological evolution in its meandering exploration of segments of the universe encounters laws, operating as selective systems, which are not described by the laws of physics and inorganic chemistry, and which will not be described by the future substitutes for the present approximations of physics and inorganic chemistry.

(4) (Downward causation) Where natural selection operates through life and death at a higher level of organisation, the laws of the higher-level selective system determine in part the distribution of lower-level events and substances. Description of an intermediate-level phenomenon is not completed by describing its possibility and implementation in lower-level terms. Its presence, prevalence or distribution (all needed for a complete explanation of biological phenomena) will often require reference to laws at a higher level of organisation as well. Paraphrasing Point 1, all processes at the lower levels of a hierarchy are restrained by and act in conformity to the laws of the higher levels.

'Downward causation' is perhaps an awkward term for point 4, and excusable only because of the shambles that philosophical analysis has revealed in our common sense meanings of 'cause'. The 'causation' is downward only if substantial extents of time, covering several reproductive generations, are lumped as one instant for purposes of analysis. In the 'instantaneous' causation of the older philosophical analyses of physics, no such direction is present. If 'causation', it is the back-handed variety of natural selection and cybernetics, causation by a selective system which edits the products of direct physical causation (Bateson, 1972; Rosenblueth, Wiener and Bigelow, 1943; Ayala, 1970; Wimsatt, 1972).

It seems worth while to try to make these points more clearly and strikingly by use of a concrete biological example. Consider the anatomy of the jaws of a worker termite or ant. The hinge surfaces and the muscle attachments agree with Archimedes' laws of levers, that is, with macromechanics. They are optimally designed to apply the maximum force at a useful distance from the hinge. A modern engineer could make little if any improvement on their design for the uses of gnawing wood, picking up seeds, etc., given the structural materials at hand. This is a kind of conformity to physics, but a different kind than is involved in the molecular, atomic, strong and weak coupling processes underlying the formation of the particular proteins of the muscle and shell of which the system is constructed. The laws of levers are one part of the complex selective system operating at the level of whole organisms. Selection at that level has optimised viability, and has thus optimised the form of parts of organisms, for the worker termite and ant and for their solitary ancestors. We need the laws of levers, *and organism-level selection* (the reductionist's translation for 'organismic purpose'), to explain the particular distribution of proteins found in the jaw and *hence* the DNA templates guiding their production. (The occasional nonfunctional mutant forms of jaws conform just as loyally to the laws of levers and biochemistry as do the more frequent functional forms.) Even the *hence* of the previous sentence implies a reverse-directional 'cause' in that, by natural selection, it is protein efficacy that determines which DNA templates are present, even though the immediate micro determination is from DNA to protein. (It is just a further complexity that this determination too is by a back-handed selective-retention process.) This is the point Dobzhansky makes in emphasising that selection is of phenotypes, not genotypes: it is the protein distribution that is selected for, not the DNA distribution directly.

If we now consider the jaw of a soldier termite or ant, a still more striking case of emergence and downward causation is encountered. In many of the highly dimorphic or polymorphic species, the soldier jaws are so specialised for piercing enemy ants and termites, huge multipronged antler-pincers, that the soldier cannot feed itself and has to be fed by workers. The soldier's jaws and the distribution of protein therein (and the particular ribonucleic acid chains that provide the templates for the proteins) require for their explanation certain laws of sociology centring around division-of-labour social organisation. The syndrome of division of labour, storable non-spoiling foodstuff such as honey or seeds, apartment house living, and professional soldiers who do no food gathering, has been repeatedly independently discovered many times among the proto-termites, as well as independently by the ants, and by six or seven separate seats of human civilisation (see Campbell (1965) for a review of this literature). This repeated convergent evolution testifies to the great selective advantage of division-of-labour social organisation, economies of cognition, mutual defence and production being some of its selective advantages.

I hope that consideration of forms so specific as the jaws of social insects has made the case for a downward determination in the relationships among the levels in social hierarchies. In Beckner's paper upward causation as a relation between the levels of a biological hierarchy is well represented and advocated, while downward determination is not mentioned even as a rejected possibility. Points 3 and 4 of my reasonable reductionist position clearly are much stronger than the merely permissive position of his 'weak organicism', in which higher levels *may* be studied independently of lower ones, and this may even lead to more efficient scientific advance. My position says that for biological systems produced by natural selection, where there is a node of selection at a higher level, the higher level laws are necessary for a complete specification of phenomena both at that higher level and also for lower levels. Scientific description is still incomplete when all the details of points 1 and 2 are solved. Yet my position is not vitalist or teleological, and is weaker than Beckner's strong organicism, since I advocate not the autonomy of higher levels, but rather the additional restraints, aspects of selective systems that these higher levels encounter.

It seems to me that Beckner needs to fill in several more levels between the weak and strong organicists, and to classify himself, Mayr, Simpson, most of our present conferees, and me, as moderate rather than weak organicists. With regard to his 'event reduction', he should recognise that it accomplishes only points 1 and 2, and leaves us lacking grounds for predicting or understanding the distribution of 'events'.

Medawar's paper is so effectively brief that it is unfair to complain about what he does not say. His presentation of the laws of higher levels as restricted versions of more general lower-level laws is quite compatible with the view presented here. But left as it is, it fails to convey his own career of scientific concern with the particular nature of the restrictions, and the understanding of why *these* restrictions rather than others. Medawar's current interest in the role of the immunity system in controlling tumour development is a clear cut case of a point 4, downward-determination hypothesis. Questions about the function of a process at one level are questions about a selective system at some higher level. For a complete scientific description of the distribution of restrictions in biological systems we need additional laws, restraints imposed by the selective systems of the highest level of selection, and affecting distributions at all lower levels. This relationship is a peculiarity of biological hierarchies not present in the mathematical order from which his conceptual model comes.

Edelman and Monod provide eloquent testimony of exquisitely interlocking processes implementing the achievement of teleonomic effects. (What new wonders of 'design' for those who still use the 'argument from design'!) Yet, in the excitement of discovering the mechanisms which implement a particular teleonomic process, they each convey the impression that the scientific puzzle is complete when the mediating process (point 2) is discovered. Basically, no doubt, they agree with points 3 and 4 as well, and recognise that the issue function, that is, of higher-order selection systems, must be answered for a complete explanation of the question of the distribution of molecular processes. But, lacking a more explicit emphasis on these points, their exciting presentations of the discovery of how teleonomic achievements are mediated tends to lead to an excessively reductionist conclusion that these mediating mechanisms provide a *complete* explanation: they do not, and for distributional problems, such as are illustrated in the protein distribution of social-insect jaws, higher-level laws encountered as attributes of a selective system operating to sustain a specific level of organisation are needed.

Worshipping man's ecological niche

This has been a reductionist translation of the very valid facts to which Michael Polanyi (1966, 1969), and others have been pointing. 'Life's irreducible structure' cannot be reduced to points 1 and 2. Can my orientation be related to his (1970) concern over the socially destructive effects of the popularised scientific reductionism of the last few centuries? Such concerns have also been well represented at this conference, as in Thorpe's closing pages, in Monod's (1971) chapter 'The kingdom and the darkness', in papers and discussions of Eccles (see also 1970), Birch, Rensch, and even Skolimowski, for all his preoccupation with our need for novel cognitive and social forms. Still others whose papers were not related to the issue expressed it in conversation, Popper among them.

Dobzhansky had devoted a book to this problem in his *The Biology of Ultimate Concern* (1967). His approach of attempting to salvage the precious values of past belief systems without compromising scientific rigour is one I would attempt to follow. As it were, where science and religion conflict we do give superior allegiance to science, but to a more complete science of the future, and not by a dogmatic adherence to every temporary scientific assertion. I also join him in distrusting that value-salvage route which lies in worshipping man's conscious experience, as by saving mind as separate from body, or making body all mind, for these too commit the ancient sin of man worshipping himself instead of purposes and forces beyond any individual's petty, transient being.

Reductionism can, of course, be destructive of social values even if 'true'. Thus my reductionism of points 1, 2, 3 and 4 could perhaps be as destructive as a narrower reductionism limited to points 1 and 2, or one denying hierarchical organisation entirely. The choice of termite jaws to make Polanyi's point is certainly iconoclastic in spirit, even if presumably done only for the purpose of emphasising the hard-headed factuality of Polanyi's descriptive claims (if not his implicit explanations). Monod's (1971) concern makes clear that we reductionists can agree with the vitalists that our reductionism, even if valid, can be destructive of useful social commitments and values. This perhaps suggests a solution through a priesthood of scientists who encourage in the public socially useful superstitions which the scientist–priests themselves no longer believe in the same terms. Some degree of such a double-standard of belief may in fact have been a part of functioning religious systems as far back as ancient Egypt. But we scientists today, vitalists and reductionists alike, reject this deception as both unworkable and, surprisingly enough, immoral (on what authority God only knows). But if these superstitions play an essential role in sustaining a social system, that is, a higher level of organisation, then they represent important truths, and in some sense are not false. If we accept a thoroughgoing neo-Darwinist evolutionary epistemology, we should also accept epistemic humility, recognising the profound indirection and presumptiveness of even our best visual perception and science. We do not know reality in the *Ding an Sich*'s own language. The meteorologist's knowledge of the Arctic climate is just as indirect (albeit more complex) as is that corresponding 'knowledge' shown in the thickness of the polar bear's coat. We do not call the polar bear's fur 'false' just because it is not self-consciously aware of its purpose and the reality it reflects. Somehow we should credit these socially useful superstitions with a comparable 'truth' (Skolimowski, 1972). Seen thus, we can recognise that the problem is in part one of translating old metaphors into newer ones (Burhoe, 1971), since one of the lessons of evolutionary epistemology is that there exists no literal language for describing truth, all languages, including mathematics, being metaphorical rather than literal or direct when used as in science for descriptive functions.

These considerations, plus an effort to specify in what sense Dobzhansky (1967) can call himself a Teilhardian, led me in our discussion to suggest that evolutionary biologists and others who are confused about the meaning of the name God can worship their Creator by worshipping the selective system that produced man, man's ecological niche. Properly approached, this could succour many aspects of man's spiritual needs now being starved under the self-worship that too often comes with loss of traditional values, and without compromising scientific scruples:

—The attitude of awe of superior unknown powers; for certainly we do not yet know much about the selective system that shaped civilised social man. (If even the niche of the termites can contain such wonders as laws of sociology, what greater wonders may not our selective system contain, eternal truths unsuspected as yet by our puny social sciences.)

—The recognition that man was shaped by purposes beyond the petty concerns of individual man.

—Recognition of a Creator that is what It is for Its own purposes, and is free to change Its purposes without regard to the previous covenants we may have felt were made in the past and with which our lusts and loyalties may be more compatible.

—A prayer and spiritual wish that we may live lives loyal to those purposes in such a way as to optimise man's survival over eternity, not just our lifetime.

—A recognition of an incomplete adaptation and an Omega Point whose outlines we can only dimly surmise, an alternative to self-inflicted extinction (Maloney, 1973).

Consideration of the problem of the genetics of altruism (Haldane, 1932; Campbell, 1973) fills out the translation of religious values, for it makes scientific sense out of the concept of temptation and sin. But it also provides grounds for doubting that the translations will be socially effective, just because of man's biologically based antisocial selfishness. Social cooperation among men, as among wolves and among apes, coexists with genetic competition among the cooperators. This sets strict limits on genetically based complex social coordination, division of labour, and self-sacrificial altruism. In the social insects, this genetic competition among cooperators was eliminated as a prior step to their evolution of complex social systems. Because of the human genetic predicament, in achieving man's complex social coordination *social* evolution has had to provide many of the relevant behavioural dispositions and has had to overcome some selfish biological dispositions in the process. As modern critiques emphasise (Olson, 1968; Hardin, 1968; Schelling, 1971) the intelligent optimising of individual utilities can lead to socially destructive outcomes. Pessimistic or not, all such considerations should give us increased respect for the belief systems we have inherited from our social evolutionary past, even though we recognise that all evolutionary adaptations are to past environments, not future ones.

References

Ayala, F. J. (1970). Teleological explanations in evolutionary biology. *Philosophy of Science*, **37**, 1–15.

Bateson, G. (1967). Cybernetic explanation. *American Behavioral Scientist*, **10**, 29–32. (Also in Bateson, G., *Steps to an Ecology of Mind*, Chandler, San Francisco.)

Burhoe, R. W. (1971). *Science and Human Values in the 21st Century*. Westminister Press, Philadelphia.

Campbell, D T. (1965). Variation and selective retention in sociocultural evolution. In *Social Change in Developing Areas: a reinterpretation of evolutionary theory* (ed. H. R. Barringer, G. I. Blanksten and R. W. Mack). Schenkman, Cambridge, Mass., 19–49.

Campbell, D. T. (1973). On the genetics of altruism and the counter-hedonic components in human culture. *Journal of Social Issues*, **28**, 21–37.

Dobzhansky, Th. (1967). *The Biology of Ultimate Concern*. New American Library, New York.

Eccles, J. C. (1970). Some implications of the scientiae for the future of mankind. *Studium Generale*, **23**, 917–24.

Haldane, J. B. S. (1932). *The Causes of Evolution*. Longman, London.

Hardin, G. (1968). The tragedy of the commons. *Science*, **162**, 1243–8.

Maloney, J. C. (1973). Man as a socioeconomic subsystem. In *Unity through Diversity* (ed. W. Grey and N. D. Rizzo) vol. II. Gordon and Breach, New York.

Monod, J. (1971). *Chance and Necessity*. Knopf, New York.

Olson, M. (1968). *The Logic of Collective Action*. Schocken, New York.

Polanyi, M. (1966). *The Tacit Dimension*. Doubleday, New York.

Polanyi, M. (1969). Life's irreducible structure. In Polanyi, *Knowing and Being*. Routledge and Kegan Paul, London, 225–39.
Polanyi, M. (1970). Why did we destroy Europe? *Studium Generale*, **23**, 909–16.
Rosenblueth, A., Wiener, N. and Bigelow, J. (1943). Behavior, purpose, and teleology. *Philosophy of Science*, **10**, 18–24.
Schelling, T. C. (1971). On the ecology of micromotives. *The Public Interest*, no. 25, 61–98.
Skolimowski, H. (1973). The twilight of scientific descriptions and the ascent of normative models. In *World Models* (ed. E. Laszlo). Brazilier, New York. (Proceedings of the Conference at Geneseo, NY, 29–30 September 1972.)
Wimsatt, W. C. (1972). Teleology and the logical structure of function statements. *Studies in the History and Philosophy of Science*, **3**, 1–80.

12. On the Relations Between Compositional and Evolutionary Theories

DUDLEY SHAPERE

On the basis of breeding ratios arrived at as a result of experiments with edible peas, Gregor Mendel concluded that certain 'differentiating characters' of the plants were paired with one another in such a way that an individual hybrid plant, though it might possess elements capable of causing either of the observable characters, nevertheless would in such a case manifest only one (the dominant as opposed to the recessive character). Crossing of plants with alternative characters of such pairs, moreover, led him to two fundamental 'laws' of heredity:

(1) *Segregation*, according to which, in such matings, the members of each pair in one parent separate to unite with a member of the corresponding pair from the other parent; if the parents each have a dominant (*A*) and a recessive (*a*) factor of the pair, the offspring may be either *AA*, *Aa*, *aA*, or *aa*; and those alternatives are equiprobable, so that the observed ratios of dominant to recessive character in the offspring of hybrids will be 3:1.

(2) *Independent assortment*, according to which any pair of elements does not affect the equiprobability of alternative combinations of any other pair.

In earlier papers,[1] I have referred to bodies of information like Mendel's peas and their characters, and the breeding ratios relating those characters, as *domains*; this concept will be further elaborated in the present paper. As a preliminary definition, a domain is a set of items of information which have come to be associated together as a unified body, and about which body there is a problem, well-defined—usually—and raised on the basis of specific considerations ('good reasons'). In addition, that problem is generally considered important (also, in the most characteristically *scientific* cases, on

[1] Shapere, Scientific theories and their domains, in *The Structure of Scientific Theories* (ed. F. Suppe), University of Illinois Press, Urbana (1974); Notes toward a post-positivistic interpretation of science, in *The Legacy of Logical Positivism* (ed. S. Barker and P. Achinstein), John Hopkins Press, Baltimore (1970).

reasonable grounds—not on the basis of some 'subjective value judgment'). Further, it must—in general, though in certain rather well-circumscribed sorts of cases not necessarily—be capable of being 'handled' at the current stage of science. Needless to say, the ideas employed in this definition—such as 'item of information', 'association' (or 'relatedness') of such items, 'problem', 'good reason', 'importance'—all need to be laid out in detail. Some such clarification will be achieved here through illustration in specific cases; but, as we will see, the primary purposes of the present paper lie elsewhere than in giving a formal elucidation of the domain concept.

The concept of a 'problem' regarding a domain will, however, be a central one for our purposes. Two major classes of such problems arise, the first having to do with elaborating the domain itself, either through improving the precision with which the items (and their interrelationships) are described, or through extending the scope of the domain. What I mean by problems of improving descriptive precision is, hopefully, intuitively clear, at least sufficiently for present purposes; but the idea of extending the scope of a domain raises a question: for what are the grounds for asserting that a certain body of information is not complete, and is thus capable of being extended? This question has different answers in different sorts of cases. Consider Mendeleev's Periodic Table of Chemical Elements, a case considered in detail in 'Scientific theories and their domains' (see footnote 1): the domain of such elements, related in a serial order according to atomic weights, and repeating with respect to other properties, was revealed to be 'incomplete' by the presence of 'gaps' in the ordering. Incompleteness of the domain was in that case indicated by the characteristics of the ordering which related the items. In the case of Mendel's peas and their breeding ratios, on the other hand, the reasons for supposing, or at least for suspecting, that the laws might apply more broadly than to peas—that, in other words, the domain could be extended—did not lie in the relations establishing the domain (bases of classifying certain objects as 'pea plants', and the breeding ratios holding between them). Rather, they lay in what we may refer to as 'background information': information which, while not incorporated as part of the domain, is taken to be knowledge applicable to the domain. Examples are: extensive and detailed similarities between peas and other plants, and specific reasons for suspecting a deep relationship between plants and animals (for example, the extensive morphological similarities between the cells of plants and animals which had accumulated since the work of Schleiden and Schwann). The existence of such 'background information' made reasonable the *suspicion*, at least, that relationships holding for peas might hold for other plants and animals, and thus made reasonable the investigation of other such organisms to see if the relationships did hold. The domain was, indeed, extended, as is well known: although Mendel himself cautiously referred to his relationships as 'the rule discovered for *Pisum*', and failed to extend them to the hawkweed, *Hieracium*, the later work of De Vries and others showed that the relationships held for a large number of other plants, while Bateson and others extended them to the animal kingdom. And although there still remained a number of formidable problems blocking the generalisation of the Mendelian laws to all (sexually reproducing) species, there was at least reason for proceeding with research aimed in that direction. But at this point a different sort of problem—the second general class of problems spoken of above, which arise with regard to domains—must be discussed.

This second class of scientific problems regarding domains calls for answers in terms of ideas different from those used in characterising the domain items and their relationships; and these ideas are expected to be, in some sense that must be explicated, 'deeper' or 'more comprehensive' than those characterising the domain itself. These new ideas, moreover, are expected to 'account for' the domain, again in a sense that requires explication. On the basis of this aspect of the domain's being 'accounted for', we may give an alternative definition of 'domain': *a domain is the total body of information for which, ideally, an answer to the problem in question is expected to account.* Problems calling for such answers will be called 'theoretical' (as opposed to the first class of problems, which may be called 'domain problems'; it must be understood, however, that theoretical problems, no less than domain problems, arise with regard to domains). Answers to theoretical problems are, of course, called 'theories'.

Beyond these very general characteristics, theoretical problems, and the theories which are or purport to be their solutions, are of very different sorts. Philosophers (and scientists also) have thus been misguided in their efforts to understand the nature of scientific theories (or theorising) by their tacit assumption that there is only *one* sort of thing (or procedure) referred to by the term 'theory' (or 'theorising'). The approach taken in 'Scientific theories and their domains' was to examine a number of cases from the history of science, with a view to unearthing some of the reasoning patterns by which theoretical problems, and the lines of research aimed at answering those problems, arise. No assumption was made that those reasoning patterns would be the same in all cases, and different types of scientific theories were defined in terms of the reasoning patterns leading to their consideration.

In that essay, two general patterns of such 'theoretical reasoning' (and therefore two general types of theories) were found; they were called *compositional* and *evolutionary*. Roughly (though a full definition—in a sense to be specified below—is given by the reasoning principles to be given shortly), a compositional *problem* is a theoretical problem about a domain calling for an answer in terms of the constituent parts of the individuals making up the domain and the laws governing the behaviour of those parts. (The parts sought need not be 'elementary'.) The Daltonian laws of combining proportions in the first half of the nineteenth century, the periodic table of chemical elements in the last quarter of the nineteenth century, the spectral characteristics of chemical elements in the same period, and the Mendelian laws of heredity in the early years of the twentieth century, are all domains which raised problems leading to the expectation of and search for compositional theories. An evolutionary problem, on the other hand, is one which calls for an answer in terms of the time development of items (whether individuals or types of individuals) in the domain. Paradigmatic examples of domains with problems calling for evolutionary theories as answers are spectral classifications of stars in the late nineteenth and twentieth centuries, the chemical elements (conceived as constituted of atomic particles) in the late nineteenth and twentieth centuries, and the biological species, particularly after the middle of the nineteenth century. There are, of course, other types of theories than compositional and evolutionary, but those others are not relevant for the purposes of the present paper.

To repeat, two major problems in the attempt to understand the nature of scientific reasoning are: first, what sorts of considerations (in the sense of 'reasons') give rise to a theoretical problem about a body of information (domain)—that is, to considering our understanding of that body of information to be inadequate in a way that suggests or requires a search for what are commonly called 'theories'? and second, what sorts of (rational) considerations lead scientists to suspect (or, sometimes, to expect or even to demand) that answers of certain specific sorts, having certain specific characteristics, should be sought for those problems?

An example dealt with in 'Scientific theories and their domains' will illustrate what is involved in this approach, particularly the concept of a 'reasoning pattern'. In that paper, a detailed examination of the historical situation with regard to the periodic table of chemical elements in the late nineteenth century led to formulation of the following 'principle of compositional reasoning':

> To the extent that a domain *D* satisfies the following conditions or some subset thereof, it is reasonable to expect (or, if the conditions are highly fulfilled, to demand) that a compositional theory be sought for *D*:
>
> (Ci) *D* is ordered [as the periodic table is, according to atomic weights];
>
> (Cii) the order is periodic [as is the periodic table];
>
> (Ciii) the order is discrete (*i.e.*, based on a property [atomic weight in the case of the chemical elements] which 'jumps' in value from one item to the succeeding one), the items having values which are (within the limits of experimental error) integral multiples of a fundamental value;
>
> (Civ) the order and periodicity are extensive, detailed and precise;
>
> (Cv) compositional explanatory theories are expected for other domains;
>
> (Cvi) compositional theories have been successful or promising in other domains;

(Cvii) there is reason to suppose that the domain under consideration is related to such other domains so as to form part of a larger domain.

This principle holds, of course, for the case from which it was extracted; but comparison with similar principles extracted from other cases that are clearly compositional in character can permit generalisation to cover a broader class of cases. For example, comparison with an equally detailed analysis of the case of chemical spectroscopy yields just such generalisation. But even intuitively, it can be seen that certain of the seven features listed are not essential to a theory's being compositional: periodicity, for example, though it tends to increase the strength of indication that a compositional theory is to be sought, need not be present at all in an ordering relation. Further, many sorts of 'ordering' are possible, and indeed, the characteristic of discreteness of values may be present without the domain-generating relation being an 'ordering' one at all in any usual sense (Mendel's laws, though they indicate the reasonableness of a search for a compositional theory, do not 'order' the breeding ratios in anything like the way in which the chemical elements are ordered in the periodic table). (Ciii) is more crucial to the domain's indicating the reasonableness of a search for a compositional theory though, even here, peculiarities of the periodic table case make the formulation given here overly restricted. This is not merely because (Ciii) is here formulated in terms of the concept of 'order', but also because there are other sorts of theories which are usefully classified as 'compositional' even though the composition they deal with is not discrete (does not consist of 'atoms', for example). There have been, in the history of science, continuum theories of the composition of individuals which have much in common with discrete theories; and these similarities are sufficiently specific to make using the term 'compositional' to cover both illuminating. Particulate explanations of the periodic table are thus cases of what may be called discrete compositional theories. (It may be added that there are—have been—also different types of discrete theories, most notably what historians of science have come to distinguish as 'atomistic' and 'corpuscular'.) It is important not only to formulate the most general principles possible, but also to note the more specific characteristics peculiar to particular theories or groups of theories within the more general classifications: for those specific characteristics may also be highly illuminating of the nature of scientific reasoning and of scientific theories. Often the specific characteristics of the domain indicate more specific characteristics of the theory to be sought (as, to those who held the periodic table to be an indication that the chemical elements had a deeper structure, the periodicity of the table was an indication of a periodicity of that structure itself).

In 'Scientific theories and their domains', examination of the case of stellar spectral classification in the nineteenth century led to the formulation of a 'principle of evolutionary reasoning' for that case. The earliest such classification systems were based on the colours of stars; and some astronomers suggested that the classes might be ordered if their colours were an indication of an evolutionary sequence, on the analogy, say, of the cooling of metals from blue or white to red hot. (This ordering was found to coincide with increasing numbers of spectral lines in the stars of the various classes; that increase could itself have been utilised for ordering purposes, though it does not seem to have played a role until after the ordering had been made on the basis of the cooling analogy.) Use of this analogy concerning colour changes of cooling bodies contributed in three ways to indicating the reasonableness of a search for an evolutionary theory for stars. First, it showed that a sequential ordering of the classes merely on the basis of increasing numbers of spectral lines could be correlated with a *temporal* order having to do with the cooling process. Secondly, it suggested a *direction* of the evolution—a direction which could not have been extracted from an ordering on the basis of lines alone, unless one made arbitrary assumptions (for example, that the evolution is from 'simple' to 'complex'). For, the final death of a star being a cold, burned-out state, the red stars should be the oldest. (Unfortunately, the suggestion was not so clear with regard to the beginning of the sequence: were the white stars like Sirius the youngest, blazing forth suddenly at their birth, and gradually burning themselves down to a red old age? Or were certain red stars young, gradually heating up to white maturity, after which they declined to a second red stage just preceding death?) And thirdly, by itself constituting the outline, at least, of a theory—an answer to the evolutionary problem regarding the domain—it suggested directions which research could take: directions in which the theory needed to be laid out in detail.

In spite of the residual ambiguity in this case, regarding the beginning of the temporal process of stellar evolution, we can see clearly at work here another principle of reasonable scientific research, this time applying to the (or rather a) way in which a problem arises, with regard to a domain, for which an *evolutionary* answer (theory) can reasonably be expected and sought:

(Ei) If a domain is ordered, and if that ordering is one which can be viewed as the increase or decrease of the factor(s) on the basis of which the ordering is made, then it is reasonable to suspect that the ordering may be the result of an evolutionary process, and it is reasonable to undertake research to find such an answer (an evolutionary theory).

(Eii) The reasonableness of such expectation is increased if there is a way (for example, by application or adaptation of some background information such as a theory from another domain, whether unrelated or—preferably—related) of viewing that sequential ordering as a temporal one, and still more if a way is provided of viewing that ordering as having a temporal direction.

Clearly, there are also analogues for evolutionary theories of conditions (Civ)–(Cvii) which were stated above for the case of compositional theories. It should be added that (Ei) alone constitutes only weak reason for undertaking research in quest of an evolutionary theory; for, without (Eii), little or no specific direction is provided for research. ('Analogy' alone is insufficient as a guide to research; analogy bolstered by other considerations can become *good* analogy, capable of furnishing specific lines of reasonable research.)

Again, this principle, extracted from a particular case, gives only a sufficient condition for seeking an evolutionary theory. Further, its formulation, based on that particular case, is too restrictive to cover many other cases of what are, on an obvious intuitive basis, to be classed as evolutionary theories. For example, in the case of the most famous instance of an evolutionary theory, *the* evolutionary theory in biology, not only is the ordering of classes (species) highly complex rather than simply linear; it is also far from clear whether there is any factor that increases (or decreases) in the ordering. And finally, the rationale behind the introduction of the temporal aspect has not been sufficiently clarified in that case, although certainly Lyell's geology played an important role in Darwin's thinking.

It is clear that compositional and evolutionary theories have very different characteristics. Compositional theories deal with behind-the-scenes entities which explain the properties and behaviour of the items of the domain; it seems natural to speak of such theories as 'explanatory'. On the other hand, one hesitates, in the case of evolutionary theories, to use the word 'explanation', and many writers have denied that such theories—Darwin's, for example—are explanatory, and have maintained that they are purely descriptive. And there is a strong tendency to think of them as being only ways of ordering 'facts'; where anything is hidden ('missing links' in the fossil record), it is something of the same *general* sort as those that are known. On the other hand, it is of the essence of an evolutionary theory that it has a 'historical' dimension, a temporal aspect, that is missing from compositional theories. In view of these (and many other) differences, it is no wonder that philosophers, attempting to give a *unitary* analysis of the concept of scientific theory, have either failed entirely, or fastened on such empty generalities that their results have given little insight into the nature of the scientific endeavour.

Sometimes when a compositional theory is available, and perhaps even accepted, for a domain, that theory comes to be regarded as itself a domain for explanation in terms of an evolutionary theory; that is, given that asserted composition, a problem arises concerning it which requires, as answer, an evolutionary theory. Under what conditions does this occur? That is, what gives rise to such evolutionary problems as legitimate scientific concerns which it is reasonable for scientists to pursue in their research? To put the question in yet another way, why do such problems seem to arise for some theories and not for others, or for some theories at a certain stage of their history and not before? (Thus the question of the evolution of the chemical elements was rarely raised—even under Prout's hypothesis of their common composition, or after Mendeleev's table—until astronomical theories of stellar evolution were applied to compositional theories of the elements.)

The converse problem also arises: under what conditions does an evolutionary theory give rise to search for a compositional theory, that is, for factors which account for the evolutionary process?

The former problem was to some extent dealt with in 'Scientific theories and their domains', where the raising of questions about the evolution of the chemical elements, seen as composed of discrete particles, was considered. The present paper will discuss the second question, that of the transition from an evolutionary to a compositional theory. The case to be dealt with will be the Darwinian theory, and the way in which a need arose for a compositional theory for it. In the course of this analysis, some discussion will also be given of the development of that compositional theory, and especially of the way in which the search for it became more specifically a search for a *discrete* compositional theory.

Earlier, two general classes of scientific problems were distinguished: domain problems and theoretical problems, the latter being problems about the domain which require 'theories' as answers. A third major class has to do with inadequacies of the theories themselves; for example, a theory may be *incomplete with respect to its domain*, that is, with respect to the total body of information for which it is expected to account. Such was the case with Darwin's theory of evolution: it is well known that Darwin did not have a satisfactory account of inheritance and variation, which was required in order to make his theory reasonably complete.

Objections against a theory because of its incompleteness (with respect to its domain) have the characteristic that they do not make reasonable the rejection of the theory (its 'falsification') unless certain further conditions are met. (The inadequacy is taken to imply the *need for supplementation*, rather than the *rejection*, of the theory.) For our purposes, the most important of these further conditions are that (1) repeated serious efforts have been made to supplement the theory so as to fill the gap represented by the incompleteness, and (2) another theory is available which does account *at least* for the item of the domain for which the original theory failed to account.[2]

[2] Even where (1) and (2) are taken to imply that the theory is *false*, it may still be *utilised*, particularly if the new 'theory' is very limited in its development, and still more if there is no new 'theory' of the sort in question (*i.e.*, if only (1) is satisfied). The terms 'rejection'

(In general, condition (2) of rejection of theories because of incompleteness comes into play only when (1) is satisfied.) Even when (1) is satisfied, there is still room for rational disagreement among scientists as to whether the theory is *incorrect* or merely *incomplete*. And even when (2) is also satisfied, there is still room for such disagreement if the new theory itself is incomplete —especially if the new theory initially accounts *only* for the item with respect to which the old theory is incomplete. In such cases, the new 'theory' hardly deserves the name; 'heuristic proposal' would be more appropriate. (Witness, as an example, Einstein's introduction of the quantum theory to account for the photoelectric effect in his 1905 paper, 'On a heuristic proposal concerning the production and transmission of light'. The accepted continuum (electromagnetic) theory had failed to account for this phenomenon, and attempts to supplement it had not been entirely successful. Einstein proposed that his quantum theory, which accounted for it quite elegantly, be pursued further in future research in order to see whether it truly deserved to replace the traditional theory.)

In the case of Darwin's theory of evolution, its vast success with a wide range of phenomena made its failure with respect to the problems of inheritance and variation appear to be due merely to incompleteness rather than to some irremediable incorrectness. Besides, it was not as though Darwin did not present *any* account of heredity and variation: he did propose a 'Provisional hypothesis of pangenesis' (1868), and he had what he considered to be good reasons for his proposals. However, his account was not a very good one, as Galton showed in his experiments with blood transfusion in rabbits. (As Darwin pointed out, of course, Galton's experiments did not refute his hypothesis, since the 'gemmules' might not be carried in the blood; but those experiments certainly increased the vagueness of the hypothesis.) Because of this unsatisfactory state of Darwin's account, his theory could reasonably be taken to be incomplete; and so it was taken by many. To others, of course, the incompleteness lay only in the degree to which the supplementary gemmule hypothesis had been developed. On the other hand, there was no lack, in the late nineteenth century, of proposals of theories of heredity and variation. However, none of them was precise

---

and 'falsification' are thus ambiguous: a theory may be considered *false* (and in that sense 'rejected') but still be useful (and in that sense not 'rejected').

In connection with these points, it might be useful to point out the main contrasts between the approach of this paper and the views—or at least the most usual interpretation of the views—of Karl Popper. Basically, there are three such contrasts. (1) The present approach emphasises that 'rationality' in science—or the distinction between science and non-science—is a matter of degree; there is no sharp 'line of demarcation' marked out by some criterion like 'falsifiability'. (2) The present approach holds that there *is* a 'rationale of inquiry', of search, and thus—in a sense which will be clarified later in this paper—denies the dictum that 'there is no logic of discovery'. (3) This approach attempts to distinguish in detail between those sorts of inadequacies or circumstances which lead to the rejection (or falsification) of those theories and those which do not.

enough, or closely enough tied to the possibility of experiment, to lead to further research and results.[3]

When Mendel's laws were brought to the attention of the scientific world in 1900, this situation was changed: here was a theory of inheritance (if not of variation) which was precise, and which could lead to further research. Did these precise relationships hold for other plants? Did they also hold for animals? What about apparent cases of continuous varitation? Did the concepts of dominance and recessiveness apply unqualifiedly to all cases? These and other questions were dealt with by De Vries, Bateson and a host of other workers in the first decade of the present century, and in the process the laws themselves—the domain they formulated—underwent modification and qualification. But it was the precision of Mendel's laws that made their study reasonable and their modification possible: it made the *sub*domain of heredity, within the more inclusive domain of biology, crucial. It made desirable the segregation, however temporary, of that subdomain from other questions—the segregation of questions of heredity from other questions with which they had traditionally (and reasonably) been associated, and indeed which had been considered more centrally important. Thus questions of development, which (partly because of the 'ontogeny recapitulates phylogeny' doctrine) had been considered more important than those of heredity, were now put aside in favour of the more tractable questions surrounding Mendel's laws. On this point, Dunn's remarks are penetrating:

> . . . that 'territory' [of the transmission mechanism of heredity] had not been recognised as such until after the break [in the continuity of ideas about that transmission system] had occurred. Heredity, with the definiteness given to it by modern genetics, now seems to be a well-marked field, but in the nineteenth century this was not so. Then, efforts to develop general theories of heredity were judged primarily by their applicability to problems of variation, evolution, and development, and we see signs today, after nearly 100 years, of a return of that attitude. Then it was due to the absence of precise questions about the transmission mechanism or to a lack of recognition of the territory. (Dunn, *A Short History of Genetics*, 34.)

> The nineteenth-century biologists, except Mendel and the 'rediscoverers', did not begin to solve the problem of transmission because they failed to recognise its real nature or even its importance. It was only when some biologists were willing to put aside the intractable problem of development and concentrate on transmission that the problem was analysed and solved. (*Ibid.*, 48.)

A domain is not always simply 'there', a matter of pure observation (whatever that is), but is—in science, and especially in science at more sophisticated stages—delineated on the basis of reasons. In the present case, a subdomain became critical partly because of its importance in dealing with the problem of inheritance (and ultimately of variation) which had arisen in connection with

3 More than a 'research programme 'is required in science; such programmes must also be precise and workable enough to be promising. And in any case, I am arguing that such programmes are more fruitfully viewed in terms of the problems and background information which produce them than in terms of some overarching approach which is accepted rather arbitrarily—as the views of Imre Lakatos ultimately imply.

Darwin's evolutionary theory, and partly because it was described in precise terms, and thus was amenable to further experimental research.

But another sort of research was also fostered by the recovery of Mendel's laws, a sort of research aimed not at delineating and extending the domain of heredity, of Mendel's laws, but rather at discovering their underlying cause. It is the reasoning by which that problem arose, and by which expectations arose regarding the character of those causes, that must now engage our attention.

It has often been remarked that Mendel's employment of statistics in determining breeding ratios was an important innovation. However, for indicating that there were hidden causal factors in heredity, statistics and breeding ratios were essentially irrelevant. For that, all that was relevant was that certain characters existing in the parental crossbred plants disappeared in the F_1 generation and reappeared in the F_2 generation. This was sufficient at least to suggest that the causal factors remain present and unchanged in their effects from generation to generation, though they will be latent, inactive, in some generations. (It is interesting that the concepts of dominance and recessiveness, though subsequently removed from their original fundamental role in genetics, nevertheless played a crucial role, indeed a decisive one, in the *discovery* process.) The breeding ratios, on the other hand—the 3:1 ratios in which the characters appeared in the F_2 generation of a crossbred grandparental stock—indicated further that these factors have the character of discrete units, combining and separating independently. The reasoning may be summarised in the following table:

Domain characteristic	Theoretical indication	Type of theory
Disappearance of character in F_1 generation, reappearance in F_2	There are hidden factors	Compositional
Breeding ratios close to integral values in character occurrence in F_2	The hidden factors are discrete units	Discrete compositional

Thus it was that the Mendelian laws, by the very character of the relations they embodied, suggested research aimed at providing a discrete compositional theory—a theory of the gene.

But the 'rationale of scientific research' is not a 'logic of discovery'. First, there are often—usually, in fact—other ways of viewing the domain that can suggest other lines of research, other general types of theories. Thus Bateson, while acknowledging the implications of discreteness in the Mendelian laws, could view that discreteness as arising not from material particles, but from motion, in his 'vortex theory' of living systems:

> . . . a living thing is not matter. It is a system-vortex, Cuvier called it, through which matter is passing. If you watch an eddy run along the dust, or through water, you will see a system —through which matter is passing—rise, *increase*, and decline. Such a system imitates the normal mechanical attributes of life fairly well. (Letter to F. A. Borradaille, 28 January 1924; quoted in Coleman, 'William Bateson and the chromosome theory', manuscript copy, 45.)

(Bateson's view, however, was never developed, and so was never a serious rival to the chromosome theory he so vehemently rejected.) This is not the place to enter into a discussion of the ways in which such alternative lines of research arise; it is sufficient here to note that the suggestions of the domain patterns are only suggestions, not logical compulsions. Any account of the notion of a 'good reason' in science must take account of the existence of alternative 'good reasons' for pursuing different lines of research with regard to the same domain.

Secondly, the 'rationale of scientific research' is not only not a *logic* of discovery; it is also not a logic of *discovery*: for there is no guarantee that the line of research indicated by the domain pattern will be successful—that the theory finally accepted will be of the form suggested and sought initially.

Thus, underlying the main line of research in genetics in the opening years of the present century were three fundamental propositions:

(1) There exists in a current accepted and important theory an inadequacy which does not imply the incorrectness of that theory. (In the present case, the incompleteness of the Darwinian theory did not imply that it was incorrect.)

(2) There is a precise relationship or set of relationships holding within the particular subdomain of that theory with respect to which it is inadequate. (Mendel's laws.)

(3) These relationships are such as to suggest that causal factors of the nature of discrete units are responsible for them, and the description of which can supplement the theory so as to remove the inadequacy in question.

Although in the case of the history of atomism there was no incomplete theory to which Dalton's views were a response—nothing, that is, corresponding to proposition 1 and the second clause of proposition 3—the parallels between the two cases are striking. (This is, of course, only to be expected in the case of two discrete compositional theories.) In the Daltonian case, the laws of combining proportions played a role analogous to that of Mendel's laws in the genetics case, relating the items of the domain in such a way as to suggest the existence of causal factors. (There were, of course, differences on this point: for example, there was nothing in the Daltonian case to correspond to the 'skipping' of the F_1 generation by a character which appeared in the parental and F_2 generations, and which was so suggestive of the existence of 'hidden' factors in the Mendelian case. Further, in calculating specific formulas of combination of his atomic units, Dalton appealed to a rule of 'simplicity' which has no analogue in Mendel, who paid little attention to speculating about the nature of his elements.)

Again, in both cases, there was a long period of dispute as to whether the newly discovered relationships, and the presumed causal factors, really were compatible with some other important scientific knowledge. In the case of genetics there was the long dispute between the 'Mendelians' and the biometricians, who maintained that the Mendelian laws, far from supplementing Darwinism and removing a critical incompleteness from it, were incompatible with it. The controversy was resolved by the 'synthetic' theory of evolution. In the Daltonian case, the question of compatibility concerned not a prior theory which was considered incomplete and which the new theory was to supplement, but rather another proposition (Avogadro's hypothesis) which seemed in rivalry with it. The dispute was resolved by Cannizzaro at the Karlsruhe Conference in 1860, by a 'synthesis' of the Avogadro and Daltonian views.

Finally, in both cases there was a positivistic opposition to considering the domain relationships as indicative of *any* sort of hidden entities, on the ground that talk about such entities was unscientific, mere 'metaphysical speculation'. The terms 'atom' and 'gene' were, according to these critics, to be taken not as referring to entities, but only as convenient summaries of those relationships themselves. (It must be remarked, however, that Carlson, in his *The Gene: A Critical History*, has seriously distorted the views of East and Johannsen in imputing to them such a positivistic attitude.[4]) Advocates of compositional theories have often been faced with such positivistic opposition; its historical record, however, has been very poor, almost invariably standing in the way of some of the most important advances in the history of science.

Nevertheless, in spite of these parallels, there was one particularly important difference between the atomistic and genetic cases: namely, that in the case of genetics, another subject was available which became unified with that subject and which localised and made possible direct study of the entity in question. This subject was cytology, and the entity it localised and studied in connection with the question of inheritance was the chromosome. The behaviour of the chromosome in mitosis and meiosis had been clarified in the early 1880s, and Roux in particular had connected their linear structure and division into equal longitudinal halves with hereditary transmission from parents. Other experiments had indicated that it was the chromosome and not any other part of the cell that was important in such transmission. Instrumental techniques (such as Caldwell's microtome) and judicious choice of organisms for study (*Ascaris*) were important in making the science 'ready' for dealing with the issues presented by genetics. The unification was indicated clearly by Sutton in a famous passage at the end of his 1902 paper, 'On the morphology of the chromosome group in *Brachystola magna*':

[4] This has been pointed out to me by L. Darden, with whom I have had many valuable discussions on the topics of this paper.

I may finally call attention to the probability that the association of paternal and maternal chromosomes in pairs and their subsequent separation during the reducing division as indicated above may constitute the physical basis of the Mendelian law of heredity.

It is unnecessary to repeat here the familiar subsequent history: the important upshot of the story is how, in the process of searching for a compositional theory for genetics, the techniques of that subject were supplemented by the more direct methods of another. Breeding ratios, the key method of the early geneticists, could provide only bases of inference to the entities in question; cytology was available and prepared to give direct access to them. And in the process, it was able to raise and answer questions that were not accessible to breeding ratio studies. (In the case of atomism, there was no analogous subject—at least not for approximately a century.) Ultimately the concepts of cytology, as well as its methods, infused the field of heredity and began to provide explanations of what had before been purely genetical concepts and relationships. (This had already been foreshadowed in Morgan's adaptation of Jannsens' chiasmatype theory to crossing over.)

In the process of delineating more and more fully a compositional theory for a domain, it often happens that the methods, and subsequently the concepts, of some other field are applied, those methods and concepts being more readily and directly applicable and appropriate at that stage. At yet a later stage of delineation, those methods and concepts may in turn be supplemented or even replaced by those of another subject (as the methods and concepts of molecular biology began in the 1950s to invade the field of cytology—to provide a deeper compositional theory for the domain now constituted by genetics and cytology).

Thus the unification of fields is often a natural by-product of search for compositional theories. And that unification can, on occasion, provide what philosophers have called the 'reduction' of one subject to another; after all, the aim of compositional theories is the explanation of a domain in terms of a deeper causative level. But the notion of 'reduction' is quite ambiguous: certainly a meaning which many philosophers have sought to give it is that the concepts of one area (or 'level') are definable in terms of those of the other, and the relationships (propositions) of the former are 'deducible' from those of the latter. In this sense, however, reduction would almost never have taken place in the history of science: for the 'deductions' involved are not strict, but involve all sorts of approximations, simplifications, idealisations, which make a precise account of any given relationship or individual event in terms of the 'lower-level' theory impossible.[5] A more useful sense of the term 'reduction' would take into account the role of such 'conceptual

[5] A detailed analysis of approximations, simplifications, idealisations, and other 'conceptual devices', as I have called them, has been given in the two papers referred to in footnote 1, above, and in my Natural science and the future of metaphysics, in *Boston Studies in the Philosophy of Science* (ed. M. Wartofsky) (forthcoming).

devices' in the reduction. And even then it would not follow that the *field* reduced would be eliminated: for not only would its laws and individual events not be strictly deducible (or predictable) from those of the reducing theory; its methods, too, might still have much to offer which is inaccessible to those associated with the reducing theory. (The subject matter of nuclear chemistry is precisely that of nuclear physics; but the methods, and the training, involved are not the same and have much of their own to offer.)

We have thus seen, in a particular case from the history of science, how a problem arising in connection with an evolutionary theory (in this case, its incompleteness with respect to its domain) called for the search for a compositional theory. And we saw at least something of the rationale of search for that theory, and the way in which the expectation arose of its being, more specifically, a discrete compositional theory. That rationale underlay a unification of scientific fields, a unification which ultimately approached what is naturally called a case of 'reduction' of one area of science to another. But the ways in which such cases have been interpreted by some philosophers have been overly restrictive, implying that 'reduction' is equivalent to 'destruction', the reduced field being eliminated entirely in favour of the reducing one. (Indeed, as we have seen, so excessively stringent is the common philosophical analysis of reduction that it makes it appear that there has been no real case of reduction in the entire history of science.) But it is not the fact of reduction that creates problems; it is only this exaggerated conception of it—and that makes the problems deserve to be called mere 'pseudoproblems'. The lines along which a more adequate conception may be developed, truer to the facts of science and its history, have been suggested here.

Discussion

Dobzhansky: Natural selection and pseudoselection

Schopenhauer said 'Philosophy is systematic misuse of a terminology devised specifically for these purposes'. Biology risks to find itself in a similar predicament. Having overcome some conceptual difficulties, biologists seem finally to have a reasonably clear idea of what selection—natural or artificial—really is. It is differential reproduction of living systems. (Cells in a body or in a tissue culture are, of course, such systems.) But some people find it tempting to use 'selection' to describe also processes of cosmic evolution, in inorganic nature, as well as in cultural evolution ('selection' of ideas, beliefs, technologies, customs and values). Perhaps by introduction of qualifying adjectives (such as 'prebiotic' 'cultural', etc.) one can avoid threatening confusion. However, 'selection' is often used in biology as a term covering both natural and artificial selection. Anyway, nothing but obfuscation can come from statements such as 'human values are products of natural selection', or simply of 'selection'. Such statements inevitably suggest that human values are determined or fixed genetically, which is doubtful to say the least.

I agree with Shapere and Edelman that 'marginal' or transitional situations will necessarily be encountered, and also that the origin of life will probably be understood some day in physicochemical terms. However, as things stand now, the processes that have led to the emergence of life are conjectural, and it is hardly useful to have basic scientific concepts founded on conjectures. I would also be reluctant to describe the processes of cosmic evolution as 'selection' of any sort. If I understand the matter correctly, what is involved is that some states of matter are more stable than others, and that in the evolutionary development of the universe stable states have predominated over unstable ones. However, if the concept of 'selection' is so diluted that it connotes only the stability or instability of an inorganic system, the 'generality' of the concept is achieved by sacrifice of its meaningfulness.

Shapere and Edelman: A note on the concept of selection

Professor Dobzhansky has suggested that *natural selection* is the result of differential reproduction of living organisms. The question arises, however, as to whether there are selective processes which do not involve self-replication and do not involve living organisms.

The question is not one of mere arbitrary definition. The very fact that there was an origin of life in a chemical environment required a selective principle, and therefore matter in general, or at least certain forms of it, has to be subject to selection, of which natural selection is a special case. Furthermore, in a number of sciences other than biology, reference is made to the 'evolution' of nonliving systems (such as stellar systems and chemical elements). More general criteria of 'selection', together with an understanding of more specific features of biological or natural selection, might help throw light on the similarities and differences between biology and those other sciences.

What is desirable is that such general criteria take into account the production of new entities from old, through the interactions of individual entities with the environment, including interaction of parts of individuals with other parts (exterior and interior milieu).

The three criteria of natural selection adduced by Professor Stebbins, suitably modified, offer the opportunity for such a useful generalisation. As stated, his criteria were:

(1) interaction between systems, as well as between each system and its environment;

(2) reproduction of those systems;

(3) differential reproduction of those systems.

The following modification of these three criteria is suggested:

(1) interaction between systems;

(2) production of new systems from old;

(3) the new systems are produced not merely as a result of the interaction of parts entering into the new system, but also as a result of interaction between those systems and their environment;

(4) self-replication with mutation (differential reproduction).

With this reconstruction, (1) and (2) serve as a general notion of 'interaction', while (3) distinguishes 'selection' as those interactions in which the environment plays a role. (It is, of course, desirable to have selection being a special type of interaction.) The addition of criterion 4 brings out what is distinctive of biological (natural) selection.

No claim is made that this set of criteria is as tight as it might be. Criterion 3, for example, could be made more complex in order to account for the role of catalysts in chemical reactions, and the distinction between 'system' and 'environment' also could be made more precise. The criteria might also be improved with respect to the following considerations. In the production of a 'new system', there is some increase of order (information, decrease of entropy). Further, the whole process generally is irreversible (that is, takes place in systems far from equilibrium). Of course, a general theory of selection which is coupled to the chemist's theories concerning forces and flows in systems far from equilibrium must await the full development of such theories. In spite of these and other questions about the above criteria, however, we believe that they point the direction which a more rigorous statement should take.

The general concept of selection, as embodied in criteria 1–3, is satisfied in other sciences than biology. For example, a racemic mixture of crystals may arise in which one form may be selected while others disappear as a result of environmental changes. This certainly fulfils the requirement of a 'prior repertoire' of individuals which differ in their properties, as well as that of differential reproduction.

There are some examples, more specifically biological, which present some difficulties of classification.

(*a*) Are there selective systems which occur among cells of an organism rather than among organisms? The immune system, which itself arose out of natural selection, may be such an example. (This case is discussed in Professor Edelman's paper.)

(*b*) Q-beta phage replicase experiments by Spiegelman are a marginal case which also may be difficult to classify as natural selection under Professor Dobzhansky's definition.

Whatever may be the difficulties of these marginal cases, the specific features of biological or natural selection are embodied in criterion 4. From the viewpoint of this conference, the important question is whether this special feature of biology implies the impossibility of its explanation in physical terms. The origin of life itself requires the appearance of molecules satisfying criterion 4, and which therefore become subject to specifically biological or natural selection. It should be added that there is nothing so far known to suggest that this transition is not perfectly understandable in physico-chemical terms.

Rensch

Your discrimination of compositional and evolutionary theories is quite convincing. I would like to mention another possibility to discriminate between two or three clearly *different types of theories*. Some theories are based purely on the laws of logic like mathematical theories, others are based on causal relations, and microphysical ones on causal relations and laws of probability. Causal theories are related to temporal succession, logical ones are not.

I believe that the need to fill the gaps in the table of chemical elements was first a logical necessity. Mendel's rules are related to a temporal succession and ultimately to causal events.

13. Problems of Rationality in Biology

HENRYK SKOLIMOWSKI

Every new form of life that appears in evolution can, with only moderate semantic licence, be regarded an artistic embodiment of a new concept of living (Dobzhansky, 1974).

Biology as a philosophical battlefield

Biology is nowadays a special science, for it has become a philosophical battlefield on which a new paradigm for all human knowledge is being established. The reign of physics as the universal paradigm is now over. Biology is aspiring to provide a new paradigm. Though its aspirations are perhaps justified, the passage to this new paradigm proves to be exceedingly rough and thorny.

Biology is also a singular science for it provides a focus for theological disputes over the place of man in this universe, and over the role of supernatural agencies in the evolution of life and man. These disputes, when carried on by biologists themselves, usually result in antitheological conclusions. Though God is denounced and declared nonexistent, his shadow still haunts the biologist turned natural philosopher.

Biology, in short, embodies some of the crucial dilemmas of the intellectual heritage of the post-Renaissance era. In the flight from the tyranny of religious orthodoxies, we have acquired the ruthless intellectual habit of denying everything that does not square with a certain conception of the empirical world. The paradox and the drama of our intellectual, cultural and indeed human development of the last three centuries lies in the dialectical reversal of once progressive attitudes into their opposites. Enlightened empiricism, which, under the banner of human dignity and individual freedom, had fought against superstition and humbug, against the supremacy of theology and an individual God, became—particularly in the twentieth century—barren positivism, which, although it adheres to the original tenets (of humanistic empiricism), is actually suppressing the search for new extensions of human dignity and new extensions of human knowledge.

Mere declarations will not suffice. Let us therefore turn to some of the present dilemmas in order to demonstrate that the difficulties of present biology are more conceptual than empirical, more rooted in our *Weltanschauung* than the actual empirical problems we investigate.

One of the most important of these dilemmas is the problem of rationality, on which hinges the whole process of understanding. I shall anticipate the outcome of my argument and say at the outset that the particular difficulties in which we now find ourselves in the realm of biology, and also in relation to the whole heritage of our scientific knowledge, stem from the restrictive harness of a rationality which is no longer adequate for the recent extensions of our knowledge and for the cognitive needs of contemporary man. The rationality developed under the auspicies of physical science is a harness, for it ties us down to a certain conceptual framework and obliges us to observe criteria of validity that are specific for this framework. The paradox of many biological explanations, particularly in evolutionary biology, consists in the fact that these explanations depart from the criteria of validity as accepted by the current scientific rationality, and yet often explain phenomena with more illumination than would have been possible within any physicalist model. The evolutionary biologist is often admonished by his molecular colleagues for not being 'scientific enough'. But by not being scientific enough he is able to shed more light on the phenomena of life than would be possible if these phenomena were to be explained merely by the interaction of chemical compounds.

In order to resolve this dilemma, in order to relieve the conscience of those biologists who are more interested in the process of life as a flow rather than in its physical–chemical components, and to reassure them that they are pursuing a valid path of inquiry; and in order to unblock roads to fruitful inquiry into the versatility of life processes on the highest level of complexity —as manifested in man—we shall have to reconstitute our notion of rationality. This will mean, in other words, establishing new criteria of the validity of accepted knowledge which will inevitably lead to a change in the scope of our knowledge. Indeed, the new paradigm for all human knowledge which biology attempts to provide cannot be successfully established unless we simultaneously reconstitute the notion of rationality and the criteria of the validity of accepted knowledge. As long as we remain in the harness of traditional rationality with its rigid criteria, our efforts to work out a new paradigm of knowledge, which would be in harmony with more recent insights into the evolutionary development of man, will be frustrated and doomed. What is at stake, therefore, is the enlargement and refinement of our concept of understanding which has been imprisoned by the canons of empiricist and positivist preconceptions.

In the post-Renaissance civilisation, physical science has been the major influence in determining the scope and the nature of our rationality. And so much so that the 'rational' has come to be used interchangeably with the 'scientific', and conversely.

In the universe of modern western man, the rational is that which is valid within a given scope of knowledge; on the other hand, the valid is that which is rational within a given scope of knowledge. Thus the validity of knowledge is interlocked with rationality. One is defined through the other. Rationality is a framework on which knowledge is based but which itself is abstracted from this knowledge. This circularity is unavoidable as long as we remain in the purely cognitive realm. If there is a non-circular justification of rationality, then this justification must be sought outside the realm of pure cognition. I shall argue later that this justification is to be found in the survival value of knowledge.

Positivist rationality and evolutionary rationality

Modern biology, and especially evolutionary biology, has proved time and again the insufficiency of the physical model of knowledge. And yet the criteria of validity characteristic of this model stubbornly persist and, indeed, are often used against some of the claims and insights of evolutionary biology, particularly in relation to the notion of evolution. Positivism dies hard. One of the reasons is that it is such a well-defined doctrine, with clear-cut boundaries, clear-cut criteria of validity and well-formulated language. The language of positivism has almost become the official language of science. In contrast, the phenomenon of life in its development might be characterised by its inherent fuzziness. How can we grasp adequately the fluidity and fuzziness of life having at our disposal only razor-sharp concepts?

By positivism I do not mean a specific philosophical doctrine of Comte or Spencer or Carnap, but I rather mean the overall approach to problems of the external world. Positivism is not only a set of doctrines. It is also a set of attitudes. It embodies a number of beliefs, of which the most important are:

(1) that all which is known can be reduced to physical laws, and that all genuinely existing phenomena have a physical foundation;

(2) that all genuine knowledge must be acquired through the scientific method which is modelled on physical science;

(3) that the pursuit of knowledge can be best conducted when we limit ourselves to simple and relatively isolated systems;

(4) that we have to give up the comprehensiveness of our inquiry for the sake of precise results; therefore, of necessity, we concern ourselves with relationships which are very clearly defined although the realm of these relationships is rather limited;

(5) that there are *a priori* rules of procedure subsumed under some such principle as the objectivity postulate which assure the validity of our cognitive pursuits;

(6) that acceptable evidence is only that which is definable and defensible in terms of physical science;

(7) that phenomena and processes which have a genuine existence are deterministic in nature or else (in evolution) are a matter of pure chance;

(8) that meaningful language must have empirical foundations and empirical consequences;

(9) that the behaviour or action of large and complex systems is the result

of the behaviour of their constitutive parts so that the total behaviour equals the sum of its parts;

(10) that the explanation of the process of knowledge must rule out any supernatural agencies which are beyond the reach and dominion of present science;

(11) that truths of science are the only truths, or at any rate the ultimate truths from which there is no appeal;

(12) that one of the most important justifications of the pursuit of knowledge lies in the idea of material progress.

In contrast, the evolutionary biologist and all who take evolution seriously, who acknowledge that evolution has produced new forms of life which can neither be reduced to the functions of older forms of life nor satisfactorily explained in terms of earlier forms of life, tend to believe a number of propositions which are in some fundamental disagreement with the tenets of positivists. The most important of these propositions are:

(1) that not all that is known can be reduced to physical laws; some knowledge is irreducible to physical knowledge;

(2) that the methods of physical science are insufficient for the study of the phenomena of life on the high level of complexity;

(3) that we cannot limit ourselves to simple and relatively isolated systems, for life systems are enormously complex and intricately interconnected;

(4) that the pursuit of knowledge, which gives us an understanding of complex life systems requires the inclusion of a large number of relationships which cannot be precisely defined but which enable us to comprehend the phenomena of life with more illumination than through a limited number of exact relationships;

(5) that there are no absolute, *a priori* rules that govern our quest for knowledge: any such principle as the objectivity postulate has its limitations; and that an *a posteriori* feedback is our best guide in the realm of cognition and elsewhere;

(6) that the acceptable evidence is evidence which explains phenomena under investigation regardless of whether or not it meets the requirements of the physical model;

(7) that the most significant phases or processes in evolution are simply beyond the dichotomy 'chance or necessity'; though not strictly determined, and in a sense unpredictable, these processes exhibit patterns of cohesion and integrity which cannot be attributed to mere chance;

(8) that meaningful language must be able to express content which is significant, both empirical and extra-empirical;

(9) that the behaviour or action of many large and complex systems is often inexplicable by the behaviour of the constituents of the system; the total behaviour often equals more than the sum of its parts or differs from this sum;

(10) that those 'agencies' or forces which go beyond the reach and scope of present science need not be feared or called 'supernatural'; and we need not

be nervous to the point of obsession or paranoia about the restitution of God and theology when we attempt to extend the reach of our present knowledge;

(11) that the truths of science are at least partially truths by convention, and that there are other kinds and sources of truth over which scientific truths have no authority;

(12) that the idea of material progress is insufficient to account either for the pursuit of knowledge or for the evolutionary processes of man.

The first set of attitudes and assumptions constitutes the rationality of hard science, or, for short, (present) *scientific rationality*. The second set of attitudes and assumptions constitutes the rationality of living systems, or, for short, the *evolutionary rationality*. The laws of falling bodies are not concerned with this or that particular falling body. Each and every particular falling body will, in actual behaviour, show some deviation from the laws which (nevertheless) govern the behaviour of this whole class of phenomena. The same is true of scientists and scientific attitudes. When one reconstructs fundamental characteristics of a school of thought or an intellectual formation, one is not obliged to make these characteristics fit *every* individual who represents this school of thought. Thus, we wish to grasp fundamental characteristics even though they do not fit uniquely any single individual. For this reason an attempt to grasp the basic features of present scientific rationality (as well as the emerging new evolutionary rationality) is as justifiable as an attempt to grasp the ideal behaviour of the gases.

Scientific and evolutionary rationalities may be seen as contrasting types of understanding. We must not think, however, that there is an unbridgeable chasm between them. Scientific rationality is a part of the evolutionary rationality, but not conversely. The characteristics of the evolutionary rationality, when simplified to the point of caricature, do become the characteristics of scientific rationality. Thus scientific rationality represents an end of the spectrum. When the variety of phenomena are reduced in number to one, when complex and open systems become simplified and closed, when language is simplified to express the physical content only, and so on, then we obtain a curious transformation of the evolutionary rationality into scientific rationality.

We must not make a fundamental mistake and consider the end of the spectrum to be the whole of the spectrum. Because this mistake lies at the foundation of many of our conceptual troubles, we must emphasise the contrast between these two intellectual attitudes which I call: scientific rationality and evolutionary rationality. In actual scientific practice these attitudes are perhaps not held in their pure forms. But we have seen enough clashes of opinion, which are not at all concerned with matter-of-fact or empirical issues, to make us perfectly aware that these contrasting attitudes are alive and competing with each other.

We must be above all aware of the historical character of *all* rationality. We do not need to go back to primitive societies in order to demonstrate the historical character of the validity of knowledge and of rationality. Was

Newton irrational when he indulged in physics and theology at the same time? worse, when he mingled physics with theology? still worse, when he used his physics in order to justify his theology? Was Kepler an irrationalist when he worked out his laws (Kepler's laws) in order to prove the harmony of the celestial spheres? We often adopt a patronising air towards these great scientists and 'forgive' them their indulgences, and we implicitly suggest that they did not know any better, whereas we do. Such an attitude is an expression of conceit on our part, and is also a misreading of the history of science.

Now the qualitative changes in the process of evolution may or may not necessitate the introduction of the concept of teleology, or at least internal teleology as Ayala (1968) suggests, but this in no sense implies the acceptance of a Master Plan; even less so the acceptance of an Omnipresent God; least of all a return to the cruelty of the Dark Ages dominated by a severely intolerant religion. For the ghost of vitalism does not threaten biology. It only threatens the atheistic theology which many scientists and natural philosophers of the new school feel obliged to subscribe to, and feel obliged to defend *per fas et nefas*. We simply seem unable to liberate ourselves from our ties with theology, even after five centuries of struggling with it. Today's situation is paradoxical for it is Newton's predicament *à rebours*. Newton through his science tried to serve theology. We through our science try to destroy theology. In trying desperately to escape the predicament of theology, we have created an antitheology which is, however, itself a theology.

The scientific spirit has attempted to eliminate every vestige of God and theology from the realm of knowledge and indeed even from our life to the extent that it has become a new theology, an anti-God theology. This is abundantly clear in biology. So very often disputes over the nature of properties, which are attributed to living cells and organisms in order to make sense of their evolution, are deviously turned around into theological disputes over the origins of life. Almost invariably in such circumstances positivists ride their high horse of antimetaphysics and antitheology, denouncing any and every trace of old theology, of Master Plans which they read into the innocuous postulates of evolutionists who are simply claiming that there is more to life than physics. It looks as though scientists have taken upon themselves the role of priests. Self-appointed vigilantes, they attempt to catch and suppress any deviance from the 'superior', secular, scientific world view. To assume such an attitude is neither tolerant nor indeed scientific, for the issues are far beyond the jurisdiction of science.

The favourite strategy which molecularists and positivist-oriented scientists assume (for instance Bronowski, 1970) is to argue that *there is no evidence* to support the claims of antireductionists. The strategy 'there is no evidence' deserves special attention for it not only signifies a semantic manoeuvre but it also signifies an important aspect of the positivist epistemology. When Bronowski and others use this phrase, or any of its variants, they mean to imply that there is no evidence in terms of physical science. Of course there *cannot* be. The essential point is that when the evolutionist exhibits the limitations of physics and wants to transcend the scope of physics, he cannot give *physical* evidence. To produce this *kind* of evidence is to support the cause of positivism (physicalism). If the evolutionist provided this kind of evidence, he would simply undermine his claim that there is more to the behaviour and function of the organism than physics can explain. Therefore, whoever argues that there is no (physical) evidence for the nonphysical manifestations or functions of organisms is either very naive or is intellectually dishonest by surreptitiously trying to force the opponent into the mould from which he seeks to liberate himself.

Complexity and hierarchy

Already in 1948 Warren Weaver, in his paper 'Science and Complexity', distinguished (i) problems of simplicity, (ii) problems of disorganised complexity, and (iii) problems of organised complexity. He argued that science, which he more or less identified with physical science, 'has succeeded in solving a bewildering number of relatively easy problems, whereas the hard problems, and the ones which perhaps promise most for man's future, lie ahead'.

Problems that are 'hard' and which perhaps promise most for man's future are, according to Weaver, problems of organised complexity which involve simultaneously '*a sizeable number of factors which are interrelated into an organic whole*' (Weaver, 1948).

Around the same time, but quite independently, Ludwig von Bertalanffy (1950, 1951) proposed the distinction between open and closed systems. A system is closed if no materials enter or leave it. A system is open if there is inflow and outflow and therefore change of the component materials. Physics and chemistry are exclusively concerned with closed systems. Biology, on the other hand, is concerned with open systems. This was a part of a larger theory of systems which in time was developed into General Systems Theory (von Bertalanffy, 1968). Some have pointed out that Bertalanffy's claims are excessive and that he proposed not so much a theory but a metaphysics. Whether this accusation was inspired by another metaphysics (held by Bertalanffy's critics) is another matter. What is undoubted is that the peculiar properties of large complex systems, especially as characteristic of biological organisms, came to be regarded as the legitimate subject matter of science.

Although the investigation of complex organised systems started a quarter of a century ago (if we look for pioneers, we can no doubt find them in antiquity), it has been vigorously pursued only during the last few years.

By a complex system we have come to recognise a large number of parts that interact in a non-simple way, that is, in such a way that 'the whole is more than the sum of the parts, not in an ultimate, metaphysical sense, but in the important pragmatic sense that, given the properties of the parts and the laws of their interaction, it is not a trivial matter to infer the properties of the whole' (Simon, 1962; 1969, 86). Most important among those complex systems are hierarchic systems. Hierarchic systems can be seen in social structures: for example, organisational hierarchy in business; in physics: for example, the atom can be seen as one end of the spectrum of a hierarchic system with the galaxy at the other end; and above all hierarchic systems can be seen in biology. Hierarchic systems have some common properties that are independent of their specific content. Hierarchy, Herbert Simon suggests, is one of the central structural schemes that the architecture of complexity uses.

One of the crucial 'objective' characteristics of hierarchic systems is that 'Hierarchic systems will evolve far more quickly than nonhierarchical systems of comparable size'. Another, and a far more subtle characteristic, bears on our comprehension of complex systems, and indeed of all systems. Simon (1969, 108) argues:

> If there are important systems in the world that are complex without being hierarchic they may to a considerable extent escape our observation and our understanding. Analysis of their behavior would involve such detailed knowledge and calculation of the interactions of their elementary parts that it would be beyond our capacities of memory or computation.

John Platt (1970) has argued that a hierarchical structure, particularly of an organism, acquires new characteristics as the result of 'hierarchical jumps'. The growth picture is non-continuous, starting from the lowest level (molecules, enzyme cycles, cells) which is stable, up to the level *i* (say, the organism). The organism grows to a new structure when it comes in touch with new and different materials or information or another organism. This is the dialectics between genotype and phenotype. Platt (1970) has distinguished five common features of hierarchical structures, or systems which develop by jumps and often as the result acquire new qualities:

(i) jumps are preceded by *anomalies* within the structure (Platt, following Gardner Quarton, talks about cognitive dissonance);

(ii) the anomalies are *all-pervading*; they are not limited to one aspect or one part of the system only;

(iii) the jump to a new level, or, in other words, the process of restructuring, is *sudden*;

(iv) after the process of restructuring is accomplished, a *simplification* takes place;

(v) the process of restructuring is often accomplished through *interactions jumping across the system level* between the old subsystems and the new supersystem that is in process of formation.

It is hard to deny Platt's contention, which he shares with Michael Polanyi and Arthur Koestler, that acts of hierarchical growth are never rationally deducible from the smaller system structures that precede them.

Another inescapable conclusion is that hierarchy and complexity in living organisms are equivalent to nonreducibility. When we *comprehend* the function of these organisms in terms of complexity and hierarchy, we *invariably* go beyond the physicochemical rudiments of these organisms.

In a similar antireductivist spirit, A. Katchalsky emphasises that function in the biological organism is not the product of structure but is another expression of living texture; a point which Dobzhansky will carry much further. Katchalsky postulates that beneath the dynamic organisation of cells, there exists in addition the hidden framework of a permanent structural set-up. It is well known, he says,

> that bacterial cells may be desiccated under the condition of high vacuum and liquid-air temperature to a dry lyophilised powder. The bacterial powder has no metabolic activity and is dead to all intents and purposes, except that when water is added and the temperature is raised, the cells revive and begin to metabolise and reproduce. Living cells are, therefore, not only loose dissipative structures in the continuous medium of a test tube, but a *dynamic pattern* superimposed on a *fixed network*, the organisation of which is dictated by the genetic code. Thus, the flow structures of cells are confined to preordained limits which represent the evolutionary history of the species (Katchalsky, 1971).

To report on the findings of others which directly or indirectly support the claim of the irreducibility of complex, hierarchical systems to their physical underpinning would take volumes. Now, if these findings are so unquestionable and important, why, then, do they have such a dubious cognitive status, why, in fact, are they constantly questioned by hard-headed molecularists and positivists in general? The answer to this question cannot be found in the realm of physical facts, but rather in the ideology of modern science.

A parenthetical remark: a variety of authors who have investigated the properties of complex hierarchical systems (David Bohm, Karl Deutsch, Ahron Katchalsky, Thomas Kuhn, John Platt, Herbert Simon and others) conspicuously avoid mentioning dialectical materialism and its canons of dialectical development, although quite often their analysis is carried on along lines very similar to those of Marxist dialecticians. Is this because we have carried an understanding of the dialectical process of nature far beyond the Marxist tenets, or is it because we still remain in the ideological straitjacket which has imposed on us certain conceptual and linguistic taboos? If the latter is the case, and I suspect this is so, then this should make us aware how vulnerable we are to the subtle pressure of ideology under which we as scientists and philosophers happen to be born and live. This should also make us aware how vulnerable we are to the pressure of the entire ideology of modern science. By ideology I do not mean crude political slogans but rather the overall climate of opinion, the predilections of society, the unwritten mores, and above all the values that inspire and control society and individuals. We are the proud inheritors and perpetuators of the scientific tradition. But perhaps also the slaves of certain modes of thinking; subjects to a conceptual tyranny which we glorify, thus being perfect slaves—slaves who enjoy their imprisonment. The uncritical attitude towards the assumptions of science often held by scientists I find quite puzzling; and the way they take for granted the ideology of science I find quite disturbing.

Orwell (1972) had written a memorable passage, intended to be a part of a preface to his *Animal Farm*, which he supressed at the time, but which was recently published in *The Times Literary Supplement*. He says:

> At any given moment there is an orthodoxy, a body of ideas which it is assumed that all right-thinking people will accept without question. It is not exactly forbidden to say this, that or the other, but it is 'not done' to say it, just as in mid-Victorian times it was 'not done' to mention trousers in the presence of a lady. Anyone who challenges the prevailing orthodoxy finds himself silenced with surprising effectiveness. A genuinely unfashionable opinion is almost never given a fair hearing, either in the popular press or in the highbrow periodicals.

Are scientists immune from this process of self-hypnotic indoctrination? Their wishful thinking that it is so is but wishful thinking.

Objectivity and its pitfalls

The ideal of objectivity is one of the most important vehicles perpetuating the ideology of modern science. Jacques Monod has been one of the most articulate spokesmen for the objectivity of science. I shall briefly examine his conception of objectivity in the context of our discussion of evolutionary biology. I shall not hold against Monod his 'limited philosophical knowledge' which many philosophical critics were so eager to point out (for instance, Toulmin (1971) and Hampshire (1972)). When one strives for novel and original conceptions one is absolved from the burden of historical reference. It is common knowledge that Wittgenstein's 'philosophical knowledge', in terms of what he knew of past philosophies, was rather limited. Yet this was hardly ever used against Wittgenstein's original contributions. We must not apply double standards: one for ourselves and the other for those we consider as intruders into the realm of philosophy.

Monod (1972) writes:

(i) 'The cornerstone of the scientific method is the postulate that nature is objective.'

(ii) 'In other words, the *systematic* denial that "true" knowledge can be reached by interpreting phenomena in terms of final causes—that is of "purpose".'

(iii) 'The postulate of objectivity is consubstantial with science',

(iv) 'and has guided the *whole* of its prodigious development for three centuries.'

(v) 'It is impossible to escape it, even provisionally or in a limited area, without departing from the domain of science.'

(vi) 'Objectivity nevertheless obliges us to recognise the teleonomic character of living organisms.'

(vii) 'Here therefore, at least in appearance, lies a profound epistemological contradiction.'

It will be a tedious task for me to show the incoherence of this position, to unravel its ambiguities, hidden assumptions and question-begging. But this task has to be performed in order, once and for all, not to be intimidated by the mere term 'objectivity', in order to show that objectivity is a normative category; in order to reveal the mythology in which the physical model is wrapped; in order to demonstrate that objectivity serves to glorify one particular approach to knowledge and thus mystifies the rest of knowledge; that it has become an instrument of cognitive intimidation and thus performs a conservative if not detrimental role in the acquisition of new knowledge.

Is nature objective (i)? In a trivial sense, yes. By assuming certain attitudes (of separating some phenomena from other phenomena, and of examining these separated phenomena microscopically) we have been able to describe quite well certain *bits* of nature. Is this all there is to nature? Obviously not. For there are integrated wholes which science, so far, cannot describe. Moreover, there *may* be some other phenomena and relationships which are quite outside the understanding of science: if they do exist, we shall never find them through the existing apparatus of science, for science *systematically* excludes that which disagrees with its universe. So the postulate of objectivity turns out to be either an unwarranted dogma—when it claims that nature is only that which science describes; or a very modest postulate indeed—when it claims that *certain* phenomena of nature and certain relationships occurring among these phenomena are the subject matter of physical science. In the latter case, our claim is so trivial that it is hardly worth mentioning.

By asserting (iii) that 'the postulate of objectivity is consubstantial with science', we only repeat a tautology or a dogma: a tautology if we define as scientific that which meets the requirements of the postulate of objectivity (whatever way we define the postulate of objectivity); a dogma if we insist that all valid knowledge is science and that all science must be based on the postulate of objectivity.

Monod seems to have given only one specific formulation of the postulate of objectivity, namely (ii) that no 'true' knowledge can be reached by interpreting phenomena in terms of final causes. So objectivity comes to as much and as little as the denial of final causes, of Master Plans, of teleology. Science is then defined in a negative way; as that which denies final causes. This is a very modest claim indeed and certainly not sufficient to make objectivity so conceived the basis of all science, present or historical. But even this modest claim is marred by some misconceptions. For it is *not* the case that no true knowledge can be acquired by resorting to final causes and teleology. True knowledge, as Popper (1963) has demonstrated, can be acquired in all sorts of ways; anything can be a source of true knowledge. We must not confuse the sources of our knowledge, which are exceedingly numerous, with the justification of knowledge, and especially with the justification of scientific knowledge. Even if we are generous to Monod and grant him that there is such a thing as objectivity, which we might identify with the ontological structure of the universe which physics *assumes* and then explores, we are bound to deny that this postulate has guided the prodigious development of science during the last three centuries. The development of science has been guided by all sorts of principles, not the least important of which was the idea of Master Plan, or the purposeful arrangement of nature by an Omnipotent Being—which we have seen in Copernicus and Newton—an idea which Monod wishes to expel from the domain of science most anxiously. And the history of science will not support Monod either in his contention that it is impossible to escape the principle of objectivity without departing from the domain of science itself. Science is a messy affair, and all kinds of principles and devices are used to get *results*; take for instance the process of the discovery of DNA as described by Watson. Monod's claim could be made valid if and only if objectivity is defined as *everything that goes on in science*. Then the postulate of objectivity is so all-embracing that it becomes meaningless.

We encounter even greater difficulties with (vi). This very 'objectivity nevertheless obliges us to recognise the teleonomic character of living organisms'. Why? In what way? From the principle of objectivity it does *not* follow logically that we have to recognise the teleonomic character of living organisms. Teleonomy does not follow from objectivity. Full stop. We have to recognise the teleonomic character of living organisms only when we recognise *living* organisms, and especially when we recognise their evolution. There is a large leap between objectivity as pursued in the domain of physics and the teleonomic character of living organisms, particularly viewed in their evolution.

The profound epistemological contradiction which Monod mentions in (vii) only arises when we accept the postulate of objectivity and the rigid deterministic physical model that goes with it, and then find ourselves (on the grounds of the accepted epistemology) unable to account for the phenomenon of life, and particularly the phenomenon of evolution. But we can easily resolve Monod's epistemological contradiction—by dissolving the alleged universality of the principle of objectivity. After such a dissolution science will not fall, nor will the universe collapse. Thus, Monod's reconstruction of his scientific methodology is defective, to use a mild term. But he is in good company: Isaac Newton was too a great scientist and a poor methodologist. If Monod is confronted with a choice—either to abandon the principle of objectivity or his actual practice of science (for one is inconsistent with the other)—he will no doubt abandon the principle of objectivity. The sooner he does so the better for him and for science.

Monod has declared himself an adherent of Popper's philosophy of science.[1] For the defence of the objectivity postulate we will derive, however, little comfort from Popper's philosophy. Popper's principle of demarcation between science and non-science is based on the idea of falsifiability. If statements and theories cannot be subjected to empirical tests which might, at least in principle, refute them, then these statements (theories) are not scientific. Falsifiability in Popper's terms is not the same as Monod's postulate of objectivity. Popper's principle of demarcation is not concerned with the denial of teleology, final causes, Master Plans, etc. Let us, therefore, not confuse the principle of demarcation with a postulate whose main purpose (as it seems) it is to escape teleology.

[1] In the discussion between Jacques Monod and Sir Peter Medawar concerning the new biology, BBC (3 June 1972), published in *The Listener* (3 August 1972).

Furthermore, Popper's views have not remained static during the last decade. Scientists (with rare exceptions such as Sir John Eccles) so far have discovered Popper of the early methodological period, the Popper of *The Logic of Scientific Discovery* and of *Conjectures and Refutations*. During the last decade Sir Karl (Popper, 1968*a*, *b*) has been developing exciting new theories which in many respects depart from his earlier philosophy of science. I say 'exciting' advisedly, for what he has proposed is a new metaphysics known as the *three worlds doctrine*, or the third world doctrine (the third world consisting of intelligibles). In this metaphysics, which is also a philosophy of science, much less emphasis is placed on refutability and falsifiability than in his earlier methodological period.[2]

Furthermore, refutability and falsifiability have undergone strange metamorphoses during the last decade, mainly under the impact of Thomas Kuhn's *The Structure of Scientific Revolutions* (1962). It is *now* admitted that theories are hardly ever refuted in actual scientific practice. We cling to them tenaciously in spite of unfavourable empirical evidence. Theories, like old soldiers, fade away rather than being killed on the scientific battlefield. Let us notice a fundamental point: if the principle of refutability is questioned, so is the idea of demarcation of science from non-science, so is the entire rationality of science as based on refutability of scientific theories; and so is in question any postulate of objectivity which attempts to derive its sustenance from the idea of refutability.

In a number of encounters and counter-encounters involving Popper, Kuhn, Feyerabend, Lakatos, J. O. Wisdom, Toulmin and others during the last ten years, philosophers of science seem to have unwittingly out-argued themselves away from the notion of rationality (see in particular Lakatos, 1970). Every known and established tenet has been so undermined that we do not know what are *scientific* theories, what are *theories*, what is *rationality*, what is *science* itself. In undermining every established orthodoxy, Feyerabend, for instance, has proposed a kind of cognitive nihilism: anything goes. Other philosophers have tried to save at least the vestiges of the rationality of science. Karl Popper has introduced his doctrine of the three worlds which justifies the rationality of science and science itself in a novel way. Thomas Kuhn (1970) has argued that existing theories of rationality are not right and that we must readjust them to explain how science actually works as it does.

[2] It seems that there are four parts to Popper's philosophy which are not quite connected together:
(i) philosophy of natural science: *The Logic of Scientific Discovery; Conjectures and Refutations*;
(ii) philosophy of social science: *The Open Society and Its Enemies; The Poverty of Historicism*;
(iii) philosophical questions of evolution: 'Evolution and the tree of knowledge' (1961); *Of Clouds and Clocks* (1965);
(iv) the Third World Doctrine.

We have argued earlier that rationality needs further justification beyond the circular definition that 'rational is that which is scientific and scientific is that which is rational'. This further, extrinsic justification lies in the value of science for the whole species. It is the *survival value* of science or of knowledge of any kind that provides the ultimate justification for all rationalities. Rationality as the property of knowledge (epistemological rationality) must be ultimately justified by the value this knowledge provides for our survival in the broad sense of the term.

Usually rationality is embedded in a system of knowledge. When this system functions well, or at least adequately, then rationality is seen and defined as a cognitive category. Only at critical moments, at moments of cognitive crisis, at times when the ultimate justification of the value of science (or knowledge of a given kind) is required, can rationality be clearly seen as a normative category.

Rationality is a function of the *Weltanschauung*. Respective world views originate respective forms of rationality which in turn justify and support these world views. The function of rationality is thus to sustain a given intellectual and ideological formation by which it was generated. As a cognitive instrument for regulating the well-being of the whole intellectual formation (which a given world view represents), rationality has a clearly normative function: *the function of rationality is the maximisation of survival value*. If, in given circumstances, the rationality of a certain type is effective in aiding us in the overall scheme of survival, then it is reasonable to retain it. If it becomes defective in the overall scheme of survival, then it is reasonable to discard it. The medieval *Weltanschauung* as a form of rationality at one time played a positive part in the overall scheme of survival. Then it had to be replaced by a secular type of rationality, which served us well for quite a while. This type of rationality proved in turn insufficient for new cognitive needs and new social circumstances of man. Hence we have to depart from the rigid and constraining rationality of hard science to a more flexible and more sustaining evolutionary rationality; and perhaps in the future to a new form of rationality. In acknowledging the need and necessity for a new rationality, we acknowledge the end of an intellectual epoch.

Beyond chance and necessity

It was already in antiquity when Zeno recognised the profound dilemma arising from our tendency towards a static and abstract comprehension of phenomena and the dynamic nature of these phenomena. The famous arrow paradox illustrates the point. Let us consider point *P* on the way of the arrow which goes from *A* to *Z*. Now, at a certain time *T*, the arrow will be at the point *P*, but since the arrow moves, once it reaches point *P* at the time *T*, it will no longer be at the point *P*; so at the time *T*, the arrow *is* and *is not* at point *P*. We have got round the difficulty this paradox poses by reinterpreting the expression 'time *T*' and the meaning of *is*. There are two different meanings of the expression 'time *T*': in the first instance it is an accent of time; in the second instance a tiny stretch of time. Secondly, instead of the ambiguous term *is*, we now say that at the time *T* the arrow *passes* point *P*. The conceptual difficulties here, though considerable, are quite trivial in comparison with those we have to confront when analysing qualitative changes within organisms directed by goals and drives.

In order to be fair to the process of evolution when we describe it on the level of *Homo sapiens*, we have to introduce into our language *open-ended* concepts, *growth* concepts and *normative* concepts. Open-ended concepts will allow us to describe without distortion the fuzziness inherent in living organisms. Growth concepts will allow us to describe without distortion living phenomena in the process of change, particularly in the process of qualitative change. Normative concepts will allow us to describe without mystification living entities guided by specific values and directed to specific goals.

The rationality of evolutionary processes is tantamount to understanding and acknowledging the properties of life. On the level of the human species, and especially on the level of *Homo symbolicus*, life is lived within the normative realm, for it embodies drives and values which are not only gratuitous elements added to the physicochemical properties of matter but are often the guiding forces of the whole process of evolution.

The evolution of our understanding has been frozen at the level of the static model of traditional physics. We have physical concepts, chemical concepts, electromagnetic concepts which we recognise. In science we do not possess concepts that attempt to grasp and depict the higher levels of the complexity of matter: matter endowed with self-consciousness and with spirituality. There is thus a great discrepancy between the dynamic units of actual biological evolution and the static and petrified units of conceptual evolution. A truly evolutionary epistemology requires matching the states of conceptual evolution with the appropriate stages of biological evolution. Ernst Mayr (1961, 1969) and others have eloquently argued that the physical interpretation of certain rudimentary biological concepts makes a caricature of these concepts.

We have some ground for our hesitation to admit open-ended, growth and normative concepts to the realm of knowledge. Open-ended concepts seem to violate the principle of the invariance of meaning. Growth concepts seem to violate the law of identity $A = A$; normative concepts seem to violate the principle of objectivity and factuality. The invariance of meaning, the static identity of process and the factuality of phenomena are all unwritten premises on which scientific knowledge is built. The admission of open-ended, growth and normative concepts together with the admission of the evolutionist rationality would amount to a new kind of knowledge, namely normative knowledge. Normative knowledge is at present anathema to scientific knowledge. From Aristotle onwards, we separated the three realms once united—truth, goodness and beauty—and insisted that science is the pursuit of truth which has nothing to do with the other two realms. The so-called objectivity of science came to epitomise our search for truth. The principle of objectivity, so it appears to many, seems to exclude such entities as growth concepts and normative concepts.

Perhaps the time has come to reexamine our whole intellectual heritage from Aristotle onwards. Perhaps the separation of truth from values was premature, or only temporary. Perhaps a compassionate attitude towards living beings, and particularly to life endowed with consciousness, is as rational as an objective attitude. Indeed the development of a compassionate attitude would allow us to accommodate all the three new sets of concepts (open-ended, growth and normative) which are necessary for the understanding of evolution in its higher stages.

The compassionate postulate demands that we study and try to understand living organisms, from the amoeba to *Homo symbolicus*, *on their own terms*, on terms that are meaningful to their *lives* and *their* values, and *not* on the terms of objectivist science, which is tantamount to subjecting living organisms to microscopic scrutiny and clinical analysis. The value element in this conception of knowledge is quite transparent. But so is it in so-called objective science. Nowadays scientists themselves admit, including arch-positivists such as Jacques Monod and moderate positivists such as Peter Medawar, that the search for truth is based on a choice of values. Science is the quest for a certain kind of values. Why not make this quest more embracing and more fulfilling?

The compassionate postulate should not be considered as a mere philosophical extravaganza, for we do not know what kind of new illumination it might render without adopting it. To be truly scientific about it, one should adopt the compassionate attitude, then practise science within the compassionate framework for, shall we say, five years and only then conclude whether this attitude yields new and illuminating results or not. Would it be a waste of time? So many scientists waste so much time on so many trivial problems that they certainly could afford a new cognitive experiment—if only they were not so imprisoned by the existing framework of science.

Secondly, and more importantly, the compassionate postulate cannot be disregarded as something whimsical simply because it clashes with the objectivity postulate. The objectivity postulate itself and the whole idiom of science based on it are in question and in need of justification, indeed show every sign of exhaustion characteristic of paradigms which are declining and fading away.

Our predicament can be reformulated as follows: what kind of new concepts, new modes of knowledge must we develop in order to account for the socio-cognitive-cultural stage of human evolution? If we capture the last stage of our evolution adequately, we shall be able to capture earlier stages; but not conversely.

We are animals of sorts. But it seems that in the animal kingdom, the more the animal knows, the better it can cope with the environment. Thus, the enlargement of the animal's knowledge is for the animal life-promoting and life-enhancing. The knowledge animals possess is life-enhancing, thus normative; not a set of abstract categories, but normative rules for acting in order to preserve and enhance life. One wonders *why* human knowledge has ceased to be of this kind. Why does knowledge often signify mere information, why does it help the individual but little in enhancing his life, why does it so often interfere with the ends of human life? Our dilemma can be once again reformulated: what kind of knowledge will be illuminating for the understanding of life processes and at the same time serve as a guide to the good life? If we accept this reformulation, then the road to a normative paradigm for knowledge is open.

Cognitive functions, Piaget (1971) argues, are an extension of organic regulations and constitute a differentiated organ for regulating exchanges with the external world. Our exchanges with the external world, including those which are carried on through the vehicle of knowledge, must serve the ends of human life.

'Objective' science, within the framework of the new normative knowledge, will not be discarded or obliterated, but integrated and sometimes dissolved into larger structures. The compassionate attitude will not eliminate the 'objective' attitude, but will only subordinate it to its proper, rather modest place. Open-ended concepts will not render useless the quest for precise meaning, but will only make us aware that 'precise' concepts are at one end of the spectrum which is not more important than the whole spectrum itself. Growth concepts will teach us that within a dynamic system in which the rate of change is close to zero, we obtain static concepts which do not epitomise the dynamic system but only a very special aspect of it. Normative concepts will not make everything relative and based on personal values, but only elucidate what is the role of descriptive concepts within the realm of knowledge which is a normative instrument.

It would be trite and trivial to repeat that man is beyond the biosphere; that the emerging new qualities of physical matter cannot be explained by resorting to the worn-out devices of 'objective' science; that, for instance, molecular biology, even when supplemented by such concepts as 'natural selection', does not even begin to explain such phenomena as human language. 'Natural selection' is one of the key phrases we use in explaining evolution. 'Natural selection' is a sort of umbrella concept, seemingly 'objective' and 'scientific', but which in fact includes normative elements. Even if we discard Higher Plans and Inner Purposes and admit that selectivity can be limited to two aspects only—(i) processing of information on the basis of feedback from the environment, and (ii) that the second source of selectivity is problem-solving in past experience—we have not escaped normative categories. On the contrary, we have multiplied them. For the ideas of *feedback*, *information*, *environment*, *past experience* are normative categories; we cannot make sense of them in any purely physical system, but only in a system that admits values, evaluations and norms.

Our riddle thus is not the riddle of *evolution per se*, but rather the riddle of *comprehending* evolution. For our dilemma arises from the incongruity between our total comprehension of evolutionary processes and this part of our comprehension which can be subsumed under the 'official' (physical) knowledge which enables us to express only a part of our total knowledge. Hence the paradox: *we know* more than 'we do know'. This paradox is only apparent, for in the first instance the *we know* refers to our total comprehension, knowledge we actually possess; in the second instance the 'we do know' is equated with the boundaries and constraints of the scientific paradigm, the knowledge we can express.

To rid ourselves of this dilemma requires a much broader concept of knowledge which, as I have argued, will be normative knowledge based on the evolutionary rationality, open-ended concepts and growth concepts. This normative paradigm of knowledge will hopefully resolve another predicament, the present crisis between science and society.

The normative paradigm will also help to handle more specific problems such as Dobzhansky's dilemma: what lies beyond chance and predestination? Dobzhansky argues that we neither arose by accident, nor were we predestined to arise. In evolution chance and destiny are not the only alternatives. He suggests that if we consider chance to be the thesis, predestination to be the antithesis, then there is still the final element of the triad which still eludes us: the synthesis. It really does not elude us. For the synthesis is creative evolution in the sense that 'every new form of life that appears in evolution can, with only moderate semantic licence, be regarded an artistic embodiment of a new conception of living' (Dobzhansky, 1974). An artistic embodiment of a new conception of living is a *par excellence* normative category. We must not shirk the intellectual responsibility that is thrust upon us; if the nature of things requires the use of normative categories for an adequate description of things, then it is our responsibility to use these categories. Otherwise we misdescribe things. We sin intellectually by ignoring aspects of phenomena which partake in the process of life.

In summary, beyond chance and predestination there is the realisation:

—that our models of understanding are frail and limited; that they mirror our knowledge as well as our ignorance and prejudices, and thus may illuminate as well as mystify phenomena for us;

—that concepts, like tools, must be suitable for the task, and that we hinder our task from the start when we try to render dynamic processes through static concepts;

—that the transformation of life into hierarchically more and more complex forms is not less of a fact (but perhaps more) than the existence of electrons;

—that 'quality' as a concept is as clear or as mysterious as 'quantity'; none can justify itself in its own terms;

—that growth concepts, which explain or attempt to account for qualitative changes, are not less rational than physical concepts;

—that the function of rationality is to illuminate and not to obscure, and that compassionate understanding, if it leads to illumination, is perfectly rational, for rationality is to be judged not by *a priori* procedures but by the outcome of the rational process;

—that evolutionary epistemology does not shun 'mysteries' and subjects that are taboo because it is aware that there is more to life than physics;

—that between the Scylla of chance and the Charybdis of necessity, there is creative evolution, which is 'objective' because making sense of any theory of evolution or any conception of knowledge presupposes a certain architecture of life, which is conscious life, which is extra-physical life, which is cognitive life;

—that to be aware as a human being is an expression of creative evolution which is 'objective' because it can be tested against innumerable individuals who embody human awareness and use it as an indespensable tool for human life, and thus initiate and embody qualitative transformations on various levels of their existence.

The architecture of physics signifies the triumph of the simple and the reducible. The architecture of human life signifies the triumph of the complex and the irreducible.

References

Ayala, F. J. (1968). Biology as an autonomous science. *American Scientist*, **56**, 207–21.
Bronowski, J. (1970). New concepts in the evolution of complexity: Stratified stability and unbound plans. *Zygon*, **5**.
Dobzhansky, Th. (1974). Chance and creativity in evolution. In this volume.
Hampshire, S. (1972). *The Observer* (May).
Katchalsky, A. (1971). The isodynamics of flow and biological organization. *Zygon*, **6**.
Kuhn, T. (1964). *The Structure of Scientific Revolutions*. Chicago University Press, Chicago.
Kuhn, T. (1970). Reflections on my critics. In *Criticism and the Growth of Knowledge* (ed. I. Lakatos and A. Musgrave). Cambridge University Press, London.
Lakatos, I. and Musgrave, A., ed. (1970). *Criticism and the Growth of Knowledge*. Cambridge University Press, London.
Mayr, E. (1961). Cause and effect in biology. *Science*, **134**.
Mayr, E. (1969). Footnotes on the philosophy of biology. *Philosophy of Science*.
Monod, J. (1972). *Chance and Necessity*. Knopf, New York.
Orwell, G. (1972). *The Times Literary Supplement* (October).
Piaget, J. (1971). *Biology and Knowledge*.
Platt, J. (1970). Hierarchical growth. *Bulletin of the Atomic Scientists*.
Popper, K. (1963). On the sources of knowledge and of ignorance. In *Conjectures and Refutations*.
Popper, K. (1968*a*). Epistemology without a knowing subject. In *Logic, Methodology, and Philosophy of Science*, vol. III (ed. B. van Rootselaar and J. F. Staal).
Popper, K. (1968*b*). On the theory of objective mind. *Proceedings of the Fourteenth International Congress of Philosophy*. Vienna.
Simon, H. A. (1962). The architecture of complexity. *Proc. Amer. Phil. Soc.*
Simon, H. A. (1969). *The Science of the Artificial*. MIT Press, Cambridge, Mass.
Toulmin, S. (1971). In *The New York Review of Books* (16 December).
von Bertalanffy, L. (1950). The theory of open systems in physics and biology. *Science*.
von Bertalanffy, L. (1951). An outline of general systems theory. *Brit. J. Phil. Sci.*, **1**.
von Bertalanffy, L. (1968). *General Systems Theory: foundations, developments, applications*.
Weaver, W. (1948). Science and complexity. *American Scientist*, **36**.

Discussion

Rensch
I believe that *rationality* has proved to be the only means to get knowledge, because we can verify and falsify scientific results only by rational thinking. In my opinion the importance of rationality is based on the fact that our thinking has been adapted to the universal laws of causality and logic in the course of our phylogeny. Even animal behaviour had necessarily to be adapted to a certain degree to the laws of causality and logic. Nearly all our scientific knowledge can be regarded as an approach to extramental reality.

Skolimowski: Rationality and survival
In answer to Professor Rensch, I want to emphasise that I do not question the validity and importance of rationality as such. Rationality has been a most important vehicle in the acquisition and extension of knowledge—thus an important factor in the process of civilising and humanising man.

What I do question is the finality of any form of rationality. Logic is embedded in rationality and we do not question the validity of logic. But rationality goes far beyond the scope of logic. Little could be proved 'rationally' if we had to rely on logic exclusively.

Knowledge is (at least to a large degree) a human invention. There are no absolute categories of the mind as Kant suggested. There is no ultimate order of nature which we can or must acknowledge. For each order which we impose on nature is only a reflection of our knowledge through which we view nature. The history of human knowledge teaches us that nature admits a variety of orders, a variety of ways of interpreting it, a variety of conceptualisations of the structure of reality. None of these ways is absolute. Even if it is, we shall not know it.

Our rationality reflects the vicissitudes of our knowledge. And our knowledge—we should never forget this—is a supreme instrument in aiding the species in the process of survival. I question present (scientific) rationality because I question the survival value of present (scientific) knowledge. If it is true that physical science has become a cognitive strait jacket and as such is impeding our progress, then it is true that the rationality characteristic of physical science needs to be replaced by a more adequate form of rationality. Only God's rationality is immutable. Human rationality reflects the socio-cognitive development of human beings, is thus alterable, emendable, capable of improvement and changes, according to the changing needs of evolving human species.

14. Chance, Necessity and Purpose

CHARLES BIRCH

I cannot think that the world as we see it is the result of chance; and yet I cannot look at each separate thing as the result of Design. . . . I am, and shall ever remain, in a hopeless muddle. (Charles Darwin in a letter to Asa Gray, 26 November 1860; see Darwin, F. (1888), 378.)

I am inclined to look at everything as resulting from designed laws, with the details, whether good or bad, left to the working out of what we may call chance. Not that this notion *at all* satisfied me. I feel most deeply that the whole subject is too profound for the human intellect. (Charles Darwin in a letter to Asa Gray, 22 May 1860; see Darwin, F. (1888), 312.)

There must be something positive limiting chance, and something more than mere matter in matter, or Darwinism fails to explain life. (Hartshorne, 1962, 210.)

Charles Darwin, with characteristic honesty, rejected the notion that the world and all that is in it was the product of a predetermined plan or design. This was an inevitable and logical conclusion from his understanding of evolution as the product of natural selection of chance variations of living organisms. In rejecting a preordained creation, Darwin ran counter to the deism of the naturalists of the seventeenth and eighteenth centuries such as John Ray who believed that the order of nature implied deterministic design with an external designer and nothing left to chance.

However, while enthroning the role of chance, Darwin found himself in a dilemma. Unlimited chance can only produce chaos, not order. Hence his description of his own state of mind as being in a 'hopeless muddle' in his letter to Asa Gray of 26 November 1860 (Darwin, F., 1888, 378). And in another letter to Asa Gray of 11 December 1861 (Darwin, F., 1888, 382) he elaborated further his feelings:

If anything is designed, certainly man must be: one's 'inner consciousness' (though a false guide) tells one so; yet I cannot admit that man's rudimentary mammae . . . were designed . . . You say you are in a haze; I am in thick mud; the orthodox would say in fetid, abominable mud; yet I cannot keep out of the question. My dear Gray, I have written a deal of nonsense.

Darwinism, and all that has followed in what some call neo-Darwinism, provides a scientific foundation for ascribing a necessity to the concept of chance in the origin and evolution of life; the chance of the appropriate environment for the assemblage of complex molecules in a primeval soup, the chance errors of replication of DNA molecules to provide a basis for diversity of living organisms, the chance that some few of these will confer survival value on their possessors, the chance that environment might be less than completely hostile to all the individuals in a population, the chance that environmental change will be sufficiently gradual that genetic mechanisms can evolve that will cope with change, the chance that success in survival and reproduction can yet lead to extinction if exponental rates of increase proceeded indefinitely, the chance that from hostile environments some few individuals may, by chance, find themselves dispersed to those few remaining places where they can survive and reproduce and act as a new focus for colonisation.

No one who studies evolution can doubt the randomness of many events both in the organism and in its environment. In the context of evolutionary theory, randomness or chance does not mean that the events have no cause but that the events have no relation to the needs of the organism at the time they occur. They are in contrast to designed events. Events which are a necessity for survival, such as the unlikely occurrence of a particular mutation, may or may not occur. The occurrence is a matter of chance. Yet, as Monod (1971, 118) says,

> once incorporated in the DNA structure, the accident—essentially unpredictable because always singular—will be mechanically and faithfully replicated and translated: that is to say, both multiplied and transposed into millions or billions of copies. Drawn out of the realm of pure chance, the accident enters into that of necessity, of the most implacable certainties.

Hence the 'chance' and 'necessity' in the title of Monod's book.

Discredited and undiscredited teleology

Monod (1971, 112) concludes from these facts that since gene mutations are accidents, and since they constitute the only known source of modifications of the genetic constitution of organisms, it necessarily follows

> that chance *alone* is at the source of every innovation, of all creation in the biosphere. Pure chance, absolutely free but blind, at the very root of the stupendous edifice of evolution: this central concept of modern biology is no longer one among other possible or even conceivable hypotheses, the only one that squares with observed and tested fact. And nothing warrants the supposition—or the hope—that on this score our position is likely ever to be revised.

It is little wonder that Darwinism and all that has been built on it, including its faithful interpretation in terms of molecular biology by Monod, has seemed to discredit completely any teleological explanations. The fact that organs and organisms fulfil 'useful' purposes or ends is explained by the natural selection of those that happen to be so constructed; those that happen not to be so constructed fall by the wayside. The 'useful' ends of organisms are their continued survival and reproduction. But this is not an end of the organism. The mating animal is not concerned with the survival of the species. In so far as it may have an objective or end in view it is probably just to mate, certainly not the survival of the species.

This way of looking at things completely discredits the sort of teleological thinking that was common in Darwin's day. We are well rid of such deterministic notions of a single design of the universe complete in all details from eternity and executed by degrees with an inevitable relentlessness. Yet Darwin repeatedly declared in his letters that he could not see how chance alone could explain the world as an ordered whole since unlimited chance is merely chaos.

What sets the limits to chance? Darwin struggled to find an answer. He even suggested (in the quotation above, Darwin, F., 1888, 312) that the 'laws of nature' are designed with the details left to chance. But he was not at all happy about that idea. The 'mud' in which he felt himself to be immersed was the opacity of a deterministic world view, whether religiously conceived or otherwise, which left no room for chance and freedom. 'But I know that I am in the same sort of muddle' he wrote to Asa Gray in 1860 (Darwin, F., 1888, 378) 'as all the world seems to be in with respect to free will, yet with everything supposed to have been foreseen or preordained'. I think he might have enjoyed Sydney Carter's satirical song 'Friday morning' about the religious determinism familiar to Darwin, particularly the verse

> You can blame it onto Adam,
> You can blame it onto Eve,
> You can blame it on the apple,
> But that I can't believe.
> It was God who made the devil and the woman and the man,
> And there wouldn't be an apple if it wasn't in the plan!

Neither monolithic design nor rigorous natural law throws any light on the randomness of events in nature. One can only accept the one by ignoring the other. But how can randomness or chance *alone* explain order?

Of course it cannot! The neo-Darwinians, such as Monod (1971) and Dobzhansky (1967), logically argue that natural selection is an ordering process which brings design out of randomness. That too is indisputable. Far from being just a sieve, natural selection is more akin to a creative process (for example, Dobzhansky (1967), and Dobzhansky's contribution to this volume). Waddington's (1960) concept of the 'genetical assimilation' of 'acquired characters' is an interpretation of natural selection which is a creative process.

Is then natural selection in all its subtleties elaborated by Dobzhansky and Waddington and others, and known so much more clearly to us than it was to Darwin, the one and only ordering process in cosmic and biological evolution? Dobzhansky (in this volume) and Monod (1971) quite clearly consider that natural selection and *no other process* turns chance into order.

It is the one and only ordering process in organic evolution, given, that is, laws of physics and chemistry that the molecules in organisms conform to.

Charles Darwin was not so sure. He confessed that his mind remained confused on the matter. It is little appreciated that Thomas Henry Huxley went through the same sort of philosophical struggle as Darwin. However, he came out of it with a much clearer resolution of the problem of chance and order. Huxley (1888) argued that the Darwinian revolution had provided his generation with the means of freeing itself from the 'tyranny of certain solutions to philosophical questions' about nature.

> The doctrine of evolution is the most formidable opponent of all the commoner and coarser forms of Teleology. . . . The Teleology which supposes that the eye, such as we see it in man or one of the higher vertebrata, was made with the precise structure it exhibits, for the purpose of enabling the animals which possess it to see, has undoubtedly received its death blow (Huxley, 1888, 201).

He then goes on to explain how he believed that Darwinism did not sweep all teleology out of the window; there remained still a meaning of teleology at another level.

> Nevertheless, it is necessary to remember that there is a wider teleology which is not touched by the doctrine of Evolution, but is actually based upon the fundamental proposition of Evolution. This proposition is that the whole world, living and non-living, is the result of mutual interaction, according to definite laws, of the powers possessed by the molecules of which the primitive nebulosity of the universe was composed. If this be true, it is no less certain that the existing world lay potentially in the cosmic vapour. . . . The teleological and the mechanical views of nature are not, necessarily, mutually exclusive. On the contrary, the more purely a mechanist the speculator is, the more firmly does he assume a primordial molecular arrangement of which all the phenomena of the universe are the consequences, and the more completely is he thereby at the mercy of the teleologist, who can always defy him to disprove that this primordial arrangement was not intended to evolve the phenomena of the universe.

The seminal idea in this quotation from Huxley is embodied in two phrases, 'the existing world lay potentially in the cosmic vapour' and 'a primordial molecular arrangement of which all the phenomena of the universe are the consequences'. This concept of the primordial nature of matter is quite consistent with a universe in which chance and necessity also reign. I would go further and say that a universe of chance and necessity demands some such view of matter.

We tend to take for granted the potentialities nascent in the unevolved cosmos from its foundations and without which the emergence of life and consciousness would not have been possible. Potentialities are unseen realities (Birch, 1971, 1972). In Huxley's statement is the idea of the potentiality of the phenomena of the universe residing in the primordial particles. Far from being a preformist view this is an epigenetic view of nature. There is no necessity that what is potentially possible must eventuate. What eventuates depends upon chance and circumstance. The chance that life might ever arise may have been remotely small. And life, having arisen, does not inevitably produce self-conscious creatures. There is no necessity that either of these events must happen.

Huxley's statement embodies the proposition that our concept of atoms and molecules has to be such as to be able to sustain *all* the phenomena of the universe known to us. One set of phenomena presents us with our major problem. What set of phenomena?

The phenomena of consciousness

Science, as ordinarily understood, is concerned with those phenomena revealed through the five senses, particularly the eyes. From a host of observations on instruments of various sorts, the physicist infers the existence of electrons, atoms and so forth. But each of us has another sort of knowledge of one special part of the universe, of one special phenomenon of the universe, namely himself. He is a conscious being, has feelings and acts purposefully. Physics deals with public data, that is to say, events that can be observed by more than one observer. But there is a second sort of data which can only be 'observed' by one observer even when others are present. How I feel when something happens to me is private data. I shall argue that what we commonly refer to as feelings, thought, consciousness, or what might be described under the category of subjective events, are described in part by physics. To categorise in this way does not imply that distinctions have not been or should not be made within the general category, but that is hardly necessary for this paper. There is, to be sure, a physical aspect of electrical impulses, ionic movements and so on to subjective events. But subjective phenomena as such are not described by physics as we know physics at present. The physical description of human events leaves out of account the private aspect which is one's own feelings. Philosophical problems arise in science when we refuse to regard ourselves merely as sentient spectators of nature and when we take seriously the fact that we are a part of nature that has to be taken into account. I find myself therefore in complete agreement with Eccles (1971) when he says 'I readily accept all that (biologists) postulate in respect of my brain, yet I find my own consciously experiencing self not satisfactorily accounted for. . . . The further we progress in research, the more each of us will realise the tremendous mystery of our personal existence as a consciously experiencing being with imagination and a sense of values and a systematisation of knowledge.' Physical explanations of human events are uncomfortably inadequate and leave unanswered a whole host of questions, some of whose data are given in our conscious experience. There is a huge gap between what physics and chemistry describe and what the person experiences. Eccles (in this volume) and Sperry (1965) recognise consciousness as a causal agent which influences physicochemical events, and which cannot be reduced to them, at least in the way physicochemical events are at present conceived. The recognition of the subjective in this context has far-reaching implications, as T. H. Huxley was hinting in the quotation above.

If, as T. H. Huxley argued, the existing world lay potentially in the primordial atoms and molecules or whatever preceded them, then that potential must include subjective factors. It must include that aspect of the universe of which I have knowledge only in a private sort of way, namely my own feelings. This is the 'teleology' that T. H. Huxley considered necessary not in spite of but because of evolutionary doctrine. The impact of the argument first became real for me also from a zoologist. It was when Professor W. E. Agar said, in a lecture to students in the University of Melbourne, 'A few thousand million years ago there was primeval chaos, and now, here we are, and I think few people can really sustain a belief that a universe which produced life and man requires no different kind of explanation than would be demanded by a universe which did not do so.' He later published his Whiteheadian interpretation of biology in a book (Agar, 1943). It seemed to me then, as it seems to me now, that biologists have tended to espouse a view of the nature of the universe which is built on the phenomena of physics alone and has not taken into account the phenomena of which we are all acutely aware in private, our own feelings. These are part of the data of the universe. Extreme materialism was thus characterised by Schopenhauer as 'the philosophy of the subject that forgets to take account of itself' (quoted in Rensch (1971), 158). Because there are feelings, this universe would require a different interpretation from one in which feelings were not possibly in the very nature of that imaginary universe.

Concentration on the study of physics has tended to result in the doctrine that human experience is merely a property of certain material combinations characteristic of living organisms, appearing with the formation of these combinations and disappearing with their dissolution. Commonly associated with these ideas is the denial of the causal efficacy of all subjective processes. When I act purposefully I know that my action is preceded by an idea of the act. But according to this other theory my idea has nothing to do with causing the action to take place. The idea is merely the accompaniment of physical processes in the brain. My idea or purpose has the same kind of relationship to my action as the rattle of a railway train has to its motion. It accompanies the motion, but is not a cause of it. The argument for this notion runs somewhat as follows. Atoms and molecules have no subjective aspect; their combining and other activities are determined solely by their objective physical properties. Since the brain is composed of atoms, everything taking place in it must also be determined by their physical properties. The argument rests on the assumption, and it is no more than an assumption, that atoms have no subjective aspect. But this is not a disclosure of physics, since physics tells us nothing about the intrinsic nature of the primary particles. On the other hand, the physicist does not hesitate to attribute to atoms the objective or physical properties he finds necessary to attribute to them to explain the physical properties of matter in bulk. Gravitation is an example. Gravity is not predictable from atomic and nuclear physics.

It is a tiny, tiny correction to all the atomic equations because the size of the gravitational force is about 10^{-38} times the size of the electromagnetic forces in the atom. The result is that if our physicists had always been operating in a space ship, doing their atomic experiments, and they had never had a big planet around to pull apples down around their heads, they might have gone on for centuries without being able to predict gravity! (Platt, 1968, 453).

Atoms also compose brains. On the same principle should not the physicist ascribe to atoms a property which will be consistent with their function as elements composing the brain. Surely the only property of atoms which could provide what we are looking for is some form of subjective activity or what Rensch (1971) calls proto-mental properties. We have no direct evidence for asserting or denying this, nor is it conceivable that we ever could have such evidence. If, however, for other reasons, a more consistent and satisfying picture of the universe and its evolution can be constructed on the assumption that in their intrinsic nature the primary particules include, like ourselves, a subjective aspect, then physics has and can have nothing to say against it. Furthermore, a major philosophical problem of evolution disappears on this assumption. It gets rid of the necessity of inserting subjects into a previously completely objective world. To argue that the subjective is a property that only 'emerges' in matter in bulk is as illogical as to say that gravitation does not exist until it is measurable between large bodies and thence to state that gravitation 'emerges' as a quality of atoms in bulk. Gravitational attraction is a property of atoms as such.

There was a time when there was no thinking. So 'thinking' did emerge from something that was not 'thinking' as such. However, it is an entirely different proposition to assume that subjectivity 'emerged' out of objectivity. That seems to me completely impossible. To think of a world in which entities are not taking account of other entities is to think of no world at all. 'Taking account of' requires both that which takes account and that which is taken account of, and that is a subject and an object. We have an insider's view only of our own taking account and even then only of its conscious aspects. But in the rest of the world as well, processes of taking account of are going on, and we should recognise that they too have the subject–object structure.

But more, the phenomenon of consciousness is quite unaccounted for by the one and only ordering process acknowledged by neo-Darwinians, namely natural selection. On the materialist view of matter, organisms might well have been nonconscious robots making all the appropriate responses to predators and the like without consciousness coming in to the picture at all. No one has demonstrated any survival value to consciousness; see also Eccles (in this volume). It is not enough to say it is there; therefore it must have survival value. That is to make of natural selection if not a Pancreston then a Procrustean bed into which all phenomena of nature have to be fitted, absence of evidence notwithstanding! On the other hand, if subjectivity is present everywhere, consciousness is an elaborated form of something germane to the primordial stuff of the universe. It does not necessarily have to have survival value. It is there, period.

Physics interpreted in terms of biology

What science describes is *not* what we experience as subjects. The physicochemical reductionist believes (in faith) that the gulf will be filled by the discovery of more and more mechanisms of physics and chemistry of the sort of physics and chemistry that we now know. On the other hand, if we turn the problem around we can derive the objective world from the subjective. The physicochemical aspects of subjective phenomena are the outer observable aspects which science can study. Experience is the inner and private reality of the experiencer. It is the view from within. We only know what it is to be ourselves. We cannot know what it is to have the experience of another person or what taking account of is for a lion, let alone for an amoeba. But that is no reason to deny any such view from within for creatures other than ourselves. There are plenty of biological phenomena which can be explained in terms of present-day physics and chemistry. There are others which cannot as yet be so explained. The strict reductionist or, in Efrom's (1967) terms, the 'hard' reductionist, asserts that all biological phenomena eventually will be accounted for and described by the laws of physics which are entirely derived from a study of inanimate entities. The 'soft' reductionist claims that this is not necessarily true but that we should accept it because there seems to be no acceptable alternative. Nagel (1968, 448) points out that no one can be sure whether phenomena as yet unexplained by present-day physics and chemistry will eventually be so explained. 'One can make guesses as to what the future will bring, but the supposition that one can establish conclusively the impossibility or the inevitability of reducing biology to physicochemistry seems to me entirely mistaken.' This leaves the door to alternatives open. One serious alternative is that we need another sort of physics and chemistry than that solely derived from the study of inanimate matter, and that is not to say that present physics and chemistry are wrong but that they are incomplete. We need a physics visited by biology, that is to say, a physics which takes into account biological phenomena and is partly derived from the study of these phenomena. It was a conviction along these lines which led John Scott Haldane to remark 'If physics and biology one day meet, and one of the two is swallowed up, that one will not be biology' (quoted in Needham (1943), 204). It is the Whiteheadian proposition 'that neither physical nature nor life can be understood unless we fuse them together as essential factors in the composition of "really real" things, whose interconnections and individual characters constitute the universe' (Whitehead, 1934, 213).

Physics has undergone and is undergoing many transformations. Its fundamental issues are much more open-ended than many biologists seem to recognise. Bohm (1969, 29), who is a physicist, considers that the question of whether the basic laws of physics are mechanical or not is of the utmost potential significance for biology. He considers that it may well be possible that natural processes cannot in general be reduced to 'automorphisms' of mechanical order and that they may contain a really 'creative movement'. If this were to be true, then the 'hard' reductionist arguments would all fall to the ground. Physics has in the past fifty years been making gigantic strides away from mechanism. In Bohm's view the main steps in that direction have taken place particularly in quantum theory and to some lesser extent in statistical mechanics. He finds it very odd that, just when physics is moving away from mechanism, biology and psychology are moving closer to it.

Yet, as Grene (1969, 62) has pointed out, the chief model of scientific method is still that of the Galilean–Newtonian philosophy, the inadequacies of which were so clearly emphasised long ago by Whitehead (1925).

Physics may undergo another transformation when its concept of the fundamental particles takes account of the existence of subjective phenomena in matter in bulk (in living organisms). Such possible transformation of modern physics has been considered by Bohm (1969), Shimony (1965), Cochran (1971), Burgers (1965).

Perception and feeling as fundamental aspects of matter

My argument seems to me to have come to this: having accepted the randomness or chance nature of events, having ruled out a direct determination of evolutionary events by any extrinsic force, having accepted natural selection as one sort of ordering process, we are driven back, not to a world of non-responsive bouncing particles whose random collisions happened to have produced ordered phenomena including conscious phenomena. We arrive at another concept of order in matter in which events analogous to subjective events in man maintain the order by responding to previous events and anticipating immediate future events. The basic order of the universe, in this view, is the ordering of the fundamental particles and their further ordering into entities, better called 'organisms', which themselves act as units that take account of or feel both the past and future possibilities. 'Organisms' are those arrangements of matter that can receive, interpret and act on 'information' as individuals. An electron and an amoeba are organisms. A sand heap is not an organism, though it is composed of organisms (the atoms). The solar system is not an organism. A tree is not an organism but a society of organisms.

The teleological aspect of organisms is subjective aim, or the urge towards self-completion within each event. It is itself a cause just as mechanical events are causal. Such a view is at least as old as Socrates, who debated the merits of the argument as he faced his last hours awaiting death in 399 BC, as reported in Plato's dialogue *The Crito*. It involves a principle of interpreting the 'lower' levels of creation (such as atoms) in terms of the higher levels of organisation (such as men). It is to look at the process of evolution not only from the electron type up but from man down. To use the analogy of a river, it is that the river is known not only by studying it at the source but by what it becomes when it empties as an estuary into the ocean. This is the approach of A. N. Whitehead, Charles Hartshorne and many other process philosophers.

At every level of organism, from electron type to man, existence is constituted by 'social' relationships. Experience or 'taking account of' has an interpretation up and down the evolutionary line. 'There is no thing in the universe.' The basic metaphysical proposition is that all is process. This is the key metaphor. As Norman Mailer has said, 'If the universe is a lock then the key to that lock is not a measure but a metaphor'.

Evolution

A common picture of cosmic evolution starts with a chaos of particles which coalesce into atoms, then molecules and, later, life appear, then later at some stage mind and responsiveness appear. On the other hand, in the view of some scientists and philosophers, the universe is both subjective and objective throughout and always has been. There has been no moment in evolution when mind first made its appearance on the earth. From the beginning evolution has consisted in the elaboration of mental–physical events together.

The evolution of matter then becomes, as Rensch (1971, 272) says, 'the evolution of consciousness from protopsychical elements to self consciousness'. Rensch (1971, 240) then quite logically develops the notion of a psychophylogenesis analogous to morphological phylogenesis. 'We must recognise' says Rensch (1971, 298)

> that all 'matter' is protopsychical in character. When atoms and molecules came into being, the protophenomena of the elementary particles acquired new relationships and in this way new properties caused by system laws emerged. Sensations and recollections, *i.e.* primary mental images, only became possible at the complex structural level at which nerve cells developed. And self consciousness and a concept of the self could only result if these sensations merged into a stream of consciousness within a larger central nervous system. Finally, human self consciousness representing the highest phylogenetic stage yet reached enabled logical thought processes to arrive at an understanding of the universe.

The connection between events is provided by the taking account of relevant previous events and the selection of anticipated future states which are possibilities of the universe not yet realised. This is not a deterministic thesis, for there is no inevitability that the stream of events will move in this direction rather than that, and at every step 'accidents' can happen, which is to say that chance plays a role. The path of evolution is not fixed in advance by the end state to which it may appear to be moving. The specific outcome of any stage may be quite indeterminate though the possibilities are not infinite. At each level of organisation in the evolutionary history, be it from helium atoms to carbon atoms or reptiles to mammals, new possibilities become realised that could not have been realised at an earlier stage. A tremendous concatenation of chance events have to occur step after step before new entities can appear. Many are called but few are chosen (or selected).

It took the insight of Whitehead (1926, 157) to see that a thoroughgoing evolutionary philosophy is inconsistent with materialism. The 'stuff' from which a materialistic philosophy starts is incapable of evolution. It is only capable of rearrangements, much as a pack of cards can be dealt in a variety of arrangements after shuffling. Any one arrangement of a pack of cards has no more significance than any other arrangement, except in so far as the card player adds significance to certain arrangements by the rules he sets for the game. The significance comes from outside the pack. It is not inherent in the pack. An evolved entity has a novel significance by virtue of the internal ties between the bits that constitute it which make it an 'organism'. Organisms are not mere rearrangements of bits of stuff of the universe. In evolution, novelty is produced. This view is the antithesis of a preformist conception of creation which Dobzhansky (in this volume) equates it with when he says that evolution on the panpsychist view is 'a rather uninspiring story'. It would be if it were simply an unfolding of some preexisting scheme. On the contrary, creation involves trial and error, uncertainty and seemingly limitless possibilities.

Conclusion

Highly speculative though these ideas are, they are nevertheless consistent with the hypothetico-deductive method as an appropriate method not only in science but also in philosophy. This point has been admirably made by Whitehead (1929, 7) in his introduction to *Process and Reality*.

> The metaphysical first principles can never fail of exemplification. We can never catch the actual world taking a holiday from their sway. Thus for the discovery of metaphysics, the method of pinning down thought to the strict systematisation of detailed discrimination, already effected by antecedent observation, breaks down. This collapse of the method of rigid empiricism is not confined to metaphysics. It occurs whenever we seek the larger generalities. In natural science this rigid method is the Baconian method of induction, a method which, if consistently pursued, would have left science where it found it. What Bacon omitted was the play of a free imagination, controlled by the requirements of coherence and logic. The true method of discovery is like the flight of an aeroplane. It starts from the ground of particular observation; it makes a flight in the thin air of imaginative generalisation; and it again lands for renewed observation rendered acute by rational interpretation. The reason for the success of this method of imaginative rationalisation is that, when the method of difference fails, factors which are constantly present may yet be observed under the influence of imaginative thought . . . the success of the imaginative experiment is always to be tested by the applicability of its results beyond the restricted locus from which it originated. . . .

It remains to mention some other biologists who have found in such a view a metaphysical framework which makes sense for them of the facts of biology. Rensch (1971, 299) gives ten facts on which his 'panpsychistic' position is based. Waddington (1969) indicates how his Whiteheadian metaphysical position has been influential in the way in which he has proceeded in his biological research. I recall him saying to me on one occasion that his decision to spend his life in research on developmental genetics was largely the outcome of having, as an undergraduate, read all the major works of A. N. Whitehead. Sewall Wright (1953, 1964), starting from the recognition of his own 'stream of consciousness', derives matter from mind, not the other way around. His earlier mechanistic viewpoint was shaken by reading Bergson's *Creative Evolution*, though he was unable to accept this as a philosophy of science. But it was a result of reading Karl Pearson's *Grammar of Science* that he emerged a convinced panpsychist, finding later strong rapport with Whitehead's *Science and the Modern World*. Agar (1943) was led to a thoroughgoing Whiteheadian position as a result of seeking to interpret what he considered to be the two major problems of biology, behaviour and development. His little-known book is an exposition of how he derives this philosophy from his consideration of biology. Rendel (1961) finds in Russell's distinction between private and public data what he prefers to call subjective and objective knowledge. He regards the physical description of mental phenomena of brain function as incomplete and argues that it is incomplete in an analogous fashion at all levels down to the physicist's fundamental particles. Teilhard de Chardin's (1960) attribution of an 'inner' aspect to all entities places him squarely in this line of thinking. Thorpe (1965, 1969) is clearly opposed to a strictly mechanistic philosophy in a way congenial to that adopted here.

My conclusions, admittedly highly speculative, are:

—that living and non-living matter do not differ in any fundamental way, there is no definite state of complexity at which life appears and no definition stage of evolution when mind appears.

—that the universe is both subjective and objective throughout, that subjective and objective are two aspects of the one phenomenon, or two ways of looking at the one phenomenon. The subjective aspect of entities is the basis of the primordial order of the universe and provides the continuity between events which prehend past events and anticipate future events in a way analogous to memory and purpose.

—that randomness or chance characterise many events; they are not completely determined by any completely deterministic external agency.

—that much of the order of living organisms is to be understood in terms of natural selection of chance events, though the possibility of such order emerging is dependent upon the more fundamental order of the physical building blocks of the universe.

—that the possibilities of the universe were nascent in the primordial cosmos before it evolved and these potentialities, though not concrete yet, are unseen realities of the universe which are part determinants of its history.

—that, as Hartshorne argues in the quotation that heads this paper, there must be something more than mere matter in matter, or Darwinism fails to explain life.

There is also the possibility, not explored in this paper, that the subjective aspects of the universe themselves constitute some sort of all inclusive mind, a concept developed in the writings of A. N. Whitehead (for example, Whitehead, 1929), Charles Hartshorne (for example, Hartshorne, 1967), J. B. Cobb (for example, Cobb, 1965, 1967, 1969), and other process philosophers, all of whom accept the role of chance and accident in the order of the universe.

Acknowledgment

I am deeply indebted to Professor J. B. Cobb, School of Theology, Claremont, California, for his criticisms and suggestions which have enabled me to express more clearly much that would otherwise have been more obscure than it is.

References

Agar, W. E. (1943). *A Contribution to the Theory of the Living Organism.* Melbourne University Press.
Birch, L. C. (1971). Purpose in the universe: a search for wholeness. *Zygon,* **6**, 4–27.
Birch, L. C. (1972). Participatory evolution: the drive of creation. *J. American Academy of Religion,* **40**, 147–63.
Bohm, D. (1969). Some remarks on the notion of order. In *Towards a Theoretical Biology* (ed. C. H. Waddington), vol. 2. *Sketches.* Edinburgh University Press, 18–40.
Burgers, J. M. (1965). *Experience and Conceptual Activity: a philosophical essay based on the writings of A. N. Whitehead.* MIT Press, Cambridge, Mass.
Cochran, A. A. (1971). Relationships between quantum physics and biology. *Foundations of Physics,* **1**, 235–49.
Cobb, John B. (1965). *A Christian Natural Theology: based on the thought of Alfred North Whitehead.* The Westminster Press, Philadelphia.
Cobb, John B. (1967). *The Structure of Christian Existence.* The Westminster Press, Philadelphia.
Cobb, John B. (1969). *God and the World.* The Westminster Press, Philadelphia.
Darwin, F. (1888). *Life and Letters of Charles Darwin,* vol. 11. John Murray, London.
Dobzhansky, Th. (1967). *The Biology of Ultimate Concern.* The New American Library, New York.
Eccles, J. C. (1971). *Facing Reality.* Longman, London.
Efrom, R. (1967). Biology without consciousness—and its consequences. *Perspectives in Biology and Medicine,* **11**, 9–36.
Grene, M. (1969). Bohm's metaphysics and biology. In *Towards a Theoretical Biology* (ed. C. H. Waddington), vol. 2, *Sketches,* 61–71.
Hartshorne, C. (1962). *The Logic of Perfection and other essays in neoclassical metaphysics.* Open Court, La Salle, Illinois.
Hartshorne, C. (1967). *A Natural Theology for our Time.* Open Court, La Salle, Illinois.
Huxley, T. H. (1888). On the reception of *The Origin of Species.* In *The Life and Letters of Charles Darwin* (ed. F. Darwin), vol. 11, 179–204.
Monod, J. (1971). *Chance and Necessity: an essay on the natural philosophy of modern biology.* Alfred A. Knopf, New York.
Nagel, E. (1968). *In* Do life processes transcend physics and chemistry? *Zygon,* **3**, 442–72.
Needham, J. (1943). *Time: the Refreshing River.* George Allen and Unwin, London.
Platt, J. R. (1968). *In* Do life processes transcend physics and chemistry? *Zygon,* **3**, 442–72.
Rendel, J. M. (1961). Consciousness. *The Australian Scientist,* **1**, 149–53.
Rensch, B. (1971). *Biophilosophy.* Columbia University Press, New York.
Shimony, A. (1965). Quantum physics and the philosophy of Whitehead. In *Boston Studies in the Philosophy of Science* (ed. R. S. Cohen and M. W. Wartofsky), vol. 2, *In honor of Philipp Frank.* Humanity Press, New York, 307–30.

Sperry, R. W. (1965). Mind, brain and humanist values. In *New Views of the Nature of Man* (ed. J. R. Platt). University of Chicago Press.
Tielhard de Chardin, P. (1960). *The Phenomenon of Man*. Collins, London.
Thorpe, W. H. (1965). *Science, Man and Morals*. Methuen, London.
Thorpe, W. H. (1969). Retrospect. In *Beyond Reductionism* (ed. A. Koestler and J. R. Smythies). Hutchinson, London.
Waddington, C. H. (1960). *The Ethical Animal*. George Allen and Unwin, London.
Whitehead, A. N. (1925). *Science and the Modern World*. Macmillan, London.
Whitehead, A. N. (1929). *Process and Reality*. Macmillan, London.
Whitehead, A. N. (1934). *Nature and Life*. University of Chicago Press.
Wright, Sewall (1953). Gene and organism. *The American Naturalist*, **87**, 5–18.
Wright, Sewall (1964). Biology and the philosophy of science. In *Process and Divinity* (ed. W. L. Riese and E. Freeman). Open Court, La Salle, Illinois.

Discussion

Beckner

It is clear from our discussions, especially of the papers of Professors Birch, Thorpe and Eccles, that the existence of consciousness is a problem of special fascination for people interested in the significance of reduction in the biological sciences. A certain consensus was reached: that no plausible reduction of consciousness has, as yet, been achieved; and that no fundamental clarification is visible on the horizon, since we know so little about what is going on in the central nervous system.

I think our collective gloom about the nature of consciousness was a little deeper than the state of the subject warrants. I would like to add a few remarks specifically on one question that recurred in discussion, namely, 'What adaptive value (if any) does consciousness have?' The question is complex, so I can do no more than provide a sketchy indication of where an answer might lie.

Consider just one case of conscious experience, my hearing the voice of a person to my right and about six feet away. It is essential to notice that the conscious experience itself bears certain relations to two aspects of myself as a biological system. The first is that the experience certainly depends upon the extraction of information from the sound that affects the auditory system, and its encoding first in the auditory system itself and then in the brain. Secondly, this auditory information is certainly encoded in a form that makes it available for the control of the voluntary motor system. No mystery so far about the pattern of the flow diagram: sensory input, central encoding and processing, motor output (but of course lots of mystery about the details). And there is no problem in seeing that certain systems of this general type would have enormous adaptive value: the input–output functions need only determine behaviour that is conducive to survival and reproduction.

But the problem seems to be that in this picture all reference to the *experience* of hearing has fallen out. Does *that* have adaptive value, or is it just a by-product of physiological functions? Now I think it is possible to reintroduce experience into our picture, and to see why it certainly does have adaptive value. But I see no way to do this short of a straight *identifica-*

tion of the experience with some physiological entity which (*a*) possesses adaptive value, and (*b*) is necessary for conscious experience.

What I propose is the identification of experience with the maintenance (by information input) in the brain of a certain information content. I think we know by introspection enough about the general features of conscious experience to describe the general physiological features of this information content. It must be a neural state which embodies information derived from all active sensory systems; information derived from past sensory input; and maintained in a form that makes it an active part of a control device for the whole motor system.

I have, no doubt, not said enough to make the identification plausible. But it should be noticed that the features of the central information state whose maintenance constitutes experience do correspond to some familiar features of consciousness, especially to such facts as that hearing a voice is only an aspect of simultaneous perception of a *world*; and that I can employ *every* discrimination in my experience as the basis of voluntary action involving *any* muscle under voluntary control. In any case, the theory (spelled out) does entail the adaptiveness of consciousness.

15. Polynomistic Determination of Biological Processes

BERNHARD RENSCH

Causality

Investigations of biological processes ultimately lead not only to problems of chemistry and physics but also to epistemology. A wealth of very heterogeneous questions is therefore involved. If one tries to treat these problems in a brief lecture, the danger of misunderstanding is very great inasmuch as everybody disposes of another realm of scientific and philosophical knowledge. This will probably render a discussion of the following comments rather difficult. (For more detailed statements, *cf.* Rensch (1971).)

Analyses of biological events are normally carried out on the assumption that all processes are *causally* determined. As soon as the cause of an event has been found, it is considered to have been 'explained'. But there is a tendency to trace the chains of causes back to the realms of molecular or even microphysical processes (as for instance in the chain of respiration enzymes or in photosynthesis). Most important biological processes have already been traced, in the course of such investigations, to chemical and physical events, and many biological laws or rules have been established which can be understood as special cases of the general law of causality. The presupposition of general causality has thus nearly always proved its worth as a *heuristic principle*. We may therefore ask if we can assume that causality is sufficient to explain all biological processes.

It will not be necessary here to discuss vitalistic hypotheses. In my opinion they are only of historical interest now. In the beginning of our century the supposition of an 'entelechy' in the sense of Driesch was still possible, although this was only a term which circumscribed unknown interrelations and did not explain anything. It was also difficult to understand how a non-causal factor could produce an 'effect'. The same holds good for similar unknown factors which should have final effects. However, as we shall see, there exist other laws besides causality, which determine natural events. But let us discuss the principle of causality first.

The metabolism of mammals has been causally analysed to a far-reaching degree, including effects of vitamins, hormones, the processes of biological oxidation with the help of respiration enzymes and the change of chemical into mechanical energy. All the processes are determined by chemical laws of composition and physical laws of diffusion, osmosis, hydrodynamics, etc. Of course, there still remain many detailed unsolved questions, but probably no specialist doubts that further progress will only be attained by causal research. The processes of excitation and conduction of excitations have also been elucidated in this successful way. The analysis of the structure of DNA led to flourishing of molecular genetics, a new field of science, by which the causal chains of gene action and gene mutation became more comprehensible. We are also able to understand how a stream of causes, induced by DNA, led us to specific synthesis of polypeptides and proteins, and to further processes determined by enzymes. In this way the problem of entropy in individual development became comprehensible in a causal manner. Among evolutionary factors, particularly selection and isolation are typical causal effects. It is important that evolutionary trends (orthogenesis), including evolutionary progress, can also be explained by causal effects of selection (Huxley, 1942; Rensch, 1947, 1959; Simpson, 1949; Dobzhansky, 1970). Our experience in many fields of biology allows to hope that nearly all open questions will eventually be solved by causal explanations.

However, we must take into consideration that chemical and biological processes, leading to more complicated stages of integration, also show the effects of *systemic* relations which often produce totally new characteristics. For example, when carbon, hydrogen and oxygen become combined, innumerable compounds can originate with new characteristics like alcohols, sugars, fatty acids, formol and so on. Most of their characteristics cannot be deduced directly from the characteristics of the three basic types of atoms, although they are doubtless causally determined. New systemic relations appear also when chemical compounds react with one another in an organism. As living beings are individuals, all alterations of structures and all developments of new organs can have far-reaching systemic consequences. Alterations of the body size, for example, change most proportions of organs. When an increase in body size continues in the course of evolution (Cope's rule) even excessive structures and vestigial organs can arise. The specificity of many biological rules and laws comes about by such systemic relations (Rensch, 1947, 1960*a*, *b*).

Laws of probability and logic

We must ask now whether there are biological processes which are determined not only by causal but also by other laws. In my opinion, we have to assume that this is the case. It is a well-known fact that in hybridisation experiments we are only able to predict the distribution of characters in a statistical manner because the pairs of chromosomes are arranged 'at random' during meiosis, and maternal and paternal chromosomes can be combined in different manner in the daughter nuclei. The terms 'at random' or 'by chance' express that we do not know the particular causal factors which determine the position of the pairs of chromosomes during the metaphase. However, we do not doubt that the position is causally determined. In such cases we can agree with the definition of Voltaire (1785): 'Le hasard c'est rien. On a inventé ce mot pour exprimer l'effet connu de toutes causes inconnues.' *Hasard* or 'chance' would only mean that it is *practically* impossible to analyse the causal determination.

This common usage of the term 'chance' is, however, appropriate to veil the important question, whether such random distributions could not be determined also by laws of probability, which cannot be detected in single cases but only in larger series of cases.

We may perhaps explain the facts best by considering a parallel example. Let us assume that we are playing roulette and that there are only equal black and white fields. Nobody will doubt, that the movements of the ball are causally determined by the force of the first push, gravitation, the friction with the substratum and the air. In each case there exists the same chance of 50 per cent that the ball will come to a stop on a black or a white field. But a greater number of trials shows that it happens very much more frequently that the ball stops on a black field successively three times than ten times or fifteen times. Thus *laws of probability* exist, which have often been discussed since the publications of De Moivre (1718) and Laplace (1814). Von Mises (1936) defined probability as the limiting value of the relative frequency in the distribution of characteristics. The convergence towards a limiting value means, however, a lawfulness (compare for instance Keynes (1921), Reichenbach (1935), Popper (1935), Carnap (1962), Juhos (1970).)

The term 'chance', used in the same sense in English, French and German, can be understood that it implies these laws of probability. The French word *hasard* or the German *Zufall*, however, seems to indicate a total unlawfulness. Monod (1970) only used *hasard* in order to express total absence of law.

Our example shows that we can say that the arrangement of the pairs of chromosomes during meiosis is also not only determined by the effect of causality but by the laws of probability as well. The same holds good for the origin of 'spontaneous' gene mutations and other processes in which the 'chance' can only be stated statistically.

It may be possible that some gene mutations originate by microphysical events. Microphysical processes, too, may show causal connections, for instance when an elementary particle hits another particle and alters its energy and its course, or when a light ray becomes reflected. In other cases it is impossible to arrive at causality; and this will probably never be possible, because of the principle of indeterminancy of Heisenberg. The impossibility to make predictions in such cases leads specialists like Bohr, Heisenberg, Born, Jordan and von Weizsäcker to conclude that an invalidation of the causal law had been achieved (Heisenberg, 1927). Jordan (1945, 1971) speaks of liberty (*Freiheit*), and even tries to trace to it the spontaneity of living beings and the free will.

Predictability, however, does not belong to the definition of causality. The principle of causality only means that nothing will happen which is not determined by preceding processes (Hartmann, 1937; Weizel, 1954). The fact that we do not know these processes does not mean that they do not exist. 'Heisenberg has apparently confounded the validity of the principle of causality with its applicability' (Hübner, 1965). Microphysicists are nevertheless able to make predictions which correspond to statistical laws. In quantum mechanics a function of probability has therefore been introduced. As quantum mechanics is not only valid in terrestrial events, but is supposed to be valid in the universe, it becomes evident that the laws of probability are *universal laws*. Therefore, Juhos (1970) emphasised that the physical laws of probability are of the same kind as the stringent laws of causality. In the realm of microphysics we have to do with a double determination by the laws of causality and the laws of probability. In many microphysical events we are able to state only the latter ones. 'Die Beschränktheit unserer Erkenntnismittel zwingt uns das Surrogat der Statistik auf und man legt sie fälschlich den Redukten zur Last' (Ziehen, 1939, 365). ['The limitation of our means of knowledge forces us to be content with the substitute of statistics, and this is erroneously charged to the "reducts" (matter, or "the last something").']

We must also consider that all macrophysical processes come about by integration of microphysical processes. How should causality originate, if it were not already valid in the realm of microphysics? The statement of probabilities is based on the number of apposite, compared with the number of possible, cases. It is a matter of *logical* relations of facts. The laws of probability can therefore be attributed to the *logical laws*, which are not only laws of human thinking, as many philosophers assumed, but universal laws valid in the whole cosmos. The basic logical statement 'If two objects equal a third object then they are also equal to each other' must have been valid with regard to three atoms anywhere in the universe, even before thinking beings originated. Otherwise our physical and cosmological conceptions could not have been developed. In the course of phylogeny animals and men have become adapted to the universal logical laws. This is evidenced by the fact that animals are capable of recognising equalities, similarities and inequalities. Human thinking and acting had also to be adapted to these laws, for illogical thinking produced erroneous actions. All elementary mathematical sentences are statements of certain universal logical relations.

This holds good also for the mathematical formulation of biological relations, for instance for cybernetic formulas.

Other physical laws

Biological processes are also determined by other laws which cannot be traced back to causal or logical laws. The law of the conservation of energy, which is so important for problems of metabolism and change of energy, can be stated in terms of causal processes, but it cannot be deduced from them. Energy could perhaps also increase, as some cosmological hypotheses suppose. The law of conservation of energy states that in a closed system the sum of kinetic and potential energy remains constant. As organisms are open systems, the intake of energy from the environment and the output of energy have therefore to be taken in account. But several intracellular processes approach processes in closed systems. In a corresponding manner the law of conservation of spin and apparently also the law of conservation of baryons (mainly protons and neutrons) is independent of the causal laws. But these laws have only some meaning for biology, in so far as they concern the microphysical basis of all processes. The same holds good for the constants of the universe, which delimit, so to speak, the causal processes. The speed of light (c), being the maximum speed for the motion of energy, Planck's constant (h) assigning the minimum possible effect, the elementary charge (e) and the constant of gravitation (G). Also not dependent on the causal laws are the microphysical principles of symmetry summarised by the PCT theorem (parity of left and right, possibility of antiparticles for all particles, possibility of a reversal of time).

As the biological processes ultimately come about by integration of microphysical processes, we may state that they are determined by a number of laws in addition to causality and laws of probability and logic. Their basis is *polynomistic*. These laws became realised in an *epigenetic manner* corresponding to the evolution of matter. As long as a celestial body consisted only of gaseous matter, the special laws of falling bodies or of hydrodynamics could not be effective. Mutations could only appear after the prestages of life-containing nucleic acids and proteins had been evolved. Mendel's rules became possible only after sexuality of organisms had been developed, and Bergmann's rule only after warmblooded animals originated. However, all these special rules and laws are implicitly involved in the general laws of causality, logic and probability and the above-mentioned laws of microphysics.

These basic laws operate on the so-called 'matter', that is to say, an integration of elementary particles. Their basic characteristics are energy, equivalent to mass, charge, spin, speed, and spatial and temporal properties. Without this ultimate irreducible basis and the above-mentioned irreducible basic laws, the origin of the multifariousness of the 'material' world, including living beings and their alterations, would not be comprehensible.

Epistemological questions

Psychophysical correspondence

It is not only very probable but almost certain that higher animals experience psychic processes like man. Nobody doubts that a dog or a monkey sees, hears, feels pain and has memory. A complete discussion of reductionism must therefore include the question whether particular psychic laws exist and how psychic processes are connected with physiological processes. Tentative solutions of these problems depend on epistemological conceptions. They are very diverse, the more so as they are determined also by religious convictions, that is to say, by non-scientific influences.

Destruction of a single forebrain region of man, and corresponding extirpations in the brains of mammals, influences by drugs as well as electrophysiological investigations, have proved that most psychic processes directly correspond to physiological brain processes. These very intimate relations have led many philosophers to speak of a unity of body and soul. Brain processes happen in a causal succession, and the same holds good for the corresponding psychic phenomena. In those cases in which the brain processes are not yet sufficiently analysed the correspondence is normally presupposed in the sense of a heuristic principle.

The opinions concerning the relations of psychic phenomena to 'material' physiological processes are very diversified. Because the excitations from the sense organs to the central sensory and associative regions, and further on to the motoric regions, apparently follow a gapless course, the conception prevails that psychic phenomena *run parallel* to physiological brain processes, like two clocks synchronised by divine influence (comparison by Geulincx and Leibniz). But such conception of a 'preestablished harmony' encounters objections.

'Running parallel' means that brain processes would take place in the same manner when accompanying sensations, and mental images would be lacking. But then psychic processes would only be epiphenomena, that is to say, additional processes which need not necessarily occur for the behaviour of animals and man. They would have no selective advantage and no survival value. It would not be understandable why they developed and were maintained during phylogeny, and why they showed evolutionary progress. Another difficulty arises when we ask where psychic phenomena should have come from, phenomena arising only in the brains of animals and man, phenomena which for billions of years did not exist on our earth and probably also not in our solar system. We must recognise that psychic phenomena cannot be regarded as a new kind of physiological acquisition; 'running parallel' means something totally different, something immaterial which is not causally connected with other processes.

On the other hand, we cannot assume that psychical phenomena could interfere with physiological brain processes, as Descartes and some other philosophers and, among biologists, Eccles and Thorpe assume. This would require additional energy, which would infringe the law of the conservation of energy, and would not be compatible with the gapless course of causal nervous events. In my opinion it is therefore more apposite to suppose a panpsychistic identism, which is also suggested by several other considerations (see below). If we accept this conception, the problem of mutual influence of soul and body no longer exists. We have only to do with physiological processes which are, however, of psychic and partly protopsychic nature.

Psychophysical parallelism would also suppose that there exist innumerable laws of correspondence (laws of parallelity in the sense of Ziehen) between physiological brain processes and psychic phenomena. For instance, a certain pattern of excitation has to be lawfully connected with the sensation 'green', another with 'red', or 'sweet', or 'painful', and so on, and the same would have to be supposed for all kinds of mental images and particular thought processes. And again we must ask from which source this multitude of special laws which have nothing to do with causal or logical laws should have been developed.

And finally we have to consider that the assumption of a psychophysical parallelism is only a hypothesis, a hypothesis which is often accepted because it can be combined with the religious conviction that a 'soul' can leave the body after death and can continue to exist. However, the assumption that physiological brain processes and the psychical phenomena are *identical* is no more or even less hypothetical. On the basis of this conception the evolution of living beings capable of perceiving, remembering and thinking is more easily understandable, and leads to a philosophical picture which is relatively free from contradiction. It can be supported by different arguments.

Of course, such conception which identifies matter and mind can not be understood in the sense of an absolute materialism, as it was assumed at the end of the eighteenth and in the nineteenth centuries, by von Holbach, La Mettrie, Cabanis and others, who equated psychological and physiological processes, and recognised only matter. On the contrary our psychic phenomena are the only indubitable facts, whereas matter is only deduced from perceptions with the help of logical operations. The different psychic phenomena are primarily 'given' facts and can therefore not be defined.

An identism must therefore recognise this psychic basis of all our physical, chemical and biological knowledge, and can only be conceived as *panpsychistic identism*. This means that what is usually called 'matter', everything transsubjective and extramental, already has a *protopsychical nature*. I use the term protopsychical and protophenomenal in order to emphasise that this shall not yet mean sensations and awareness in the sense of self-awareness, but only prestages. By means of complicated integration of this protopsychic 'matter' in neurones and brains, sensations can arise by systemic conditions.

Protopsychic characteristics are therefore as dissimilar to real phenomena as the characteristics of carbon, hydrogen and oxygen are dissimilar to the characteristics of sugars, alcohols, fatty acids and other compounds coming about by their systemic integration.

Identistic conceptions have often been proposed in very different versions by great philosophers like Spinoza, Berkeley, Fechner, von Hartmann, Wundt, Bain, Spaulding, Riehl, Mach, Erdmann, Schuppe, Ziehen, Feigl and many others. They are speculative but, as already mentioned, by no means more speculative than psychophysical parallelism, and they can be supported by deductions from biological statements.

Psychophylogenesis

The existence of psychic phenomena in animals can only be deduced by conclusions of analogy. These conclusions are very sure with regard to higher animals. The structure of their sense organs and brains is very similar to those of man, and the neurones function in the same manner. There is no reason to doubt that a monkey or a dog would have visual, auditive, tactile and painful sensations, and that they would not have a memory. In former times the assumption prevailed that these phenomena corresponded to physiological processes of the cortex of the forebrain. However, this is very improbable because birds which do not have a cortex like mammals show no less sensation and brain accomplishments than do small carnivores or rats. In fish the main visual associations are connected with the midbrain. But their eyes function in a similar manner as those of higher vertebrates. They can learn to master a number of visual tasks simultaneously, they have a long-term memory and are capable of generalisation, that is to say, they act as if they have formed averbal concepts. They are subjected to optical illusions and simultaneous colour contrast like man. It is therefore highly probable that they have perceptions, mental images, memory and feelings.

But bees, bumblebees and Octopus also have well-functioning eyes and a brain with some hundred-thousand neurones (bees about 800 000; Witthöft (1967)), and the neurones work in more or less the same manner as in vertebrates. They can learn up to three or four visual tasks and master them simultaneously. They are capable of retaining learned tasks for at least several days. Bees are also subjected to the simultaneous colour contrast. Although the central nervous systems of these higher invertebrates have a structure which is totally different from brains of vertebrates, the neuronal functions, particularly the synaptic connections producing associative learning, are very similar. We have therefore no reason to deny that these higher invertebrates would not have sensations and memory (Rensch, 1970, 1973).

In lower invertebrates sense cells and neurones function in a similar manner as in higher animals. In annelids primitive associative learning has been proved, and they can show a learned reaction after one day. Turbellaria can learn at least in the sense of conditioned reflexes. Although conclusions of analogy are very much vaguer here, it seems at least possible to suppose that these animals still have sensations which must, however, not be combined in a continuous stream of perceptions.

The responses of unicellular organisms to light, touch, chemical stimuli and so on are also often called 'sensory reactions', because particular regions of the cell, for instance stigmata, tactile processes in Dileptus or Lacrymaria, and the region around the cytostome, react like sense organs. Ciliates also show 'habituation', but apparently no 'memory' (Machemer, 1966). If we take into consideration that sense cells and nerve cells of multicellular animals apparently originated by increase of reactivity to particular stimuli, the possibility exists to ascribe at least prestages of sensation to unicellular organisms. Such speculation can only be justified by the fact that animal phylogenesis was a continuous process and that psychic phenomena of higher animals have probably been developed from prestages. If we would assume that psychic phenomena are something immaterial, running parallel to physiological brain processes, we must ask once again from which root it could originate. If, however, we ascribe a protopsychical nature to all 'matter', it would become understandable that sensations could be developed from prestages by systemic integration, after neurones and nervous centres had been developed. This conception allows us to regard the whole animal phylogeny as a continuous process beginning with the causal development of prestages of life. Such speculation can be justified by the fact that other biological processes allow corresponding conclusions.

Psychontogenesis

We have to do with a very similar problem when we consider the development of psychic phenomena during ontogeny. The development of a human embryo begins with a fertilised egg cell showing no signs of anything 'psychical'. And the same holds good for the next multicellular stages. Only after sense cells, neurones and a brain have been developed, some weeks before birth at the earliest, we may suppose the existence of first sensations. If these phenomena would be something immaterial, only running parallel to the beginning function of the brain, from which root should they develop? Should we assume that at a certain stage an immaterial 'soul' has been inserted—a 'soul' which, however, shows characteristics of parents and grandparents? It seems to me to be more probable that protopsychical characteristics of the organic 'matter' have been integrated during the development of the nervous system and produced sensations and other psychic phenomena by systemic relations.

Genetical investigations and particularly twin research have clearly proved that many psychic characteristics are inherited, mental endowments as well as weaknesses, individual tempo, reserve, some types of homosexuality, etc. (*cf.* for instance Juda (1934, 1935), Dubitscher (1937), Newman *et al.* (1937), Skodak and Skeels (1949), von Verschuer (1954), Stern (1960), Shields (1962),

Erlenmeyer-Kimling and Jarvik (1963)). The investigations of Shields are of special interest because he compared monozygotic twins brought up apart and brought up together. He could clearly prove that intelligence, sociability, neuroticism and a variety of personal characteristics are inherited to a high degree. All these characteristics are transmitted by DNA. As molecules are capable of transmitting psychic particularities from one generation to the next one, it is obvious to suppose that 'matter' already has a protopsychic nature.

Psychophysical substance

The so-called 'matter' composing our body seems to be divided into two different kinds. The functions of a large part of our brain cells correspond to mental processes, whereas this is not the case with all other brain and body cells. The 'matter' which composes the psychophysical cells only consists, however, of compounds and atoms, which are known also in the nonliving state. All material which builds up our brain cells finally derives from the food—first from the nourishment which we receive by the placenta, later the components of our own blood. Besides, most compounds of our brain cells show a rather rapid turnover. As the chemical and electrophysiological functions of these compounds, ions and elementary particles correspond to psychic processes, it seems to me to be most easily understandable when we suppose that all matter already has a protopsychic nature. But psychic phenomena like sensations originate only in highly complicated stages of integration in neurones and brains, and they can only be experienced in coherent awareness.

Characteristics of matter

It is necessary to ask now whether and how far the panpsychistic and identistic conception which I have developed is compatible with the concept of matter in modern physics. Already Democritus, Plato (in the *Theaitetos*) and St Augustine have suggested, and Locke (1690) first proved in a more detailed manner, that matter has no 'secondary qualities', that is to say, no colour, smell, taste, hardness. All these characteristics only come about by processes in our sense organs and brains. Most men find it hard to free themselves from the naive realistic conception that all these qualities (particularly hardness and solidity) which seem to be the main characteristics of matter belong to matter itself. Whereas two revolutions of our picture of the world, started by Copernicus and by Lamarck and Darwin, have altered the conceptions of educated men rather rapidly, the epistemological revolution, mainly started by Locke three hundred years ago, has not influenced the thinking even of scientists very much. The reason is perhaps that matter, lacking colour, smell, solidity and so on, seems to become an unreal concept.

But now physics also denies that matter is something 'solid' and 'substantial' in the former sense. Mass is equivalent to energy. Mass can become radiation, and *vice versa*. Matter can only be defined as a complex of relations between energy, speed, charge, spin, and spatial and temporal characteristics. It is not necessary to assume a 'carrier' of these characteristics. All alterations of matter are determined by the laws of causality, logic, the laws of conservation, the universal constants and principles of symmetry.

Furthermore, we must consider that only our sensations and thought processes are indubitable primary reality. The physicist attains to a concept of matter, that is to say, of extramental reality, only by a process of reduction. He abstracts from secondary qualities and feeling tones and he pays attention to the possibility that the primary qualities, the spatial and temporal characteristics can probably be attributed to material processes only with certain restrictions. As optical illusions and hallucinations are also primary reality he will also exclude deductions from those phenomena which contradict the results from too many other sensations and thought processes. In this process of reduction from his psychic phenomena he does not reduce, however, from consciousness in a general sense. His concept of matter, which is rather far from common sense, is primarily still a psychic phenomenon, and it is a *hypothesis* when he concludes that extramental reality is not psychical or protopsychical. It is *as well* possible that extramental matter has a protopsychic nature and that consciousness results when a very complex integration of matter in neurones and brains leads to new systemic relations, to sensations and mental images. Only a stream of consciousness can be experienced by an individual. This means that there is no difference in principle between phenomenal reality and being (Berkeley: *esse* = *percipi*), but only between two systems of relationship: those which enter a stream of consciousness and those which remain extramental and are mainly protopsychical (Rensch, 1971, 298–9). This conception makes it easier to insert the physical knowledge in an identistic picture of the world. Already Eddington (1939) and Jeans (1943) had made similar suggestions.

Final remarks

The reduction of biological processes leads to an increasing degree to processes in the molecular and even microphysical realm. We see that all biological events, like all 'material' processes, are based ultimately on a relatively very small number of irreducible components, the alterations of which are determined by a limited number of basic laws. This means that the enormous multitude of inorganic and organic chemical compounds and biological structures, and the multitude of cosmological, geological, chemical and biological events, have developed in an *epigenetic manner*. The ultimate basis of all events is *polynomistic*. This conclusion is independent of special philosophical conceptions, whether they correspond to a functional materialism, a psychophysical parallelism or an identism.

If we base our conception on the indubitable facts of our own psychic phenomena, the analysis of which leads to scientific knowledge, we may come to a panpsychistic identism which in my opinion yields a picture of the world relatively free from contradiction. Such conception avoids the objections which arise when we assume an interplay between 'soul' and 'matter' (violation of the law of the conservation of energy), or a psychophysical parallelism (unintelligible development of the 'prestabilised harmony' of physiological brain processes and conscious processes; special difficulty in explaining inherited mental characteristics; evolution of psychic phenomena being epiphenomena without selection value). Besides, panpsychistic identism may have an important meaning because the analysis of physiological brain processes also explains the corresponding psychological processes.

The identification of mind and matter corresponds to the *realism* of many biologists who are not concerned with epistemological considerations and to philosophers who hold a more or less materialistic view. It is instructive to consider, for instance, the conception of the well-known Australian philosopher Smart (1963, 1964). I share most of his opinions, but he ignores too much the indubitable primary reality of psychic phenomena. A 'materialist' who does not deny that he experiences psychic phenomena is an identist!

Among modern biologists Wright (1964) is perhaps one of the few who holds a clear panpsychistic view. After discussing the question that 'the human mind is not to appear by magic', but must be a product of the development from the fertilised egg to the thinking individual, he wrote: 'The only satisfactory solution of the dilemma would seem to be that mind is universal, present not only in all organisms and in their cells, but in molecules, atoms and elementary particles'.

Eccles (1970), discussing Popper's theory of the 'three worlds' (1968), seems to hold a dualistic view. However, he also stresses intimate relations and a gradual development of 'world I', the world of matter, and 'world II', the world of consciousness. In chapter X, he writes: 'It is my contention that, just as in biology there are new emergent properties of matter, so at the extreme level of organised complexity of the cerebral cortex, there arises still further emergence, namely the property of being associated with conscious experiences. In some such manner we may eventually account for the emergence of self-awareness in Neanderthal men—and in all subsequent men.' In my opinion this view is not incompatible with panpsychistic identism, when we understand emergence as a new integration of components which leads to entirely new characteristics by systemic relations in a similar manner to the integration of the light metal sodium and the gas chlorine to produce the entirely new characteristics of salt.

But it is not probable that self-awareness is characteristic only of the human species, as Eccles believes. Dobzhansky also seems to hold this view. He wrote (1967, 68) that self-awareness is 'the most fundamental characteristic of the human species'. The findings of animal psychology, however, bring evidence that higher animals, particularly apes, not only have consciousness but apparently form a concept of their own self in a fairly accomplished manner. This can be deduced from different observations. Higher animals distinguish their body from the environment by proprioreceptive and reciprocous sensations, by seeing it, hearing their own voice, and by intense feelings which accompany some sensations, mainly pain, hunger, thirst and sexual feelings. They learn their rank in a family or in a society. Domesticated animals can learn their name. Apes can help one another, and they even learn to use gestures or material symbols for their own self (Gardner and Gardner, 1969; Premack, 1971). Chimpanzees can also plan most complicated successions of their personal actions which presuppose a certain concept of their own self (Köhler, 1925; Rensch and Döhl, 1968; Rensch, 1973).

During the phylogeny of animals the mental characteristics have surely been as gradually developed as morphological characters, particularly those of the brain. But this development led to entirely new abilities by systemic relations, mainly by the development of speech and the new vast possibilities of generalisation and complicated and logical thought processes. I believe that it is possible to assume that the whole evolution beginning with the development of our planet and leading to prestages of life, an apparently very improbable but possible event, further on to true organisms and at last to man, another very improbable but possible event, as he developed a language and created a culture, was a continuous process governed by eternal laws. During this development the main levels—inanimate matter, life, man—surely differ on principle. But they came about gradually by new systemic integrations.

Summary

(1) The enormous progress of biological knowledge, in many cases traced back to molecular and even microphysical processes, came about by presupposing the principle of causality, which always proved to be valid. It was, however, necessary to take into consideration that processes leading to more complicated stages of integration normally show effects of systemic relations which often produce totally new and often unpredictable characteristics.

(2) All events in the universe come about by a series of irreducible ultimate characteristics and laws. Biological processes are not only determined by causal laws, but also by laws of probability and logic, laws of conservation, the universal law of natural selection, and the lawfully effective microphysical constants. Most special causal laws became realised in an epigenetic manner corresponding to the succession of cosmological, geological and biological evolutionary stages.

(3) The immediate correspondence of psychic phenomena and physiological brain processes can be explained best by an identistic and panpsychistic conception, that is to say, by an identification of matter and mind. A psychophysical parallelism encounters the difficulty that psychic phenomena had to

be regarded as epiphenomena which would not have any selective value and would therefore not have been evolved, maintained and led to psychic progress. The assumption of a functional effect between psychic and physiological processes meets with the difficulty that then the law of the conservation of energy would be infringed upon.

(4) Panpsychistic identism presupposes a protopsychical nature of 'matter'. This assumption can be supported by deductions from statements of the phylogenetic and ontogenetic development of psychic phenomena and the effect of the psychophysiological substance in the brain which is composed of the same atoms as nonliving matter. The characterisation of matter by modern physics as a system of relations between basic constants and laws is compatible with a panpsychistic identism.

(5) Scientists who believe they are materialists, but do not deny that they experience psychic phenomena, are identists.

References

Carnap, R. (1962). *Logical Foundations of Probability*. Chicago.

Dobzhansky, Th. (1967). *The Biology of Ultimate Concern*. New American Library, New York.

Dobzhansky, Th. (1970). *Genetics of the Evolutionary Process*. Columbia University Press, New York, London.

Dubitscher, F. (1937). Der Schwachsinn. In A. Gütt, *Handb. Erbkrankheiten*, vol. 1. Leipzig (Thieme).

Eccles, J. C. (1970). *Facing Reality: philosophical adventures of a brain scientist*. Springer, New York, Heidelberg, Berlin.

Eddington, A. (1939). *The Philosophy of Physical Science*. Cambridge University Press.

Erlenmeyer-Kimling, L. and Jarvik, L. F. (1963). Genetics and intelligence: a review. *Science*, **142**, 1477–9.

Feigl, H. (1958). The 'mental' and the 'physical'. *Minnesota Studies in the Philosophy of Science*, vol. II.

Gardner, R. A. and Gardner, B. T. (1969). Teaching sign language to a chimpanzee. *Science*, **165**, 664–72.

Hartmann, M. (1937). *Philosophie der Naturwissenschaften*. Springer, Berlin.

Heisenberg, W. (1927). Über den anschaulichen Inhalt der quantentheoretischen Kinematik und Mechanik. *Z. f. Physik*, **43**.

Hübner, K. (1965). Zur gegenwärtigen philosophischen Diskussion der Quantenmechanik. *Philosophia naturalis*, **9**, 3–21.

Huxley, J. S. (1942). *Evolution: the modern synthesis*. Allen and Unwin, New York, London (2nd edition, 1963).

Jeans, G. (1943). *Physics and Philosophy*. Macmillan, New York.

Jordan, P. (1945). *Die Physik und das Geheimnis des organischen Lebens* (4th edition). Vieweg, Braunschweig (1st edition, 1941).

Jordan, P. (1971). *Wie frei sind wir? Naturgesetz und Zufall*. Fromm, Osnabrück.

Juda, A. (1934). Über Anzahl und psychische Beschaffenheit der Nachkommen von schwachsinnigen und normalen Schülern. *Z. Neurol.*, **151**.

Juda, A. (1935). Über Fruchtbarkeit und Belastung bei den Seitenverwandten von schwachsinnigen und normalen Schülern und deren Nachkommen. *Z. Neurol.*, **154**.

Juhos, B. (1970). *Wahrscheinlichkeit als Erkenntnisform*. (*Erfahrung und Denken* Bd. 34). Duncker und Humblot, Berlin.

Keynes, J. M. (1921). *A Treatise on Probability*. London, New York.

Köhler, W. (1925). *The Mentality of Apes*. Harcourt, Brace and World, New York.

Laplace, P. S. (1814). *Essai philosophique sur les Probabilités*. Paris.

Locke, J. (1690). *An Essay concerning Human Understanding*. In *Philosophical Works* (ed. J. A. John), vol. 1, London (1877).
Machemer, H. (1966). Versuche zur Frage nach der Dressierbarkeit hypotricher Ciliaten unter Einsatz hoher Individuenzahlen. *Z.f. Tierpsychol.*, **23**, 641–54.
Mises, R. von (1928). *Wahrscheinlichkeit, Statistik und Wahrheit*. Wien.
Mises, R. von (1931). *Wahrscheinlichkeitsrechnung und ihre Anwendung in der Statistik und theoretischen Physik*. Wien.
Moivre, A. de (1718). *The Doctrine of Chances, or a Method of Calculating the Probabilities of Events in Play* (cit. after Juhos, 1970). London.
Monod, J. (1970). *Le Hasard et la Nécessité*. Seuil, Paris.
Newman, H. H., Freeman, F. N. and Holzinger, K. J. (1937). *Twins: a study of heredity and environment*, Chicago University Press, Chicago.
Popper, K. R. (1935). *Logik der Forschung: zur Erkenntnistheorie der modernen Naturwissenschaft*. Springer, Wien.
Popper, K. R. (1968). On the theory of the objective mind. *Akten d. XIV. Interm. Kongr. Philos.*, vol. 1. Wien.
Premack, D. (1971). Language in chimpanzee? *Science*, **172**, 808–22.
Reichenbach, H. (1935). *Wahrscheinlichkeitslehre*. Leiden.
Rensch, B. (1959). *Evolution Above the Species Level*. Methuen, London; (1960*a*) Columbia University Press, New York.
Rensch, B. (1960*b*). The laws of evolution. In *Evolution After Darwin* (ed. S. Tax). Chicago University Press, Chicago, vol. 1, 95–116.
Rensch, B. (1967). The evolution of brain achievements. In *Evolutionary Biology* (ed. Th. Dobzhansky, M. K. Hecht and W. C. Steere). Appleton-Century-Crofts, New York, vol. I, 26–68.
Rensch, B. (1968). *Biophilosophie auf erkenntnistheoretischer Grundlage*. G. Fischer, Stuttgart.
Rensch, B. (1969). Die fünffache Wurzel des panpsychistischen Identismus. *Philosophia Naturalis*, **11**, 129–50.
Rensch, B. (1970). Evolution of matter and consciousness and its relation to panpsychistic identism. In *Essays in Evolution and Genetics in Honor of Theodosius Dobzhansky* (ed. M. K. Hecht and W. C. Steere). Meredith Co., New York, 97–119.
Rensch, B. (1971). *Biophilosophy*. Columbia University Press, New York, London.
Rensch, B. (1973). *Gedächtnis, Begriffsbildung und Planhandlungen bei Tieren*. Parey, Berlin.
Rensch, B. and Döhl, J. (1968). Wahlen zwischen zwei überschaubaren Labyrinthwegen durch einen Schimpansen. *Z.f. Tierpsychol.*, **25**, 216–31.
Shields, J. (1962). *Monozygotic Twins, brought up apart and brought up together*. Oxford University Press, London, New York, Toronto.
Simpson, G. G. (1949). *The Meaning of Evolution*. Yale University Press, New Haven.
Smart, J. J. C. (1963). *Philosophy and Scientific Realism*. Routledge and Kegan Paul, London.
Smart, J. J. C. (1968). *Between Science and Philosophy: an introduction to the philosophy of science*. Random House, New York.
Skodak, M. and Skeels, H. M. (1949). A final follow-up study of one-hundred adopted children. *J. Genet. Psychol.*, **75**, 85–125.
Stern, C. (1960). *Principles of Human Genetics* (2nd edition). Freeman, San Francisco.
Thorpe, W. H. (1965). *Science, Man and Morals*. Cornell University Press, Ithaca, New York.
Verschuer, O. von (1954). *Wirksame Faktoren im Leben des Menschen: Beobachtungen an ein- und zweieiigen Zwillingen durch 25 Jahre*. Steiner, Wiesbaden.
Voltaire (1785). *Philosophie générale: Metaphysique, Morale et Théologie*. In *Oeuvres complètes*, vol. 40. Soc. Littér. Typograph., Paris.
Weizel, W. (1954). Das Problem der Kausalität in der Physik. *Arbeitsgemeinschaft f. Forsch. Nordrhein-Westf. H.*, **43**, 37–62.
Witthöft, W. (1967). Zahl und Verteilung der Zellen im Hirn der Honigbiene. *Z. Morph. Ökol. Tiere*, **61**, 160–84.
Wright, S. (1964). Biology and the philosophy of science. In *Process and Divinity* (ed. R. Reese and E. Freeman) (The Hartshorne *Festschrift*). Open Court, La Salle, Ill., 101–25.
Ziehen, Th. (1934, 1939). *Erkenntnistheorie* (2nd edition). G. Fischer, Jena.

Discussion

Shapere

Contemporary physics attributes certain intrinsic properties to the elementary material particles—for example, mass, charge, spin, strangeness, baryon number. In the case of all these properties, their behaviour in particle decays and interactions is precisely specified, each particle which has the property being assigned a numerical value of that property, which is then governed in decays and interactions by conservation principles such as conservation of baryon number. (Of course, not all particles have all the fundamental properties or obey all the conservation principles.)

What Birch and Rensch are saying seems to amount to this: that the list of such intrinsic properties as given by contemporary physics is incomplete—that there is another property of all matter, which they call 'consciousness'. Now there is no doubt that the list of properties attributed to matter at the present or any other time might be incomplete; thus most of the presently-assigned properties have been added in the twentieth century in order to account for various observed aspects of particle behaviour. And it might be that, at some future date, it might be found appropriate, on grounds of its particular explanatory function, to christen some such newly-found property 'consciousness'.

But the suggestion of Birch and Rensch fails to satisfy two conditions which it must meet if it is to be entertained as a serious scientific hypothesis. First: the *need* of admitting consciousness as a fundamental property of matter has not been established. It may yet turn out that consciousness (conscious behaviour) will be explained in terms of more fundamental properties of matter, perhaps in terms of those already known, just as have many widespread properties such as colour, wetness, solidity, etc. And secondly, if and when it does become appropriate to introduce 'consciousness' as a fundamental property of matter, we will have to know—as we do in the case of other such properties—how to operate with it precisely in calculations. For example, if physical explanation continues to follow its present pattern, a 'consciousness number' would presumably be assigned to each type of particle, and the behaviour of that number would be governed by an appropriate conservation principle. Until these two conditions, of explanatory need and precision, are satisfied, the suggestion that the present list of fundamental properties is incomplete because it omits 'consciousness' is purely gratuitous speculation.

It is futile to argue that scientific method, conceived as 'hypothetico-deductive', allows the proposal of any hypothesis whatever, or in particular of this one. For—many contemporary philosophers of science notwithstanding—there are constraints on the kinds of hypotheses that are scientifically admissible as worthy of serious consideration, and among these are the two conditions mentioned here, which the hypothesis of Birch and Rensch fails to satisfy.

Rensch

I believe that there may be some misunderstanding. I did not speak of normal 'consciousness' of matter, but of a 'protopsychical nature' of matter. This protopsychical essence is not assumed as a property *besides* other properties, and we can not make 'precise calculations' about it, because it is supposed to be the fundamental *nature* of matter. As I suppose—like many philosophers—that mind and matter must be regarded as being identical, specific protopsychic properties are assumed to be the same as the physical properties of elementary particles like energy, charge, spin, etc. Different combinations of elementary particles yield new protopyschical characteristics by new systemic relations.

Although we have a certain inductive basis by statements of phylogeny and ontogeny of psychic phenomena and of brain functions, a panpsychistic identism must mainly be found in a deductive manner, because we can make exact statements only about properties of consciousness in the last stage of their phylogenetic development, in *Homo sapiens*. We can therefore only speculate about their phylogenetic and ontogenetic development on the base of conclusions by analogy. But as we are absolutely sure that such developments took place during phylogeny and takes place during ontogeny, we are forced to consider this problem when we want to get a satisfying ontological picture. Physical matter, on the other hand, can be analysed from molecules 'down' to elementary particles and their properties, and we can make exact statements about an upward development to atoms, molecules, aggregates of macromolecules and living matter.

Of course, panpsychistic identism is 'only' an epistemological theory, but it is in no way more hypothetical than a psychophysical parallelism or functionalism. But whereas substantiated objections against these two dualistic theories can be raised, panpsychistic identism is free from such contradictions. It supposes a continuous development of the basically protopsychic matter from inorganic compounds to life, and to higher animals and man, guided by universal laws. The materialistic view does not differ from panpsychistic identism very much, provided that their advocates do not deny that they experience psychic phenomena.

Birch

Nowhere do I say, nor does Rensch, that 'consciousness' is a property of all matter. It is a characteristic of matter organised in man. That is the subjective aspect of human life. My paper embodies the proposal that it is necessary to posit a subjective aspect to all entities that can be called individual entities such as atoms and electrons. The subjective aspect of an electron is not consciousness for that is probably associated only with a complex nervous system. By subjective aspect I mean what the entity is to itself in its relation to other entities. It has nothing to do with adding to the list of physical properties in the way Shapere suggests; that is to confuse the subjective and the objective. The question Shapere has to answer is: could anything exist if everything did not have a subjective aspect as well as the objective properties described by physics? I think not. That is the explanatory need and it has nothing to do with objective physical properties as such.

Shapere

Birch and Rensch speak of a 'subjective aspect' of all entities (Birch), a 'protopsychical nature' of matter (Rensch). That they did not *identify* this 'subjective aspect' or 'protopsychical nature' with 'consciousness' is irrelevant to my point. (I might better have given the name 'protoconsciousness' to the 'aspect' or 'nature' of which they spoke.) The point is that, whatever we call this 'aspect' or 'nature' that is ultimately responsible for consciousness, it must either be one of the fundamental properties of matter or it must follow from one or more of those properties as a causal consequence. Professor Rensch says that his protopsychical property 'is not assumed as a property *besides* other properties . . . because it is supposed to be the fundamental *nature* of matter'. But what does this mean? He does declare it to be a property. Presumably it does not follow from the other (known or unknown) properties of matter, since he considers it to be the 'fundamental nature' of matter. Then do the other properties, or some of them, follow from it? Such an assumption is completely gratuitous. Attempts to rely on the concept of the 'natures' of things as an explanatory concept have long been rejected by most philosophers not only as being obscure, but as being non-explanatory. Professor Rensch's proposal adds one more chapter to that dismal history.

To Professor Birch, this 'subjective aspect' has to do with 'what the entity is to itself in its relation to other entities'. This seems to me quite obscure; in so far as I can make any sense at all out of it, I would suppose that even if the 'subjective aspect' is or arises from relations between things, the precise character of that relation must have its ground in the things related. What then is the ground (or are the grounds) for such relations? Either they reside in properties of the entities related (and therefore either those currently known, or an additional one or ones which we might call 'protoconsciousness'), or else they are effects thereof.

But if the protoconsciousness property, on either Birch's or Rensch's view, is (or is grounded in) one or more of the currently known physical properties, or else is a property in addition to these, then the consequences to which I referred in my original comment follow.

16. Scientific Reduction and the Essential Incompleteness of All Science

K. R. POPPER

I

The thesis from which I start* is that, for a conference convened by biologists, the outstanding questions of reduction are three:

(1) Can we reduce, or hope to reduce, biology to physics, or to physics and chemistry?

(2) Can we reduce to biology, or hope to reduce to biology, those subjective conscious experiences which we may ascribe to animals and, if question (1) is answered in the affirmative, can we reduce them further to physics and chemistry?

(3) Can we reduce, or hope to reduce, the consciousness of self and the creativeness of the human mind to animal experience, and thus, if questions (1) and (2) are answered in the affirmative, to physics and chemistry?

It is obvious that the replies to these three questions (to which I shall turn later in the paper) will partly depend on the meaning of the word 'reduce'. But for reasons which I have given elsewhere (1945, vol. II, 9–21) I am opposed to the method of meaning analysis and to the attempt to solve serious problems by definitions. What I propose to do instead is this.

I will begin by discussing some examples of successful and unsuccessful reductions in the various sciences, and especially the reduction of chemistry to physics; and also the residues left by these reductions.

In the course of this discussion, I will defend three theses. First, I will suggest that scientists have to be reductionists in the sense that nothing is as great a success in science as a successful reduction (such as Newton's reduction

* I am greatly indebted to David Miller and Jeremy Shearmur for their comments on an earlier draft of this paper.

—or rather explanation[1]—of Kepler's and Galileo's laws to his theory of gravity, and his correction of them; see my (1957)). A successful reduction is, perhaps, the most successful form conceivable of all scientific explanations, since it achieves what Meyerson (1908, 1930) stressed: an identification of the unknown with the known. Let me mention however that by contrast with a reduction, an explanation with the help of a new theory explains the known—the known problem—by something unknown: a new conjecture (see my (1963), 63, 102, 174).

Secondly, I will suggest that scientists, whatever their philosophical attitude towards holism, *have* to welcome reductionism as a *method*: they have to be either naive or else more or less critical reductionists; indeed, somewhat desperate critical reductionists, I shall argue, because hardly any major reduction in science has ever been *completely* successful: there is almost always an unresolved residue left by even the most successful attempts at reduction.

Thirdly, I shall contend that there do not seem to be any good arguments in favour of *philosophical* reductionism, while, on the contrary, there are good arguments against essentialism, with which philosophical reductionism seems to be closely allied. But I shall also suggest that we should, nevertheless, on methodological grounds, continue to attempt reductions. The reason is that we can learn an immense amount even from unsuccessful or incomplete attempts at reduction, and that problems left open in this way belong to the most valuable intellectual possessions of science: I suggest that a greater emphasis upon what are often regarded as our scientific failures (or, in other words, upon the great open problems of science) can do us a lot of good.

II

Apart from Newton's, one of the very few of the reductions known to me which have been almost completely successful is the reduction of rational fractions to ordered pairs of natural numbers. (That is, to relations or ratios between them.) It was achieved by the Greeks, although one might say that even this reduction left a *residue* which was dealt with only in the twentieth century (with the successful reduction, by Wiener (1914) and Kuratowski (1920), of the ordered pair to an unordered pair of unordered pairs; moreover, one should be aware that the reduction is one to *sets* of equivalent pairs, rather than to pairs themselves). It encouraged the Pythagorean cosmological research programme of arithmetisation which, however, broke down with the

[Note added in proof] In the text of this paper I have disregarded—perhaps carelessly, or because I dislike terminological minutiae—the distinction that can well be made between explanation in general, and reduction in the sense of an explanation by way of an established or more 'fundamental' theory. A distinction of major interest, I suppose, would be that between an explanation of something known by a new (unknown) theory on the one hand, and a reduction to an old (known) theory on the other. I have added an allusion to this distinction here to the text, and also the footnotes and Postscript, in the hope of avoiding possible misunderstandings.

proof of the existence of irrationals such as the square roots of 2, 3, or 5 (*cf.* my (1950), vol. I, ch. 6, n. 9; and (1963), ch. 2, 75–92). As I have suggested (*loc. cit.*) Plato replaced the cosmological research programme of arithmetisation by one of geometrisation, and this programme was carried on successfully from Euclid to Einstein. However, the invention of the calculus by Newton and Leibniz (and the problem of excluding the paradoxical results which their own intuitive methods failed to exclude) created the need for a new arithmetisation—a new reduction to natural numbers. And in spite of the most spectacular successes of the nineteenth and early twentieth centuries, we can say now, I believe, that this reduction has not been fully successful.

To mention only one unresolved residue, a reduction to a sequence of natural numbers or to a set in the sense of modern set theory is not the same as, or even similar to, a reduction to a set of equivalent ordered pairs of natural numbers. As long as the idea of a set was used naively and purely intuitively (as by Cantor) this was perhaps not obvious. But the paradoxes of infinite sets (discussed by Bolzano, Cantor and Russell) and the need to axiomatise set theory showed, to say the very least, that the reduction achieved was not a straightforward arithmetisation—a reduction to natural numbers—but a reduction to axiomatic set theory; and this turned out to be a highly sophisticated and somewhat perilous enterprise.

To sum up this example, the programme of arithmetisation—that is, of the reduction of geometry and the irrationals to natural numbers—has partly failed. But the number of unexpected problems and the amount of unexpected knowledge brought about by this failure are overwhelming. This, I shall contend, may be generalised: even where we do not succeed as reductionists, the number of interesting and unexpected results we may acquire on the way to our failure can be of the greatest value.

III

I have briefly hinted at the failure of the attempted reduction of the irrationals to natural numbers, and I have also indicated that attempts at reduction are part of the activities of scientific and mathematical explanation, simplification and understanding.

I will now discuss in a little more detail the successes and failures of attempted reductions in physics, and in particular the partial successes of the reduction of macrophysics to microphysics and of chemistry to both microphysics and macrophysics.

IV

I have elsewhere (1956, 365–72; 1963, ch. 3, 103–7) given the name 'ultimate explanation' to the attempt to explain or reduce things by an appeal to something that is neither in need of, nor capable of, further explanation, more especially an 'essence' or a 'substance' (*ousia*).

A striking example is the Cartesian reduction of the whole of the physics of inanimate bodies to the idea of an *extended substance*; a substance (matter) with only one essential property; that is, spatial extension.

This attempt to reduce the whole of physics to the one apparently essential property of matter was highly successful in so far as it gave rise to an understandable picture of the physical universe. The Cartesian physical universe was a moving clockwork of vortices in which each 'body' or 'part of matter' pushed its neighbouring part along, and was pushed along by its neighbour on the other side. Matter alone was to be found in the physical world, and all space was filled by it. In fact, space too was reduced to matter, since there was no empty space but only the essential spatial extension of matter. And there was only one purely physical mode of causation: *all causation was push*, or action by contact.

This way of looking at the world was found satisfactory even by Newton, though he felt compelled to introduce by his theory of gravity a new kind of causation: *attraction*, or action at a distance.

It was the almost incredible explanatory and predictive success of Newton's theory which destroyed the Cartesian reduction programme. Newton himself, I have elsewhere conjectured (1969, 107, n. 21), attempted to carry out the Cartesian reduction programme by explaining gravitational attraction by the 'impulse' (radiation pressure combined with an umbrella effect) of a cosmic particle bombardment (the attempt is usually linked with the name of Le Sage). But I also conjectured that Newton became aware of the fatal objection to this theory. Admittedly it would reduce attraction and action at a distance to push and to action by contact; but it would also mean that all moving bodies would move in a resisting medium which would act as a brake on their movement (consider the excess push of rain on the windscreen of a car over that on the rear window) and which would thus invalidate Newton's use of the law of inertia.

Thus, in spite of its intuitive attractiveness, and in spite of Newton's own rejection as 'absurd' of the view that attraction at a distance could be an essential property of matter, the attempt at an ultimate reduction of attraction to push breaks down.

V

We have here our first and very simple example of a promising scientific reduction and its failure, and of how much one can learn by attempting a reduction and discovering that it fails.

(I conjecture that this failure was the immediate reason why Newton described space as the sensorium of God. Space was 'aware', so to speak, of the distribution of all bodies: it was, in a sense, omniscient. It was also omnipresent, for it transmitted this knowledge with infinite velocity to all locations at every moment of time. Thus space, sharing at least two characteristic properties of the divine essence, was itself part of the divine essence.

This, I suggest, was another attempt by Newton at an essentialist ultimate explanation.)

The Cartesian reduction may be taken as an illustration of my remark that for methodological reasons we have to attempt reductions. But it may also give an indication of the reason why I suggest that as reductionists we must not be sanguine but can be only somewhat despairing concerning the complete success of our attempted reductions.

VI

It is clear, I think, that the Cartesian attempt (which, if I am right in my historical conjecture, was also a Newtonian attempt) to reduce everything in the physical world to extension and push became a failure when it was judged against the success of Newton's theory of gravity. And the success was so great that Newtonians, beginning with Roger Cotes, began to look upon Newtonian theory itself as an ultimate explanation and thus at *gravitational attraction* as an essential property of matter, in spite of Newton's own views to the contrary. But Newton had seen no reason why *extension* (of his atoms) and *inertia* should not be essential properties of mass (*cf.* my (1956), 370, or (1963), 106*f.*). Thus we can say that Newton was clearly aware of the distinction, later stressed by Einstein, between inertial and gravitational mass, and of the problem opened by their proportionality (or equality); a problem which, because of the obscurantism of the essentialist approach, was almost lost sight of between Newton and Eötvös or even Einstein.

Einstein's Special Relativity theory destroyed the essentialist identity of inertial and gravitational mass, and this is the reason why he tried to explain it, somewhat *ad hoc*, by his principle of equivalence. But when it was discovered (first by Cornelius Lanczos) that Einstein's equations of gravitation led by themselves to the principle, previously separately assumed, that gravitating bodies move on a space–time geodesic, the principle of inertia was in fact reduced to the equations of gravitation and thus inertial mass to gravitational mass. (I believe that Einstein, though strongly impressed by the importance of this result, did not fully accept that it solved Mach's central problem—the explanation of inertia—in a more satisfactory way than the famous but far from unambiguous 'Mach principle': the principle that the inertia of each body is due to the combined effect of all the other bodies in the universe. To Einstein's disappointment, this principle was, at least in some of its interpretations, incompatible with General Relativity which, for a space empty of all bodies, yields Special Relativity, in which the law of inertia, contrary to Mach's suggestion, is still valid.)

Here we have what I regard as a most satisfying example of a successful reduction: the reduction of a generalised principle of inertia to a generalised principle of gravitation. But it has been rarely considered in this light; not even by Einstein, though he strongly felt the significance of a result which, from a purely mathematical point of view, could be regarded as elegant but not as particularly important. For the dependence or independence of an axiom within a system of axioms is in general not of more than formal interest. Why should it matter, therefore, whether the law of motion on a geodesic had to be assumed as a separate axiom or could be derived from the rest of gravitational theory? The answer is that by its derivation, the identity of inertial and gravitational mass was *explained*, and the former reduced to the latter.

In this way one might say that Newton's great problem of action at a distance (couched in the phraseology of essentialism) was solved not so much by the finite velocity of Einsteinian gravitational action as by the reduction of inertial matter to gravitational matter.

VII

Newton and the Newtonians knew, of course, about the existence of magnetic and electrical forces; and until at least the beginning of the twentieth century, attempts were made to reduce electromagnetic theory to Newtonian mechanics, or to a modified form of it.

The outstanding problem in this development was the reduction of *prima facie* non-central forces (Oersted forces) to central forces, the only ones which seemed to fit into even a modified Newtonian theory. The outstanding names in this development were Ampère and Weber.

Maxwell too began by trying to reduce Faraday's electromagnetic field of (lines of) forces to a Newtonian mechanism or model of the luminiferous ether. But he gave up the attempt (though not the luminiferous ether as the carrier of the electromagnetic field). Helmholtz also was attracted by a Newtonian and partly Cartesian reduction programme, and when he suggested to his pupil, Heinrich Hertz, that he should work on this problem, Helmholtz seems to have done so in the hope of saving the research programme of mechanics. But he accepted Hertz's confirmation of Maxwell's equations as crucial. After Hertz and J. J. Thomson, precisely the opposite research programme became more attractive—the programme of reducing mechanics to electromagnetic theory.

VIII

The electromagnetic theory of matter—that is, the reduction of both mechanics and chemistry to an electromagnetic theory of atomism—was strikingly successful from at least 1912, the year of Rutherford's planetary or nuclear atom model, until about 1932.

In fact, quantum mechanics (or 'the new quantum theory', as it was once called) was, until at least 1935, simply another name for what was then regarded as the final form of the reduction of mechanics to the new *electromagnetic theory of matter*.

In order to realise how important this reduction appeared to leading physicists even shortly before quantum mechanics, I may quote Einstein who wrote (1920; 1922, 24; see also my (1967) where I discuss the same point): '. . . according to our present conceptions the elementary particles [that is, electrons and protons] are . . . *nothing else* than condensations of the electromagnetic field . . . , our . . . view of the universe presents two realities . . . , namely, gravitational ether and electromagnetic field, or—as they might also be called—space and matter.'

Note the 'nothing else' which I have italicised because it is characteristic of reduction in the grand style. Indeed, to the end of his life, Einstein tried to unify the gravitational and the electromagnetic fields in a unified field theory, even after his view of 1920 had been superseded—or rather, had broken down (especially owing to the discovery of nuclear forces).

What amounts, essentially, to the same reductionist view was accepted at that time (1932) by almost all leading physicists: Eddington and Dirac in England and, besides Einstein, Bohr, de Broglie, Schrödinger, Heisenberg, Born and Pauli on the continent of Europe. And a very impressive statement of the view was given by Robert A. Millikan (1932, 46), then of the Californian Institute of Technology:

> Indeed, nothing more beautifully simplifying has ever happened in the history of science than the whole series of discoveries culminating about 1914 which finally brought practically universal acceptance to the theory that the material world contains but two fundamental entities, namely, positive and negative electrons, exactly alike in charge, but differing widely in mass, the positive electron—now usually called a proton—being 1850 times heavier than the negative, now usually called simply the electron.

This reductionist passage was written in the very nick of time: it was in the same year that Chadwick (1932) published his discovery of the neutron, and that Anderson (1933) first discovered the positron. Yet some of the greatest physicists, such as Eddington (1936), continued to believe, even after Yukawa's suggestion of the existence of what was to be called the meson (1935), that with the advent of quantum mechanics the electromagnetic theory of matter had entered into its final state and that all matter consisted of electrons and protons.

IX

Indeed, the reduction of mechanics and of chemistry to the electromagnetic theory of matter seemed almost perfect. What had appeared to Descartes and Newton as the space-filling essence of matter, and as Cartesian push had been reduced (as Leibniz had demanded long ago) to *repulsive forces*—the forces exerted by negative electrons upon negative electrons. The electrical neutrality of matter was explained by the equal number of positive protons and negative electrons; and the electrification (ionisation) of matter was explained by a loss of electrons from (or excess of electrons in) the planetary electron shell of the atom.

Chemistry had been reduced to physics (or so it seemed) by Bohr's quantum theory of the periodic system of elements, a theory which was ingeniously perfected by the use of Pauli's exclusion principle; and the theory of chemical composition, and of the nature of covalent chemical bonds, was reduced by Heitler and London (1927) to a theory of (homeopolar) valency which also made use of Pauli's principle.

Although matter was revealed to be a complex structure rather than an irreducible substance, there had never before been such unity in the universe of physics, or such a degree of reduction.

Nor has it ever been achieved again since.

True, we still believe in the reduction of Cartesian push to electromagnetic forces; and Bohr's theory of the periodic system of elements, though considerably changed by the introduction of isotopes, has largely survived. But everything else in this beautiful reduction of the universe to an electromagnetic universe with two particles as stable building blocks has by now disintegrated. Emphatically, we have learned an immense number of new facts in the process of this disintegration: this is one of my main theses. But the simplicity of the reduction has disappeared.

This process, which started with the discovery of neutrons and of positrons, has continued with the discovery of new elementary particles ever since. But particle theory is not even the main difficulty. The real disruption is due to the discovery of new kinds of forces, especially of short-range nuclear forces, irreducible to electromagnetic and gravitational forces.

Gravitational forces did not trouble the physicists very much in those days, because they had just been explained away by General Relativity, and it was hoped that the dualism of gravitational and electromagnetic forces would be superseded by a unified field theory. But now we have at least four very different and still irreducible kinds of forces in physics: gravitation, weak decay interaction, electromagnetic forces and nuclear forces.

X

Thus Cartesian mechanics—once regarded by Descartes and Newton as the basis to which all else was to be reduced—was, and still is, successfully reduced to electromagnetism. But what about the admittedly most impressive reduction of chemistry to quantum physics?

Let us assume for argument's sake that we have a fully satisfactory reduction to quantum theory of chemical bonds (both of covalent or twin electron bonds and of non-covalent, for example plug-and-hole, bonds), in spite of the telling remark of Pauling (1959), author of *The Nature of the Chemical Bond*, that he was unable to 'define' (or state precisely) what the nature of the chemical bond was. Let us further assume for argument's sake that we have a fully satisfactory theory of nuclear forces, of the periodic system of the elements and their isotopes, and especially of the stability and instability of the heavier nuclei. Does this constitute a fully satisfactory reduction of chemistry to quantum mechanics?

I do not think it does. An entirely new idea has to be brought in, an idea which is somewhat foreign to physical theory: the idea of evolution, of the history of our universe, of cosmogony.

This is so because the periodic table of the elements and the (reformulated) Bohr theory of the periodic system explain the heavier nuclei as being composed of lighter ones; ultimately as being composed of hydrogen nuclei (protons) and neutrons (which in turn *might* be regarded as a kind of composition of protons and electrons). And this theory assumes that the heavier elements have a history—that the properties of their nuclei actually result from a rare process which makes several hydrogen nuclei fuse into heavier nuclei, under conditions which are only rarely encountered in the cosmos.

We have much evidence in favour of the view that this really happened and still happens; that the heavier elements have an evolutionary history and that the fusion process by which heavy hydrogen is transformed into helium is the main source of the energy of our own sun and also of the hydrogen bomb. Thus helium and all the heavier elements are the result of cosmological evolution. Their history, and especially the history of the heavier elements, is, according to present cosmological views, a strange one. The heavier elements are at present regarded as the products of supernovae explosions. Since helium, according to some recent estimates, forms twenty-five per cent of all matter by mass and hydrogen two-thirds or three-quarters of all matter by mass, all the heavier nuclei appear to be extremely rare (together perhaps one or two per cent by mass). Thus the earth and presumably the other planets of our solar system are made mainly of very rare (and I should say very precious) materials.

At present the most widely accepted theory of the origin of the universe[2]—that of the hot big bang—claims that most of the helium is the product of the big bang itself: that it was produced within the very first minute of the existence of the expanding universe. The precariousness of the scientific status of this speculation (originally due to Gamow) need not be stressed. And since we have to appeal to theories of this kind in our attempts to reduce chemistry to quantum mechanics, it can hardly be claimed that this reduction has been carried out without residue.

The truth is that we have reduced chemistry, at least in part, to cosmology rather than to physical theory. Admittedly, modern classical relativistic cosmology started as an applied physical theory; but, as Bondi has stressed, these times seem now to be over and we must face the fact that some of our ideas (for example, those that started with Dirac and Jordan) could almost be described as attempts to reduce physical theory to cosmogony. And both cosmology and cosmogony, though immensely fascinating parts of physics, and though they are becoming better testable, are still almost borderline

[2] [Added after the conference]: This theory may now be threatened by the new theory of redshifts proposed by J. C. Pecker, A. P. Roberts and J. P. Vigier, Non-velocity redshifts and photon–photon interactions, *Nature*, **237** (1972), 227–9.

cases of physical science, and hardly yet mature enough to serve as the bases of the reduction of chemistry to physics. This is one reason why I regard the so-called reduction of chemistry to physics as incomplete and somewhat problematic; but of course I welcome all these new problems.

XI

But there is a second residue of the reduction of chemistry to physics. Our present view is that hydrogen alone, and especially its nucleus, is the building material of all the other nuclei. We believe that the positive nuclei strongly repel each other electrically down to very short distances, but that for still shorter distances (achievable only if the repulsion is overcome by tremendous velocities) they attract each other by nuclear forces.

But this means that we attribute to the hydrogen nucleus relational properties which are inoperative in the overwhelming majority of the conditions in which hydrogen nuclei exist in our universe. That is to say, these nuclear forces are potentialities that become operative only under conditions which are extremely rare: under tremendous temperatures and pressures. But this means that the theory of the evolution of the periodic table looks very much like a theory of essential properties which have the character of *predestination, or of a preestablished harmony.*[3] At any rate, a solar system like ours depends, according to present theories, on the preexistence of these properties, or rather, potentialities.

Moreover, the theory of the origin of the heavier elements in explosions of supernovae introduces *a second kind of predestination or preestablished harmony*. For it amounts to the assertion that gravitational forces (apparently the weakest of all, and so far unconnected with nuclear or electromagnetic forces) can, in big accumulations of hydrogen, become so powerful as to overcome the tremendous electrical repulsion between the nuclei, and to make them fuse due to the action of the nuclear forces. Here the harmony is between the inherent potentialities of nuclear forces and of gravitation. I do not want to assert the untruth of any philosophy of preestablished harmony. But I do not think that a preestablished harmony can be regarded as a satisfactory reduction; and I suggest that the appeal to it is an admission of the failure of the method of reducing one thing to another.

Thus the reduction of chemistry to physics is far from complete, even if we admit somewhat unrealistically favourable assumptions. Rather, this reduction assumes a theory of cosmic evolution or cosmogony, and in addition two kinds of preestablished harmony, in order to allow sleeping potentialities, or relative propensities of low probability built into the hydrogen atom, to become activated. It appears, I suggest, that we should

[3] I have used the term 'preestablished harmony' here to stress that our explanation is not in terms of the manifest physical properties of the hydrogen atom. Rather, a hitherto unknown and unsuspected property of the hydrogen nucleus was postulated, and used as an explanation.

recognise that we are operating with the ideas of *emergence* and of *emergent properties.*[4] In this way we see that this very interesting reduction has left us with a strange picture of the universe—strange, at any rate to the reductionist; which is the point I wanted to make in this section.

XII

To sum up what has been said so far: I have tried to make the problem of reduction clear with the help of examples, and I have tried to show that some of the most impressive reductions in the history of the physical sciences are far from completely successful, and leave a residue. One might claim (but see footnote 1 above) that Newton's theory was a complete successful reduction of Kepler's and Galileo's. But even if we assume that we know much more physics than we do, and that we have a unified field theory which yields with high approximation General Relativity, quantum theory and the four kinds of forces as special cases (this is perhaps a claim implicit in Mendel Sachs's unified field theory), even then we can say that chemistry has not been reduced without residue to physics. In fact the so-called reduction of chemistry is to a physics that assumes evolution, cosmology and cosmogony, and the existence of emergent properties.

On the other hand, in our not fully successful attempts at reduction, especially of chemistry to physics, we have learned an incredible amount. New problems have given rise to new conjectural theories, and some of these, such as nuclear fusion, have not only led to corroborating experiments, but to a new technology. Thus from the point of view of method, our attempted reductions have led to tremendous successes, even though it can be said that the attempted reductions have, as such, usually failed.

XIII

The story here told and the lesson here drawn from it will hardly strike a biologist as unexpected. In biology too, reductionism (in the form of physicalism or materialism) has been extremely successful, though not fully successful. But even where it has not succeeded, it has led to new problems and to new solutions.

I might perhaps express my view as follows. As a philosophy, reductionism is a failure. From the point of view of method, the attempts at detailed reductions have led to one staggering success after another, and its failures have also been most fruitful for science.

It is perhaps understandable that some of those who have achieved these scientific successes have not been struck by the failure of the philosophy. Perhaps my analysis of the success and of the failure of the attempt to reduce chemistry completely to quantum physics may give them pause, and may make them look at the problem again.

[4] I use here the term 'emergent' to indicate an apparently unforeseeable evolutionary step.

XIV

The main points made so far may be regarded as an elaboration of a brief remark made by Jacques Monod in the Preface to his *Chance and Necessity* (1970; 1971, xii): 'Nor can everything in chemistry be predicted or resolved by means of the quantum theory [or reduced to quantum theory] which, beyond any question, underlies all chemistry.' In the same book Monod also puts forward a suggestion (not an assertion, to be sure) concerning the origin of life, which is very striking, and which we may consider from the point of view reached here. Monod's suggestion is that life emerged from inanimate matter by an extremely improbable combination of chance circumstances, and that this may not merely have been an event of low probability but of zero probability—in fact, a *unique* event.

This suggestion is experimentally testable (as Monod pointed out in a recent discussion with Eccles). Should we succeed in producing life under certain well-defined experimental conditions, then the hypothesis of the uniqueness of the origin of life would be refuted. Thus the suggestion is a testable scientific hypothesis, even though it may not look like one at first sight.

What, besides, makes Monod's suggestion plausible? There is the fact of the uniqueness of the genetic code, but this could be, as Monod points out, the result of natural selection. What makes the origin of life and of the genetic code a disturbing riddle is this: the genetic code is without any biological function unless it is translated; that is, unless it leads to the synthesis of the proteins whose structure is laid down by the code. But, as Monod points out, the machinery by which the cell (at least the non-primitive cell which is the only one we know) translates the code 'consists of at least fifty macromolecular components *which are themselves coded in DNA*' (Monod, 1970; 1971, 143). Thus the code cannot be translated except by using certain products of its translation. This constitutes a really baffling circle: a vicious circle, it seems, for any attempt to form a model, or a theory, of the genesis of the genetic code.

Thus we may be faced with the possibility that the origin of life (like the origin of the universe) becomes an impenetrable barrier to science, and a residue to all attempts to reduce biology to chemistry and physics. For even though Monod's suggestion of the uniqueness of life's origin is refutable—by attempts at reduction, to be sure—if true, it would amount to a denial of any fully successful reduction. With this suggestion Monod, who is a reductionist for reasons of method, arrives at the position which, I believe, is the one forced upon us all in the light of our earlier discussion of the reduction of chemistry to physics. It is the position of a critical reductionist who continues with attempted reductions even if he despairs of any ultimate success. Yet it is in going forward with attempted reductions, as Monod stresses elsewhere in his book, rather than in any replacement of reductionist methods by 'holistic' ones, that our main hope lies—our hope of learning more about old problems and of discovering new problems, which in turn may lead to new solutions, to new discoveries.

I do not want to discuss holism in any detail here, but a few words may be needed. The use of holistic experimental methods (such as cell transplantation in embryos), though inspired by holistic thought, may well be claimed to be methodologically reductionist. Holistic theories are, on the other hand, trivially needed in the description of even an atom or a molecule, not to speak of an organism or of a gene population. There is no limit to the variety of possibly fruitful conjectures, whether holistic or not.[5] In view of my main thesis, doubt arises only about the character of experimental methods in biology: whether they are not all, more or less, of a reductionist character. (A similar situation arises, incidentally, as David Miller reminds me, with regard to deterministic and indeterministic theories. Though we must, I think, be metaphysical *indeterminists*, methodologically we should still search for deterministic or causal laws—except where the problems to be solved are themselves of a probabilistic character.)

XV

I should like to point out that even if Monod's suggestion of the uniqueness of the origin of life should be refuted by the production of life from inanimate matter under definite experimental conditions, this would not amount to a complete reduction. I do not wish to argue *a priori* that a reduction is impossible; but we have produced life from life for a long time without understanding what we have been doing, and before we had even an inkling of molecular biology or the genetic code. Thus it is certainly possible that we may produce life from inanimate matter without a full physicochemical understanding of what we are actually doing; for example, how we managed to break the vicious circle inherent in the translation of the code.

At any rate we can say that the undreamt-of breakthrough of molecular biology has made the problem of the origin of life a greater riddle than it was before: we have acquired new and deeper problems.

XVI

As I have tried to show, the attempt to reduce chemistry to physics demands the introduction of a theory of evolution into physics; that is, a recourse to the history of our cosmos. A theory of evolution is, it appears, even more indispensable in biology. And so is, in addition, the idea of purpose or teleology or (to use Monod's term) of teleonomy, or the very similar idea of problem solving; an idea which is quite foreign to the subject matter of the non-biological sciences (even though the role played in these sciences by maxima and minima and by the calculus of variations has been regarded as remotely analogous).

[5] This is now stressed in the second point of the Postscript to the present paper.

It was of course the great achievement of Darwin to show that there is a possibility of explaining teleology in non-teleological or ordinary causal terms. Darwinism is the best explanation we have. There are not, at the moment, any seriously competing hypotheses (*cf.* my (1961) and (1966*a*)).

XVII

Problems and problem solving seem to emerge together with life (see my (1966*a*)). Even though there is something like natural selection at work prior to the origin of life—for example, a selection of the more stable elements owing to the radioactive destruction of the less stable ones—we cannot say that for atomic nuclei, survival is a 'problem' in any sense of this term. And the close analogy between crystals and microorganisms and their molecular parts (organelles) breaks down here too. Crystals have no problems of growth or of propagation or of survival. But life is faced with the problems of survival from the very beginning. Indeed, we can describe life, if we like, as problem solving, and living organisms as the only problem solving complexes in the universe. (Computers are *instrumental in* problem solving but not, in this sense, problem solvers.)

This does not mean that we have to ascribe to all life a *consciousness* of the problems to be solved: even on the human level we constantly solve many problems, such as keeping our balance, without becoming aware of them.

XVIII

There can be little doubt that animals possess consciousness and that, at times, they can even be conscious of a problem. But the emergence of consciousness in the animal kingdom is perhaps as great a mystery as is the origin of life itself.

I do not want to say more about this than that panpsychism, or hylozoism, or the thesis that matter is, generally, endowed with consciousness (of a low degree), does not seem to me to help in the least. It is, if taken at all seriously, another theory of predestination or of a preestablished harmony. (It was of course part of Leibniz's original form of his theory of preestablished harmony.) For in nonliving matter, consciousness has no function at all; and if (with Leibniz, Diderot, Buffon, Haeckel and many others) we attribute consciousness to nonliving particles (monads, atoms) then we do so in the vain hope that it will help to explain the presence of those forms of consciousness which have some function in animals.

For there can be little doubt that consciousness in animals has some function, and can be looked at as if it were a bodily organ. We have to assume, difficult as this may be, that it is a product of evolution, of natural selection.

Although this might constitute a programme for a reduction, it is not itself a reduction, and the situation for the reductionist looks somewhat desperate; which explains why reductionists have either adopted the hypothesis of panpsychism or why, more recently, they have denied the existence of consciousness (the consciousness say, of a toothache) altogether.

Though this behaviourist philosophy is quite fashionable at present, a theory of the nonexistence of consciousness cannot be taken any more seriously, I suggest, than a theory of the nonexistence of matter. Both theories 'solve' the problem of the relationship between body and mind. The solution is in both cases a radical simplification: it is the denial either of body or of mind. But in my opinion it is too cheap (see my (1970), 7–9). I shall say a little more about this second 'outstanding question' and especially about panpsychism in Section XXI where I criticise psychophysical parallelism.

XIX

Of the three 'outstanding questions of reduction' listed at the beginning of this paper I have briefly touched upon two. I am now coming to the third one, the question of the reduction of the human consciousness of self and the creativeness of the human mind.

As Sir John Eccles has often stressed, this third question is the problem of the 'mind–brain liaison'; and Jacques Monod calls the problem of the human central nervous system the 'second frontier', comparing its difficulty with the 'first frontier', the problem of the origin of life.

No doubt this second frontier is a dangerous region to dwell in, especially for a lay biologist; nevertheless I may say that the attempts at a partial reduction seem to me more hopeful in this region than in that of the second question. As in the region of the first question, it seems to me that more new problems can be discovered here with reductionist methods, and perhaps even solved, than in the region of the second question—a region which looks to me comparatively sterile. I hardly need to stress that a completely successful reduction in any of the three regions seems to me most unlikely, if not impossible.

With this, it may perhaps be said, I have fulfilled my promise to discuss, or at any rate mention, those three outstanding questions of reduction listed at the beginning of this paper. But I wish to say a little more about the third of them—about the body–mind problem, or mind–body problem—before proceeding to my thesis of the incompletability of all science.

XX

I regard the problem of the emergence of consciousness in animals (question 2), of understanding it and, perhaps, of reducing it to physiology, as most likely insoluble; and I feel similarly about the further problem of the emergence of the specifically human consciousness of self (question 3)—that is, the body–mind problem. But I do think that we can throw at least some light upon the problem of the human self.

I am, in many ways, a Cartesian dualist (see my (1953)), even though I should prefer to describe myself as a pluralist; and of course I do not subscribe to either of Descartes's two substances. Matter, we have seen, is no ultimate substance with the essential property of extension, but consists of complex structures about whose constitution we know a great deal, including an explanation of its 'extension': that it takes up space by electrical repulsion.

My first thesis is that the human consciousness of self, with its apparently irreducible unity, is highly complex, and that it may perhaps be, in part, explicable.

In a course of lectures given at Emory University in May 1969 I suggested (as I had done some years before in lectures at the London School of Economics) that the higher human consciousness, or consciousness of selfhood, is absent in animals. I also suggested that Descartes's conjecture that locates the human soul in the pineal gland may not be as absurd as it has often been represented, and that, in view of Sperry's results with divided brain hemispheres (1964; see also Eccles (1970), 73–9), the location is to be looked for in the speech centre, in the left hemisphere of the brain. As Eccles has more recently informed me (1972), Sperry's later experiments (not known to me at the time) support this guess to a degree: the right brain may be described as that of a very clever animal while only the left brain appears to be human, and aware of selfhood.

I had based my guess upon the role which I ascribe to the development of a specifically human language.

All animal language—indeed, almost all animal behaviour—has an *expressive* (or symptomatic) and a *communicative* (or signalling) function, as Karl Bühler has pointed out. But human language has, besides, some further functions, which are characteristic of it and make it a 'language' in a narrower and more important sense of the word. Bühler drew attention to the basic *descriptive* function of human language, and I pointed out later (1949, 1953) that there are further functions (such as prescriptive, advisory and so on) of which the most important and characteristic one for human beings is the *argumentative* function. (Professor Alf Ross (1972) points out that many other functions may be added, for example, those of giving orders or making requests or promises.)

I do not think (and I never did think) that any of these functions are reducible to any of the others, least of all the two higher functions (description and argument) to the two lower ones (expression and communication). These, incidentally, are always present, which may perhaps be the reason why many philosophers mistake them for properties which are characteristic of human language.

My thesis is that, with the higher functions of the human language a new world emerges: the world of the products of the human mind. I have called it 'world 3' (following a suggestion of Sir John Eccles: originally I called it the 'third world'). I call the world of physical matter, fields of force, and so on, 'world 1'; the world of conscious and perhaps also subconscious experience 'world 2'; and 'world 3' especially the world of spoken (written or printed) language, like story telling, myth making, theories, theoretical problems, mistakes and arguments. (The worlds of artistic products and of social institutions may either be subsumed under world 3 or be called 'world 4' and 'world 5': this is just a matter of taste.)

I introduce the terms 'world 1', 'world 2' and 'world 3' in order to emphasise the (limited) *autonomy* of these regions. Most materialists or physicalists or reductionists assert that, of these three worlds, only world 1 really exists, and that it is therefore autonomous. They replace world 2 by behaviour, and world 3, more particularly, by verbal behaviour. (This, as indicated above, is just one of those all too easy ways of solving the body–mind problem: the way of denying the existence of the human mind and of a human consciousness of self—that is, of those things which I regard as some of the most remarkable and astonishing in the universe; the other equally easy way out is Berkeley's and Mach's immaterialism: the thesis that only sensations exist, and that matter is just a 'construct' out of sensations.)

XXI

There are in the main four positions with respect to the interrelationship between the body, or the brain, and the mind.

(1) A denial of the existence of the world 1 of physical states; that is, immaterialism, as held by Berkeley and Mach.

(2) A denial of the existence of the world 2 of mental states or events, a view common to certain materialists, physicalists and philosophical behaviourists, or philosophers upholding the identity of brain and mind.

(3) An assertion of a thoroughgoing parallelism between mental states and states of the brain; a position that is called 'psychophysical parallelism'. This was first introduced in the Cartesian school by Geulincx, Spinoza, Malebranche and Leibniz, mainly in order to avoid certain difficulties in the Cartesian view. (Like epiphenomenalism, it robs consciousness of any biological function.)

(4) An assertion that mental states can interact with physical states. This was the view of Descartes which, it is widely believed, was superseded by (3).

My own position is that a brain–mind parallelism is almost bound to exist *up to a point*. Certain reflexes, such as blinking when seeing a suddenly approaching object, are to all appearances of a more or less parallelistic character: the muscular reaction (in which no doubt the central nervous system is involved) repeats itself with regularity when the visual impression is repeated. If our attention is drawn to it we may be conscious of its happening, and so with some (but of course not all) other reflexes.

Nevertheless, I believe that the thesis of a *complete* psychophysical parallelism—position (3)—is a mistake, probably even in some cases where mere reflexes are involved. *I thus propose a form of psychophysical interactionism.* This involves (as was seen by Descartes) *the thesis that the physical world 1 is not causally closed*, but open to the world 2 of mental states and events; a somewhat unattractive thesis for the physicist, but I think one that is supported by the fact that world 3 (including its autonomous regions) acts upon world 1 *via* world 2.

I am quite willing to accept the view that whenever anything goes on in world 2, something connected with it goes on in world 1 (in the brain). But in order to speak of a complete or thoroughgoing parallelism, we would have to be able to assert that 'the same' mental state or event is always accompanied by an exactly corresponding physiological state, and *vice versa*.

As indicated, I am prepared to admit that there is something correct in this assertion, and that for example the electrical stimulation of certain brain regions may regularly give rise to certain characteristic movements or sensations. But I ask whether, as a universal rule about all mental states, the assertion has any content; whether it is not an empty assertion. For we can have a parallelism between world 2 elements and brain processes, or between world 2 *Gestalten* and brain processes, but we can hardly speak of a parallelism between a highly complex, unique and unanalysable world 2 process and some brain process. And there are many world 2 events in our lives which are unique. Even if we disregard creative novelty, hearing a melody twice and recognising that it is the same melody is not a repetition of the same world 2 event, just because the second hearing of the melody is connected with an act of *recognising* the melody, which was absent the first time. It is the world 1 object (in this case the melody) which is repeated, but not the world 2 event. Only if we could accept a kind of world 2 theory which, like associationist psychology, looks upon world 2 events as composed of atom-like elements could we make a clear distinction between the repeated part of the world 2 experience—the *hearing* of the same melody—and the non-repeated part, the *recognition* that it is the same melody (where the recognition experience in its turn is capable of recurrence in other contexts). But I think that it is clear that such an atomistic or analytical psychology is quite incapable of carrying us far.

World 2 is highly complex. While if we attend only to such fields as sense perception (that is, perception of world 1 objects) we may think that we can analyse world 2 by atomic or molecular methods, for instance *Gestalt* methods (methods which, I think, are all unrewarding as compared with the biological or functional methods of Egon Brunswik or Richard Gregory), the application of such methods turns out to be quite inadequate if we consider our unique attempts to invent, and to understand, a world 3 object, such as a problem or a theory.

The way in which our thinking or our understanding interacts with attempts at linguistic formulation and is influenced by it; the way in which we have first a vague feeling for a problem or a theory which becomes clearer when we try to formulate it, and still clearer when we write it down and criticise our attempts to solve it; the way in which a problem may change and still be in a sense the old problem; the way in which a train of thought is on the one hand interconnected and on the other hand, articulated: all this seems to me to be beyond analytical or atomistic methods, including the interesting molecular methods of *Gestalt* psychology. There is a unique history of unique world 2 events involved in all of these attempts, and as a consequence, the talk about (strictly) *parallel* physiological processes loses all content.

Besides, we have reason to believe that often, if one region of the brain is destroyed, another region can 'take over', with very little or perhaps no interference with world 2—another argument against parallelism, and this time based on experiments in world 1 rather than on the necessarily vague consideration of the more complex world 2 experiences.

All this sounds, of course, very antireductionist; and as a philosopher who looks at this world of ours, with us in it, I indeed despair of any ultimate reduction. But as a methodologist this does not lead me to an antireductionist research programme. It only leads to the prediction that with the growth of our attempted reductions, our knowledge, and our universe of unsolved problems, will expand.

XXII

Let us return now to the problem of the specifically human consciousness of self; my suggestion was that it emerges in interaction (feedback, if you like) between world 2 and the worlds 1 and 3. My arguments for the role played by world 3 are as follows.

The human consciousness of self is based, among other things, upon a number of highly abstract *theories*. Animals and even plants have, no doubt, a sense of time, and temporal expectations. But it needs an almost explicit *theory* of time (*pace* Benjamin Lee Whorf) to look upon oneself as possessing a past, a present and a future; as having a personal history; and as being aware of one's personal identity (linked to the identity of one's body) throughout this history. Thus it is a *theory* that, during the period of sleep, when we lose the continuity of consciousness, we—our bodies—remain essentially the same; and it is on the basis of this theory that we can consciously recall past events (instead of merely being influenced by them in our expectations and reactions which, I suggest, is the more primitive form which the memory of animals takes).

Some animals, no doubt, have personalities; they have something closely analogous to pride and ambition, and they learn to respond to a name. But the human consciousness of self is anchored in language and (both explicitly and implicitly) in formulated theories. A child learns to use his name of himself, and ultimately a word like 'ego' or 'I', and he learns to use it with the consciousness of the continuity of his body, and of himself; he also combines it with the knowledge that consciousness is not always unbroken. The great complexity and nonsubstantial character of the human soul, or the human self, become particularly clear if we remember that there are cases where men have forgotten who they are; they have forgotten part or the whole of their past history, but they have retained, or perhaps recovered, at least part of their selfhood. In a sense, their memory has not been lost, for they *remember how* to walk, to eat, and even to speak. But they do not *remember that* they come from, say, Bristol, or what their names and addresses are. In so far as they do not find their way home (which animals normally do) their consciousness of self is affected even beyond the normal level of animal memory. But if they have not lost the power of speech, some human consciousness is left that goes beyond animal memory.

I am not a great friend of psychoanalysis, but its findings seem to support the view of the complexity of the human self, in contrast to any Cartesian appeal to a thinking substance. My main point is that the consciousness of the human self involves, at the very least, an awareness of the (highly theoretical) temporal or historical continuity of one's body; an awareness of the connection between one's conscious memory and the single, unique body which is one's own; and the consciousness of the normal and periodical interruption of one's consciousness by sleep (which, again, involves a theory of time and temporal periodicity). Moreover, it involves the consciousness of belonging locally and socially to a certain place and circle of people. No doubt much of this has an instinctive basis and is shared by animals. My thesis is that in raising it even to the level of unspoken human consciousness, human language or interaction between worlds 2 and 3 plays an important role.

It is clear that the unity of the human self is largely due to memory, and that memory can be ascribed not only to animals but also to plants (and even perhaps, in some sense, to non-organic structures such as magnets). It is therefore most important to see that the appeal to memory as such is not enough to explain the unity of the human self. What is needed is not so much the 'ordinary' memory (of past events), but a memory of theories that link the consciousness of having a body to world 3 theories about bodies (that is, to physics); a memory which is of the character of a 'grasp' of world 3 theories. It comprises the dispositions which enable us to fall back on explicit world 3 theories if we need to, with the feeling that we possess such dispositions and that we can make use of them in order to articulate those theories if we need to. (This would, of course, explain to a certain extent the difference between the human consciousness of self with its dependence on human language, and animal consciousness.)

XXIII

These facts seem to me to establish the impossibility of any reduction of the human world 2, the world of human consciousness, to the human world 1, that is, essentially, to brain physiology. For world 3 is, at least in part, autonomous of the two other worlds. If the autonomous part of world 3 can interact with world 2, then world 2, or so it seems to me, cannot be reducible to world 1.

My standard examples of the partial autonomy of world 3 are taken from arithmetic.

I suggest that the infinite series of natural numbers is an invention, a product, of the human mind, and a part of developed human language. (There are, it appears, primitive languages in which one can count only 'one, two, many' and others in which one can count only to 'five'.) But once a method of counting without end has been invented, distinctions and problems arise autonomously: even and odd numbers are not *invented* but *discovered* in the series of natural numbers, and so are prime numbers, and the many solved and unsolved problems connected with them.

These problems, and the theorems which solve them (such as Euclid's theorem that there does not exist a greatest prime) arise autonomously; they arise as part of the internal structure of the man-created series of natural numbers, and independently of what we think or fail to think. But we can *grasp* or *understand* or *discover* these problems, and solve some of them. Thus our thinking, which belongs to world 2, depends in part on the autonomous problems and on the objective truth of theorems which belong to world 3: world 2 not only creates world 3, it is partly created by world 3 in a kind of feedback process.

My argument now runs as follows: world 3, and especially its autonomous part, are clearly irreducible to the physical world 1. But since world 2 depends, in part, upon world 3, it is also irreducible to world 1.

Physicalists, or philosophical reductionists as I called them (1970), are thus reduced to denying the existence of worlds 2 and 3. But with this, the whole of human technology (especially the existence of computers), which makes so much use of world 3 theorems, becomes incomprehensible; and we must assume that such violent changes in world 1 as are produced by the builders of airports or skyscrapers are ultimately produced, without the invention of world 3 theories or world 2 plans based on them, by the physical world 1 itself: they are predestined; they are part of a preestablished harmony built, ultimately, into hydrogen nuclei.

These results seem to me absurd; and philosophical behaviourism or physicalism (or the philosophy of the identity of mind and body) appears to me to be reduced to this absurdity. It seems to me to stray too far from common sense.

XXIV

Philosophical reductionism is, I believe, a mistake. It is due to the wish to reduce everything to an ultimate explanation in terms of essences and substances, that is, to an explanation which is neither capable of, nor in need of, any further explanation.

Once we give up the theory of ultimate explanation we realise that we can always continue to ask: 'Why?'. Why-questions never lead to an ultimate answer. Intelligent children seem to know this, but give way to the adults who, indeed, cannot possibly have time enough to answer what is in principle an endless series of questions.

XXV

The worlds 1, 2 and 3, though partly autonomous, belong to the same universe: they interact. But it can easily be shown that knowledge of the universe, if this knowledge itself forms part of the universe, as it does, must be incompletable.

Take a man who draws a detailed map of the room in which he is working. Let him try to include in his drawing the map which he is drawing. It is clear that he cannot complete the task, which includes an infinity of smaller and smaller maps within each map: every time he adds a new line to the map, he creates a new object to be drawn, but not yet drawn. The map which is supposed to contain a map of itself is incompletable.

The story of the map shows the incompleteness and openness of a universe that contains world 3 objects of knowledge. Incidentally, it can also be used as an argument for the view that our universe is indeterministic. For while, admittedly, each of the different 'last' strokes actually entered into the map determines, within the infinite sequences of maps to be drawn, a dependent stroke, the determinacy of the strokes holds only if we do not consider the fallibility of all human knowledge (a fallibility which plays a considerable role in the problems, theories and mistakes of world 3). Taking this into account, each of these 'last' strokes entered into the map constitutes a *problem* for the draughtsman, a problem of entering a further stroke which depicts the last stroke *precisely*. Because of the fallibility that characterises all human knowledge, this problem cannot possibly be solved by the draughtsman with absolute precision; and the smaller the strokes to which the draughtsman proceeds, the greater will be the relative imprecision, which in principle will be unpredictable and indeterminate and will constantly increase. In this way, the story of the map shows how the fallibility which affects objective human knowledge contributes also to the essential indeterminism of our universe, apart from showing the openness and unknowability of a universe that contains human knowledge as a part of itself.

This example can help us to see why all explanatory science is incompletable; for to be complete it would have to give an explanatory account of itself.

An even stronger result is implicit in Gödel's famous theorem of the incompletability of formalised arithmetics (though it has to be admitted that to use Gödel's theorem and other metamathematical incompleteness theorems in this context is to use heavy armament against a comparatively weak position). Since all physical science uses arithmetic (and since for a reductionist only science formulated in physical symbols has any reality), Gödel's incompleteness theorem renders all physical science incomplete; which to the reductionist should show that all science is incomplete. For the nonreductionist, who does not believe in the reducibility of all science to physically formulated science, science is incomplete anyway.

Not only is philosophical reductionism a mistake, but the belief that the method of reduction can achieve complete reductions is, it seems, mistaken too. We live, it appears, in a world of emergent evolution; of problems whose solutions, if they are solved, beget new and deeper problems. Thus we live in a universe of emergent novelty; of a novelty which, as a rule, is not completely reducible to any of the preceding stages.

Nevertheless, the method of attempting reductions is most fruitful, not only because we learn a great deal by its partial successes, by partial reductions, but also because we learn from our partial failures, from the new problems which our failures reveal. Open problems are almost as interesting as their solutions; indeed they would be just as interesting but for the fact that almost every solution opens up in its turn a whole new world of open problems.

References

Anderson, C. D. (1933). Cosmic ray bursts. *Physical Review*, **43**, 368–9.

Anderson, C. D. (1933). The positive electron. *Physical Review*, **43**, 491–4.

Chadwick, J. (1932). Possible existence of a neutron. *Nature*, **129**, 132.

Eccles, J. C. (1970). *Facing Reality*. Springer-Verlag, Berlin, Heidelberg, New York.

Eccles, J. C. (1972). Unconscious actions emanating from the human cerebral cortex (unpublished).

Eddington, A. (1936). *Relativity Theory of Protons and Electrons*. Cambridge University Press.

Einstein, A. (1920). *Äther und Relativitätstheorie*. Springer, Berlin. (Translated as (Einstein, 1922).)

Einstein, A. (1922). *Sidelights on Relativity*. Methuen, London.

Heitler, W. and London, F. (1927). Wechselwirkung neutraler Atome und homöopolare Bindung nach der Quantenmechanik. *Zeitschrift für Physik*, **44**, 455–72.

Kuratowski, C. (1920). Sur la notion de l'ordre dans la théorie des ensembles. *Fundamenta mathematica*, **2**, 154–66.

Meyerson, É. (1908). *Identité et Réalité*. F. Alcan, Paris. (Translated as (Meyerson, 1930).)

Meyerson, É. (1930). *Identity and Reality*. Allen and Unwin, London.

Millikan, R. A. (1932). *Time, Matter and Values*. University of North Carolina Press, Chapel Hill.

Monod, J. (1970). *Le Hasard et la Nécessité*. Editions du Seuil, Paris. (Translated as (Monod, 1971).)

Monod, J. (1971). *Chance and Necessity*. Knopf, New York.

Pauling, L. (1959). [Discussion remark.] In *The Origin of Life on the Earth* (proceedings of the First International Symposium on The Origin of Life on the Earth, Moscow, 19–24 August 1957, ed. A. I. Oparin and others). Ed. F. Clark and R. L. M. Synge. Pergamon Press, London, 119.

Popper, K. R. (1945). *The Open Society and its Enemies*, 2 volumes. Routledge and Kegan Paul, London. (See now (Popper, 1966*b*).

Popper, K. R. (1949). Towards a rational theory of tradition. *The Rationalist Annual*, London, 36ff. (Now ch. 4 of (Popper, 1963).)

Popper, K. R. (1950). *The Open Society and its Enemies*, revised edition. Princeton University Press. (See now (Popper, 1966*b*).)

Popper, K. R. (1953). Language and the body–mind problem. In *Proceedings of the XIth International Congress of Philosophy*, **7**, 101–7. (Now ch. 12 of (Popper, 1963).)

Popper, K. R. (1956). Three views concerning human knowledge. In *Contemporary British Philosophy* (ed. H. D. Lewis). Allen and Unwin, London, 355–88. (Now ch. 3 of (Popper, 1963).)

Popper, K. R. (1957). The aim of science. *Ratio* [Oxford], **1**, 24–35. (Now ch. 5 of (Popper, 1972).)

Popper, K. R. (1961). Evolution and the tree of knowledge (Herbert Spencer Lecture). Published for the first time as ch. 7 of (Popper, 1972).

Popper, K. R. (1963). *Conjectures and Refutations*. Routledge and Kegan Paul, London; Basic Books, New York. (See now also the third and fourth editions (Popper, 1969, 1972).)

Popper, K. R. (1966*a*). Of clouds and clocks (the second Compton Memorial Lecture, delivered 1965). Washington University Press, St Louis, Missouri. (Now ch. 6 of (Popper, 1972).)

Popper, K. R. (1966*b*). *The Open Society and its Enemies*, fifth edition, two volumes. Routledge and Kegan Paul, London.

Popper, K. R. (1967). Quantum mechanics without 'the observer'. In *Quantum Theory and Reality* (*Studies in the Foundations, Methodology, and Philosophy of Science*, volume 2) (ed. Mario Bunge). Springer-Verlag, Berlin, Heidelberg, New York, 7–44.

Popper, K. R. (1969). *Conjectures and Refutations*, third (revised) edition, Routledge and Kegan Paul, London. (Fourth edition, 1972.)

Popper, K. R. (1970). A realist view of logic, physics, and history. In *Physics, Logic, and History* (ed. W. Yourgrau and A. D. Breck). Plenum Press, New York and London, 1–30. (Now ch. 8 of (Popper, 1972).)

Popper, K. R. (1972). *Objective Knowledge: an evolutionary approach*. Clarendon Press, Oxford.

Ross, A. (1972). The rise and fall of the doctrine of performatives. In *Contemporary Philosophy in Scandinavia* (ed. R. E. Olsen and A. M. Paul). Johns Hopkins Press, Baltimore, 197–212.

Sperry, R. W. (1964). The great cerebral commissure. *Scientific American*, **210**, 42–52.

Wiener, N. (1914). A simplification of the logic of relations. *Proceedings of the Cambridge Philosophical Society*, **17**, 387–90.

Yukawa, H. (1935). On the interaction of elementary particles, I. *Proceedings of the Physico-Mathematical Society of Japan*, 3rd series, **17**, 48–57.

Postscript

Except for minor revisions and a reference or two to this postscript, I have left the paper as originally prepared. But before it was discussed at the conference, I criticised it there myself, pointing out the first two of the following four important omissions.

(1) The first of these omissions is that in the paper there is no mention of the attempts to reduce thermodynamics to mechanics. This is an important example of a reduction, and an interesting one from the point of view of my thesis. For while the results of the attempted reduction have been important, there has not been anything like a complete reduction without remainder.

(2) There is a second and more important omission—a point which in the paper I took more or less for granted (I mentioned it only briefly in Section XIV; see text to footnote 5, above). It is this. Before we can even attempt a reduction, we need as great and as detailed a knowledge as possible of whatever it may be that we are trying to reduce. Thus before we can attempt

a reduction, we need to work on the level of the thing to be reduced (that is, the level of 'wholes'). I had pointed this out previously.[6]

(3) A third omission (not mentioned at the conference) is connected with the distinction (indicated at the beginning of the paper: see text to footnote 1) between a *reduction* which explains some theory by an existing theory and an *explanation with the help of a new theory*: though I will not quarrel about words I should now be disinclined to call an explanation with the help of a new theory a 'reduction'. Yet if this terminology is adopted the explanation of the wave theory of the propagation of light by Maxwell's theory of electromagnetism could be claimed as an example of a completely successful reduction (perhaps the only example of a completely successful reduction). However, it may be better not to describe this as a reduction of one theory to another, or one part of physics to another, but rather as a radically new theory which succeeded in unifying two parts of physics.

(4) Without wishing to advocate what one might call an antireductionistic research programme for biology, the following seems to be a reasonable comment on the situation.

The Newtonian mechanistic programme for physics broke down over the attempt to include electricity and magnetism, or, more precisely, over Faraday's introduction of non-central forces. (Maxwell's attempt to reduce these non-central forces to Newtonian theory by constructing a mechanical model of the ether proved extremely fruitful in suggesting to him his field equations, but nevertheless was unsuccessful and had to be dropped.) Einstein's realisation that Newton's and Maxwell's theories are incompatible led to Special Relativity. So physicists had to accept a radically new theory rather than a reduction. A similar fate befell physics when both mechanics and electromagnetic theory in the unified form due to Lorentz and Einstein were applied to new and largely statistical problems of the microstructure of matter. This led to quantum mechanics. We cannot rule out the possibility that the inclusion of biological problems may lead to a further expansion and revision of physics.

Discussion

Rensch

I still have some difficulties in understanding your world 3. In my opinion the human culture is the practical effect of your world 2, that is to say, of our mental abilities, or capability to use a language, to think in generalised symbols and to draw conclusions in conformity with universal logical laws. Being an identist I would regard your worlds 1, 2 and 3 as three evolutionary levels. If your world 3 is also characterised by the fact that culture originated through the influence of or the adaptation to causal and logical universal

6 See my (1972), 285–318, esp. 297.

laws, then it would contain also elements of world 1. I put this question because you mentioned in your paper that you did not invent your classification but only discovered its existence.

Popper

What Professor Rensch calls 'your world 3' is indeed, as I always emphasise, the product of world 2. This agrees with Professor Rensch's remark that it is 'the practical effect' of world 2. I also think, like Professor Rensch, that we can 'regard . . . worlds 1, 2 and 3 as three evolutionary levels'. But all this does not mean that world 3 is part of world 2, or world 2 part of world 1, or anything of the sort. If I have a very bad toothache then, no doubt, it may be regarded as the product of my very bad tooth. Even though I feel the pain 'in' the tooth, few of us doubt that a nervous signal has to be transmitted from the tooth to the brain before I can feel it. Similarly, a book which I have written is the product of my thought processes, of my world 2; but the book, the product of my world 2, can be bought and read, while no part of my thought processes, of my world 2, can be bought and read. Thus the products of world 2 do not necessarily belong to world 2.

Look at the sentence 'This book is for sale in the bookshops, and it can be read'. Here the word 'book' is used for a world 1 object, located in physical space, and possessing mass and weight; but it is here also used for a world 3 object. For reading a book involves more than merely observing black marks on white paper: it involves the grasping (a world 2 process) of an *objective thought content*, a typical world 3 object.

Only thought contents can stand in the objective logical relationship of compatibility or incompatibility, or in the relationship of premise and conclusion. We can distinguish and, I suggest, we ought to distinguish, between Euclid's *discovery*—a world 2 object—of the theorem that there is no greatest prime number, and *that theorem itself*, which is a world 3 object.

This distinction (I do not think I spoke of classification) is very clearly made in the works of Bolzano and Frege. However, neither Bolzano nor Frege attributed to what I call world 3 objects the power of interacting with world 1 (*via* world 2).

17. Adaptive Shifts and Evolutionary Novelty: A Compositionist Approach

G. LEDYARD STEBBINS

With respect to the topic of this conference, reductionism in biology, the present contribution is intended to provide an example to show how the alternation between methodological reduction and synthesis or composition can clarify a problem that is of basic significance in biology. By methodological synthesis I mean systematic comparisons between examples of factual data that have been chosen either because they have been very well analysed, are particularly significant for answering the questions that have been asked, or for both of these reasons.

In our era, the study of organic evolution has become polarised around two widely distant foci. One of these is evolution in the broad sense; a succession of events that took place over billions of years of time, and gave rise successively to living matter, organised cells, multicellular organisms, animals having instincts and intelligence, and finally to man, the only organism that is conscious of his intelligence and is therefore capable of guiding the future course of evolution. The other focus is the study of evolution at the level of populations which seeks to recognise, characterise and reproduce experimentally those evolutionary changes that can be observed by a scientist in his own life time, even though they may appear to be minuscule compared to the grand sweep of evolution as a whole.

These two foci differ from each other not only with respect to the subject material that they include, but also with respect to methods employed, relationships to other scientific disciplines and public interest. The population biology focus is quantitative, experimental and pragmatic. Descriptive information is obtained in the field, and is analysed quantitatively on the basis of rigidly defined parameters. Experiments are designed in order to reproduce and explain short-term sequences of events that are inferred from field data. Theories and hypotheses are largely confined to models that are often highly mathematical and make use of computer simulation. The purpose of these models is to explain population dynamics over short periods of time, and the limited amount of evolutionary change that is responsible for the diversification of races and species. The origin of major evolutionary innovations is usually ignored entirely by evolutionists who are concentrating upon the population biology focus of our discipline. On the other hand, the scientist whose major interest is the entire course of evolution recognises population biology as only one corner of his field of interest. It is a bridgehead from which, by extrapolation, he must reach out into remote areas of scientific knowledge. He can extrapolate successfully only by acquiring a broad knowledge of apparently disparate facts. On the basis of a carefully considered and clearly expressed philosophy he must put these facts together, like the pieces of a jigsaw puzzle, into a rational, harmonious, synthetic whole.

With respect to other disciplines, the evolutionist whose focus is the contemporary differentiation of populations has much in common with geneticists not primarily interested in evolution, with statistical mathematicians, and in recent years with enzymologists. His niche in the area of experimental science is becoming increasingly secure. As more information becomes available about the nature of gene action and of genetic control systems, this focus of evolutionary study is destined to develop closer ties with molecular biology. By contrast, an integrated concept of evolution as a whole must be based not only upon population genetics, but also upon systematics which alone can reveal the vast diversity of the products of evolution; and upon palæontology, which is the chief source of direct information about the origin of organismic diversity. Moreover, evolutionists concerned with macroevolutionary syntheses must be familiar with world philosophies and the principles of logic, since their productions must stand comparison with other world philosophies. On the other hand, the events with which they deal have occurred over such long periods of time that they cannot be experimentally reproduced. Moreover, realistic mathematical models dealing with these events are impossible to construct, since not enough parameters are known with sufficient precision.

The difference between the foci with respect to public interest is equally profound. The study of population differentiation is the 'evolutionist's evolution'. Its appeal is to professional scientists who prefer the out of doors to the laboratory, but who nevertheless share with other scientists a devotion to intellectual precision, to the elegance of a well-designed experiment, and to deductions that are based upon hard facts. The outsider, even the biologist who is not well versed in the field, may have difficulty in understanding these experiments and deductions. Even if he does understand them, he may reply to the evolutionist with that most frustrating of comments, 'so what?' On the other hand, the macroevolutionary focus inevitably provides generalisations which, whether true or false, have wide public appeal. The public, however, both scientists and nonscientists, does not have enough background to discriminate between broadly based, well integrated hypotheses and wild speculations that extrapolate from a few facts that the theorist happens to regard as important. In the field of general evolutionary theory, acceptance by the intellectual public of a point of view often appears to depend more upon the way in which it is packaged in catchy phrases and titillating examples than upon the solid worth of the package's contents.

How can better connections be established between the subspecific and the transspecific focus?

Obviously, progress towards a complete understanding of evolution will be made only by establishing and maintaining close connections between these two major foci. Their advocates and practitioners must understand each other fully, and maintain open and continuous lines of communication with each other. The best way of doing this is through the medium of some principle which is common to both foci; one that can be applied with equal significance to the differentiation of populations as to the major trends of evolution.

One does not have far to seek in order to find such a principle. Expressed in general terms already by Darwin, and in precise modern scientific language by Simpson (1953), it is the following. At every level, from the differentiation of populations into races and species up to the origin and diversification of the major phyla, the rate and direction of evolution depend upon interactions between the gene pools of populations and their environments.

Emphasis upon interactions between populations and their environments is the only way of making sense out of the various processes that control evolution. Considered in the context of a constant environment, natural selection is tautological. It says only that the best genotypes are preserved, and that the frequency of genes contributing to these genotypes will increase at the expense of unfavourable alleles. Operationally, these superior genes and genotypes are recognised by the fact of their preservation and increase. On the other hand, when attention is paid to the variation of environments in space and time, natural selection appears as the mediating factor which makes possible the differentiation of populations in response to these environmental changes. Given an environment that remains constant over long periods of time, and the absence of open niches into which segments of the population can migrate, stabilising selection holds sway, and inhibits evolutionary change of the population as a whole, no matter how many mutations or other genetic changes may be occurring in its separate individuals. If, however, significant changes in the environment are taking place, those populations having the genetic equipment for doing so will respond to these changes. They will differentiate by means of adaptive shifts. If a previously stable environment becomes suddenly diversified, opening up several new ecological niches almost simultaneously, or if members of an old and stable population are able to enter a new area containing several unoccupied niches, simultaneous adaptive shifts in several directions become possible, and adaptive radiation based upon diversifying selection (Dobzhansky, 1970) is the result. If several successive environmental changes in the same general direction take place over time spans that are measured in millions of years, successive adaptive shifts may produce the succession of similar adaptive shifts that have been collectively characterised as directional selection (Simpson, 1953).

Similarly, mutation as a major evolutionary process is insignificant with respect to a constant population–environment interaction because of the deleterious effects of mutations upon harmonious, well-integrated genotypes. On the other hand, given environmental changes that reduce the effectiveness of existing population–environment interactions, mutations that correct this unbalance or that contribute to new harmonious gene combinations will have an enhanced adaptive value. A realistic conception of the significance of mutation in evolution requires that all models based upon constant adaptive values of alternative alleles be rejected and replaced by models in which adaptiveness is regarded as relative. It is a function of interactions between individual alleles and the genotypic composition of the entire population, as well as with the available environments.

On the other hand, evolutionists who confine their attention to the environment and to descriptive accounts of visible changes in populations will also fail to understand evolution as a whole. Mistaken concepts about how evolutionary changes are brought about can be recognised and rejected only on the basis of a thorough understanding of population genetics. The last fortress of Lamarckian theory, adaptive shifts in microorganisms, was overcome by experiments in the 1940s and 1950s on population dynamics (Luria and Delbrück, 1943; Cavalli-Sforza and Maccacaro, 1952).

Evolution by means of some kind of internal direction, variously characterised as 'orthogenesis', 'aristogenesis', 'telefinalism', etc., although rendered extremely unlikely by careful analyses of palæontological data (Simpson, 1953), nevertheless was a tenable hypothesis until molecular biology revealed the extreme indirectness between alterations by mutation of proteins, the primary products of genes, and any visible changes in morphology that can be recognised by taxonomists or palæontologists. In the future, all general theories about evolution will have to be based chiefly upon established facts of population and molecular genetics.

The spectrum of adaptive shifts

The unity of the evolutionary process can be seen most easily from a comparative survey of adaptive shifts. Those having the simplest genetic basis are shifts in the frequency of particular genes that promote survival in an altered habitat. The examples of industrial melanism and the acquisition of DDT resistance in flies are familiar to everyone. Hardly more complex are adaptive shifts based upon a single phenotypic character that is controlled by many genes. Examples are the evolution of mineral tolerant races of plants on mine tailings (Jowett, 1964; Bradshaw *et al.*, 1965; Allen and Sheppard, 1971). Essentially similar but involving a large geographic area and, presumably, long periods of time is the differentiation of races of plants having different photoperiodic requirements and adapted to different latitudes, as in the goldenrod, *Solidago maritima* (Goodwin, 1944; Stebbins, 1950).

The origin of races, subspecies and ecotypes belonging to the same species differs from the above examples only in that the adaptive shifts are more complex, involving simultaneous alterations in the frequency of genes that affect many different characteristics. If adaptive shifts of this kind are accompanied by changes in the frequency of genes that control reproductive isolating mechanisms, restriction of gene flow can make possible the divergence of new evolutionary lines, and sets the stage for evolution above the species level.

Adaptive shifts that provide transitions between examples of speciation and those that give rise to higher categories have been reviewed by Hecht (1965). If the new habitat entered is so different from the one previously occupied that the animal can survive only by altering completely its entire way of life, then the evolutionary line is said to have entered into a new adaptive zone. The origin of the polar bear (*Ursus maritimus*) from an ancestor that resembled the brown bear is a good example. In this case, relatively few anatomical changes were necessary, and alterations of the genotype were relatively slight, as is evident from the fact that polar bears can hybridise with brown bears, and the hybrid is partly fertile. In another example, the fish-eating ichthyomyine cricetid rodents of South America, the adaptive shift that gave rise to them was of about the same order of magnitude, but it was followed by adaptive radiation in the new zone, so that five genera and sixteen species of these rodents can be recognised (Walker *et al.*, 1968).

For the most part, families of mammals and birds are distinguished from each other on the basis of morphological and ecological characteristics that differ in magnitude but not in kind from those which distinguish the Ichthyomyinae from other cricetid rodents. The working taxonomist generally recognises families as compared to subfamilies not on the basis of the magnitude of the morphological and ecological differences that separate the family from its neighbours, but on the basis of distinctness, that is, of marked discontinuities between groups of genera that are placed in different families. At this level of the taxonomic hierarchy, therefore, the origin of a new category depends as much upon the extinction of intermediate, bridging species and genera as it does upon further evolutionary divergence.

The importance of this fact lies in the existence at the family level, and to a lesser extent at the level of genera, of organisms which have successively invaded similar adaptive zones and have exploited these zones at increasing levels of efficiency and complexity of organisation. Consider, for instance, the adaptive zone of large, herbivorous mammals. According to the fossil record as outlined by Romer (1966), this zone was invaded at the beginning of the Tertiary period by condylarths, and at various subsequent epochs by pantodonts, uintatheres, titanotheres, oreodonts, rhinoceroses, horses, camels, elephants and bovids (antelopes, bison and cattle). Some of these groups persisted until modern times and shared the habitat with later invaders, while others became extinct. The important fact is that with the possible exception of the origin of pantodonts and uintatheres from condylarths, none of these later evolving groups was derived from preexisting occupants of the same habitat. They represent immigrations of increasingly efficient and successful animals, rather than the evolution of increased efficiency *in situ.* This and similar examples lead to the following generalisation: the ability to exploit a particular adaptive zone at a higher level of efficiency and organisation is rarely or never acquired by an evolutionary line that is occupying continuously the zone in question. Higher levels of organisation are usually acquired by means of successive adaptive shifts, from one adaptive zone to another zone, where the environmental challenge is particularly great, and then back to the original zone at a higher level.

This generalisation was recognised by Simpson (1960) as the 'relay phenomenon' of replacement. He pointed out that extinction is an essential part of it. Unless the older group is becoming reduced in scope, the relay replacement cannot occur. In many instances, the older group has been entirely eliminated. When this occurs, the expansion of the younger group usually is in part responsible, via direct competition.

Adaptive shifts, evolutionary novelty, and major evolutionary advances

The relay phenomenon is of such basic importance that it needs to be documented as fully as possible. In particular, the question must be asked: is this the way in which major classes and phyla came into being? Unfortunately, this question cannot be answered for most organisms, since the fossil record that might document the origin of classes and phyla is either very scanty or completely lacking. Nevertheless, several examples point towards an affirmative answer.

One of these is the evolution of bony fishes, lungfishes and amphibia from primitive fishes, as described by von Wahlert (1965) and Thomson (1969). The primitive Acanthodii were free-swimming fishes of fresh or salt water that lacked swim bladders. Consequently they could not stay motionless, suspended in the water, as do most modern fishes. When not swimming, they had to rest on the bottom. In the early part of the Devonian period, the Acanthodii or forms similar to them gave rise to one or several adaptive radiants that lived in shallow fresh water in climates having seasonal drought. Under these conditions, the oxygen content of the water became so low that these fishes could survive only by coming to the surface and gulping air. Hence modifications of the digestive tract that produced oxygen-storing pouches, or primitive lungs, acquired a high adaptive value.

These earliest lung fishes radiated in three directions, probably based upon differences in food preferences and in ways of food getting. One direction, which probably took place in several different lines, was towards active predation on smaller fishes and other free swimming animals. In fishes that adopted this way of life, beginning with the Actinopterygii, the lungs became modified into a single swim bladder, and ushered in the modern, more efficient way of exploiting open water on the part of bony fishes. The second direction was taken by fishes which became adapted to feeding upon molluscs and other sedentary prey. They developed crushing teeth, became more sluggish in their habits, and their lateral fins became long and slender, so that they were either useless or served as prehensile organs for clinging to aquatic vegetation while the fish was feeding. These fishes retained lungs, and evolved the ability to aestivate, becoming encased in mud and completely dormant when the water that they inhabited dried up. They became extinct or reached evolutionary stabilisation by the middle of the Mesozoic era, their modern survivors being the three species of lungfishes now existing in Australia, Africa and South America.

The third line of radiation, consisting of the rhipidistian suborder of crossopterygian fishes, continued to seek actively moving prey, but may have adopted a different mode of predation (Szarski, 1963; Schaeffer, 1965). Instead of overtaking their prey by rapid swimming, they probably lay in wait, hidden among the aquatic vegetation, and captured fast-moving fishes or arthropods by sudden lunges. In connection with such habits, two kinds of modifications would have had a high adaptive value. One was the elongation of the snout, which was characteristic of the later rhipidistians, and provides the strongest resemblance between them and the earliest amphibians. The second was the thickening of the lateral fins, the strengthening of their bones, and the development of stronger muscles in them, particularly those attached to the pectoral girdle (Schmalhausen, 1964). Such strong fins would have provided a purchase on the bottom of the water for the sudden lunges after prey. Even greater efficiency for this purpose could be acquired by modifying the ends of the fins into feet having movable digits. These deductions suggest the hypothesis that the earliest amphibians acquired feet not primarily as a means of invading the land, but as a more efficient way of executing their highly specialised method of food getting in the water. Once acquired, their feet could easily have been modified for overland locomotion.

Following the same line of reasoning, I would like to suggest that the first invasion of the land by amphibia was not as a means of moving from a body of water that was beginning to dry up to another that was more favourable, as Romer (1966) has suggested. If these amphibia were completely adapted to preying upon arthropods, some of their species could have occupied a new niche by ascending to the marshy land that surrounded the pools of water that were their original home, and preying upon the terrestrial arthropods that had already evolved. The simultaneous adaptive radiation and great expansion of both primitive insects and amphibians during the upper Carboniferous (Romer, 1966; Lanham, 1964) may have been not merely a coincidence, but in association with a relationship between predators and prey.

A further suggestion that the lunging method of capturing prey was characteristic of early amphibians is provided by the fact that their ancestors, the rhipidistians, had particularly strong pectoral fins and large pectoral girdles. An early amphibian, *Ichthyostega*, retained the latter condition to some extent (Szarski, 1962; Schmalhausen, 1964). In later amphibia and reptiles, which evolved different methods of capturing prey, the limbs were all about equally developed, or the hind limbs became more strongly developed than the forelimbs.

This example illustrates three completely different ways by which adaptive radiation at the level of species or genera can affect major trends of evolution. The rhipidistian–amphibian transition suggests that a new class can arise from a highly specialised offshoot of a preexisting class, but that this specialisation includes preadaptive structures that, with only slight modifications, can become major adaptations to a new way of life. The evolution of lungfishes supports the hypothesis of Schmalhausen (1949) that evolutionary stabilisation or stagnation is most likely to result from adaptive radiation into a mode of life in which a passive defence against extreme environmental pressures, in this case aestivation as a means of avoiding drought, has become a key factor for survival. The origin of primitive bony fishes from primitive lung-bearing fishes illustrates the indirect way in which increasing efficiency and organisational complexity is usually acquired.

The origin of mammals is an even more compelling example of the latter principle. This is because, as Crompton and Jenkins (1968) have pointed out, specialised reptiles, therapsids, that in some respects were already transitional towards mammals, existed for a hundred million years and for much of this time were the dominant land animals. Late in the evolution of therapsids there appeared small or medium-sized forms having particularly high levels of specialisation, the ictidosaurs and tritylodonts. These animals inhabited niches that were peripheral to the adaptive zone of therapsids in general, and they may have represented the evolution of nondominant, marginally adapted forms from more completely dominant ancestors. They were most probably nocturnal, and many of them may have been arboreal. For a nocturnal form, the acquisition of homoiothermy has a particularly high adaptive value, since it permits the animal to be active throughout the night. The adaptive syndrome for homoiothermy would have included also the acquisition of hair as a more efficient insulation than reptilian scales, the four-chambered heart along with other changes that increased the efficiency of circulation, and various alterations of enzyme systems that would increase the rate and efficiency of cellular metabolism. In an arboreal form, egg laying is a hazardous way of producing young, unless it is accompanied by nest building. This habit is not very difficult to acquire by a bird that can fly to and from the sources of nest building material, but would be very difficult for a small running animal, for which vivipary is a more efficient solution. Hence the shift from egg-laying to vivipary could well have evolved first in arboreal therapsids or early mammals. The shift in ear bones that palæontologists use as diagnostic for the transition from reptiles to mammals probably evolved in response to strong selective pressure for more efficient action of the jaws as well as for increased hearing ability in a nocturnal animal. This increased sensitivity of hearing would have been made possible only by a simultaneous development of certain centres of the brain, in this way laying the foundation for the increase in intelligence that accompanied the later evolution of mammals.

Of equal importance with the origin of mammals from small, highly specialised therapsids was their long period of existence and evolution as a minor element of the world's fauna until, after the extinction of the dinosaurs at the end of the Cretaceous period, they underwent their explosive phase of evolution at the beginning of the Tertiary. No clearer example could be desired of the course of evolution postulated by Bock (1965); a major ascent in grade of organisation that resulted from an initial radiation into a peripheral adaptive zone, a long period of successive adjustments to various peripheral zones, and a final return to a major zone as its dominant inhabitants after the major ascent in grade had been completed.

The two examples just presented can also be used to illustrate how specific novel structures and functions can be acquired as a result of certain kinds of adaptive radiations. This subject has been well reviewed by Mayr (1960), so that a detailed discussion of it is not necessary here. He pointed out that novelties are usually acquired through a transfer of function, the transitional conditions being adaptive in an intermediate habitat because either (1) two different structures can serve the same function or (2) a single organ can perform two functions. The first situation is illustrated by lungfishes as well as many amphibians, which receive oxygen via gills if the water is well aerated, but via the lungs if the water has a low oxygen content. During the transition from paired lungs to a single dorsal swim bladder, one can imagine with little difficulty the existence of an air sac that could have performed either function, depending upon the condition and depth of the water. According to the hypothesis of amphibian origins presented above, the appendages that were intermediate between fins and legs could have helped the animal's locomotion when it was not seeking prey, and have been useful at other times for the predatory lunge. Many authors have suggested that during the reptilian–mammalian transition, the small, partly detached articular and quadrate bones of the lower jaw could have amplified sound vibrations even before they became incorporated as accessory bones in the ear. Consequently, the best palæontological examples that are available to us indicate that no great difficulties are involved in assuming that specific new and more efficient functions can arise as byproducts of certain kinds of adaptive shifts.

Preconditioning as a basis for the origin of evolutionary novelty

The review of vertebrate palæontology that was undertaken for this discussion amply confirms the statement of Simpson that by far the commonest fate of evolutionary lines, even those that have initiated novel ways of life, has been extinction. Moreover, among those lines that have persisted into modern times, a high proportion have remained essentially stable for long periods of time, and are now represented by only a few taxonomically isolated forms. A review of mammalian and reptilian families as outlined by Romer indicates that among the 186 families of reptiles listed, 85 per cent are now extinct, 8 per cent are represented by only one or a few stabilised modern genera, and only 7 per cent contain groups that have undergone adaptive radiation in recent times. For the 273 families of mammals, the figures are 62 per cent extinct, 17 per cent represented only by stabilised relics, and 21 per cent containing some examples of recent adaptive radiation. Since even among the latter group about 75 per cent contain more extinct than recent genera, the proportion of mammalian genera that have undergone successful adaptive radiations is not more than 5 per cent of those that have evolved. Since mammals are generally recognised as having evolved far more actively during the Tertiary period than any other class of organisms, this percentage is probably much higher than that for animals in general.

Equally significant is the fact that if both mammals and reptiles are considered, the examples of families that have been reduced to evolutionary stabilisation are nearly as numerous (64) as those in which some active evolution has continued (74). Apparently, even if a group does not become extinct, its chances of becoming stabilised at a particular organisational level are just as great as those for future divergent evolution. This being the case, one must ask the question: what factors favour continued evolution as opposed to those that lead towards stabilisation and stagnation?

This question was answered some time ago by both Simpson (1953) and Schmalhausen (1949). According to Simpson, evolutionary stabilisation or 'bradytely' is characteristic of organisms that occupy broad, continuously available adaptive zones, such as the ocean and the great forest belts. Within these zones, bradytely is favoured by a very general kind of organism–environment interaction, of such a nature that many factors of the environment can change without materially affecting the ability of the available gene pool of the bradytelic organism to exploit it successfully.

Schmalhausen specifies for animals certain kinds of population–environment interaction that promote evolutionary stability. The principal ones are the reliance of the population on a high reproductive rate, and the evolution of a very generalised kind of protection from predators, such as the carapace of the horseshoe crab (*Limulus*), and the shells of turtles and tortoises.

Examples of this sort could be multiplied indefinitely. They show clearly that the long-term fate of an evolutionary line—whether it becomes extinct, stabilised, or undergoes evolutionary progress—depends not upon any biological laws, similar to the laws of chemistry and physics, but upon specific *ad hoc* interactions between populations and their environment. The evolutionist may wish to systematise his study of these interactions by making certain generalisations, but if he does so he must recognise a basic axiom of biology. This is that the only generalisation about evolutionary biology which has no exceptions is the following one: all generalisations that the evolutionary biologist may wish to make are subject to some exceptions. Contrary to the physical sciences, at least as they are regarded by those not specialists in them, the science of evolutionary biology is not governed by universal laws, to which no exceptions are permitted. The appearance of regularity that some observers may detect actually results from an emphasis upon a highly selected succession of events, and a disregard for those events that do not contribute to the regularity that is being described.

The origin of eucaryotes and of multicellular organisms

The examples already presented have been from vertebrates, partly because these organisms are the most familiar to non-biologists and biologists alike, and partly because they have the best fossil record. Nevertheless, the point of view expressed above can be regarded as fully valid only if it is supported by evidence from all kinds of organisms. In the present section, I shall examine in this respect three major advances in biological organisation: the origin of eucaryotes, of multicellular plants, and of Metazoa, or multicellular animals.

Most biologists now agree that the most important single advance in biological organisation was the origin of the eucaryote cell: one in which the bulk of the DNA is incorporated into chromosomes that are flexible organelles containing protein and enclosed between divisions in a nuclear membrane, and in which cell division is by mitosis and the cytoplasm is equipped with a complex system of organelles, such as mitochondria, endoplasmic reticulum, lysosomes, Golgi bodies and, in autotrophic organisms, plastids (Stanier, Adelberg and Doudoroff, 1970). Two principal hypotheses exist to account for the origin of eucaryote cells. One is that the various organelles originated in association with the enlargement of a simple procaryotic cell like that of bacteria and the smaller blue-green algae (Allsopp, 1969). The other is that the eucaryote cell originated via a series of symbiotic inclusions of various kinds of procaryotic cells within a single large cell that originally was also procaryotic (Margulis, 1970, and earlier authors cited there).

Whichever hypothesis one adopts, the necessary prelude to the origin of the eucaryotic cell is the existence of a relatively large procaryotic cell having a highly flexible cellular membrane. If one adopts the single cell theory, one must postulate that internal membranes and organelles were derived by invagination of the surrounding cellular membrane. The symbiotic theory must postulate the initial existence of a procaryote cell that was capable of ingesting other cells, amoeboid fashion, via the activity of its cellular membrane.

Consequently, the most logical intermediate stage between procaryotes and eucaryotes is a large procaryotic cell having a flexible, highly extensible cellular membrane, and capable of ingesting other procaryotes. When it arose, this method of feeding must have been unique, representing a high degree of specialisation. The evolution of a membrane that could become extended in response to tactile stimuli would have required a whole succession of genetic modifications, for the origin of which one would have to postulate a course of directional selection similar to that which was responsible for the origin of lungs, homoiothermy and other specialisations that in vertebrates represented fundamental advances in grade of organisation. Once perfected, it made possible the occupation of a broad adaptive zone on the basis of a more efficient way of getting food. The various structures of the eucaryote cell, regardless of the way in which they were acquired, could be interpreted adaptively as increasingly efficient ways of exploiting this zone.

On the basis of either the single cell or the symbiont theory, the origin of autotrophic eucaryotes from autotrophic procaryotes must be regarded as an example of the indirect acquisition of a new grade of organisation, through multiple and reversed adaptive radiations. If the single cell theory is favoured, the intermediate stage could be visualised as a unicellular organism which was ecologically similar to the modern *Euglena*. The permanent plastids of the ancestral procaryote became modified into membrane-bound proplastids, similar to those that exist in the embryonic cells of a higher plant. Given the presence of light, these proplastids could develop into photosynthetic organelles, and the organism would become autotrophic. If, on the other hand, the cell should exist in a habitat having insufficient light, the plastids would fail to develop, and the organism would lead a heterotrophic existence by means of ingesting its food, amoeboid fashion. Such an organism could occupy a particularly broad adaptive zone. Its evolutionary line could eventually give rise in one direction to a *Chlamydomonas*-like autotrophic organism having a rigid cellulose wall, and in another direction to a heterotrophic ingester, similar to an amoeba. On this hypothesis, the intermediate stage between an autotrophic procaryote and an autotrophic eucaryote would be an ambivalent organism capable of either autotrophic or heterotrophic existence.

As Margulis (1970) has pointed out, the symbiont hypothesis requires that autotrophic eucaryotes evolved from heterotrophic eucaryotes via the ingestion of small autotrophic procaryotic cells, which were then converted into self-reproducing plastids. On the basis of this hypothesis, therefore, the rise in grade from the procaryotic to the eucaryotic conditions was achieved by autotrophs via a heterotrophic intermediate.

For the origin of multicellular plants, similar sequences of *ad hoc* adaptive advantages can be postulated. The simplest kind of multicellular plant that is capable of evolving further into an archegoniate or a large alga is an unbranched, polarised filament attached at one end to a substrate. The green algae *Ulothrix* and *Oedogonium* are familiar examples (see Fritsch (1935) for the source of this and the subsequent descriptive material). For such a plant to evolve from a unicellular autotroph such as *Chlamydomonas*, three new characteristics must be acquired. The first is the ability of a cell to become polarised in such a way that rhizoids are formed in contact with the substrate, and new cells are added to the distal end. The second is the ability to form separate cytoplasms around the two daughter nuclei at mitotic telophase, and their eventual separation by a cell wall, so that two firmly connected cells are produced by each mitotic nuclear division. Finally, in the aquatic plants that preceded terrestrial plant life, at least some cells of mature plants must have been capable of giving rise to motile zoospores or gametes.

Of these three characteristics, only the formation of firmly connected cells had to be acquired *de novo*. In *Ulothrix* and other filamentous algae, the polarity of the filament is a direct modification of the polarity of the zoospore or of the zygote, since the motile cell becomes attached to the substrate by its 'front' or flagellated end. The polarity which eventually made possible the anchoring and distal growth of multicellular plants was acquired through gradual modification of the preexisting cellular polarity of unicellular motile autotrophs, associated with their ability to move in a particular direction. Furthermore, many unicellular autotrophs have a non-flagellate resting stage, from which motile flagellate cells emerge at the end of the dormant period. The alterations of organelles and of cell physiology that give rise to the motile cells of multicellular algae may, therefore, be homologous to and derived from similar alterations in their motile unicellular ancestors.

The adaptive basis of the sequence of events from an autotrophic, amoeboid or nonmotile unicellular organism to a primitive filamentous alga can be postulated as follows. First, in association with more rapid cell proliferation, a reduction in cell size would have an adaptive value. This would, however, reduce the cells to a size range capable of being ingested by the ever-present amoeboid organisms. To escape such predation, the most immediately successful evolutionary strategy would be rapid motility with the aid of flagella. If, however, the organisms were in small bodies of water heavily populated with amoeboid predators, this strategy would become increasingly ineffective. Four alternative strategies would then be possible for these autotrophs. One would be the acquisition of a hard outer cell wall, as in *Chlorella*, diatoms and desmids. The second would be the formation of colonies of semiindependent cells, as in the Volvocales. The third would be anchorage to the bottom of quiet water, followed by rapid growth in cell size, and many nuclear mitoses unaccompanied by cell division. This would give rise to coenocytic algae, such as *Acetabularia* and *Vaucheria*. The fourth strategy would be anchorage to swiftly moving water, either streams or tidal pools, which the amoeboid organisms could not enter without being destroyed. In such habitats, both anchorage and the firmness that would result from the evolution of cell wall formation and the formation of filaments of uninucleate cells would have a great adaptive advantage. Under these conditions of maximum selective pressure exerted by the nonbiotic environment, accompanied by minimal pressure from predators, the ability to form polarised, anchored filaments could evolve to its maximum condition of efficiency. Once this condition had been acquired, the filamentous algae could either reinvade quiet water at a higher level of organisation, or could enter moist terrestrial habitats, as is true of some modern filamentous algae (*Trentepohlia*).

The analogy between this postulated sequence of events and that which was described for primitive fishes, bony fishes, lungfishes and amphibians is obvious. This analogy can be carried further by considering the further evolutionary consequences of the three strategies described above. The cellular forms that became encased in thick cell walls either became stabilised (*Chlorella*) or went through extensive evolution at the unicellular level (diatoms, desmids). Possibly, desmid-like unicellular forms gave rise to the multicellular apolar Conjugales, but these were almost certainly a side line that evolved no further. The strategy of increase in size through acquisition of the coenocytic condition gave rise to several different groups of algae, some of them of moderate size (*Caulerpa*, *Hydrodictyon*), but the bulk of the larger and dominant groups of algae, as well as all autotrophic land plants except for lichens are most probably derived originally from polarised, uninuclear filamentous algae. The evolutionary future of plants, therefore, depended upon successful passage through a stage in which an extreme and in one way inhospitable habitat, flowing water, was conquered and colonised.

This postulated sequence of events is, I admit, highly speculative, and I doubt that firm evidence in favour of it can ever be acquired. Nevertheless, it is plausible, and shows that the origin of a major kingdom of organisms can be explained on the basis of adaptive shifts similar to those that give rise to species and genera.

The origin of Metazoa

The problem of the origin of multicellular animals, or Metazoa, is even more difficult and speculative than any of the other problems connected with the early evolution of eucaryotes. This is because not only is there a complete lack of fossil evidence but, in addition, living intermediate forms, which point towards analogous sequences leading to the origin of multicellular algae and fungi, are completely lacking with respect to all phyla of Metazoa, with the possible exception of sponges (Hyman, 1940; Greenberg, 1959). Even the question of mono *vs* polyphyletic origin of Metazoa is by no means settled. The separate origin of sponges (Porifera) from choanoflagellate Protozoa is highly probable and widely accepted. The hypothesis of Greenberg (1959), that Cnidaria (Coelenterata) and acoel flatworms (Platyhelminthes) were separately derived from different groups of Protozoa, appears to be highly plausible, but zoologists are by no means in agreement on this point.

Given this uncertainty, the only thing that the general evolutionist can do is to consider the structural and physiological characteristics that are common to all Metazoa but are not general properties of all multicellular organisms, and to ask whether any immediate adaptive advantages can be postulated for these characteristics in any habitats in which primitive Metazoa might have lived.

These distinctive metazoan characters are as follows (Greenberg, 1959):

(1) The presence of from three to five different kinds of cells, including specialised eggs and sperm, and two layers of body cells that are differentiated from each other.

(2) Ability to form hollow or solid colonies.

(3) Gastrulation or some similar form of cellular infolding or inversion.

(4) Ingestion of food rather than photosynthesis, as in algae, or external digestion, as in fungi.

Since ingestion is postulated as the original condition in eucaryotes, this ability was very likely retained throughout the ancestral line or lines that led to Metazoa. Along with it, the highly flexible cell membrane characteristic of these organisms was also retained. On the other hand, the general appearance of metazoan cells suggests that they did not evolve directly from ameboid protozoa but from flagellates. Consequently, evolutionary reduction in size and the acquisition of high motility through the action of flagella is postulated for the ancestors of Metazoa just as for those of plants and fungi. Adaptations for escape from ameboid predators, therefore, could have also followed the sequence, increased motility followed by colony formation, which was postulated for some autotrophic organisms, particularly the Volvocales. The line or lines of evolution leading to Metazoa, following Greenberg (1959), are postulated as similar and analogous to, but not homologous with, the line that led from *Chlamydomonas*-like unicellular organisms to *Volvox*.

Based upon this analogy, a reasonable postulate is that the differentiation of eggs and sperm is based upon the adaptive value in seasonally unfavourable habitats of a resting cell that contains a large amount of food and is protected by a heavy wall. This adaptive value would be considerably higher in an organism that obtained its food by ingestion than in either an autotrophic organism or one that subsisted by external digestion. The unicellular organism growing from the egg would have to become big enough to carry out ingestion before it could obtain any food except for that with which it was equipped during dormancy. On the other hand, autotrophic and externally digesting organisms could obtain food almost as soon as they emerged from a dormant condition, and very little new growth of cellular structures would be necessary.

In *Volvox*, an infolding process similar to gastrulation has evolved apparently as a means of protecting the young, growing colony. This suggests that in the earliest Metazoa gastrulation may have evolved as a method of placing the growing oöcyte in the inside of the colony, and thus protecting it. A fundamental feature of primitive Metazoa which did not evolve in *Volvox* is the presence of two differentiated layers of cells. Such division of labour must have helped greatly the ability for ingestion, and at the same time have rendered possible the differentiation of specialised locomotor perceptive and protective cells (Hyman, 1940).

This line of reasoning suggests that the principal selective pressures that led towards Metazoa were all connected directly or indirectly with retention and increase in efficiency of the ability to capture and ingest food. Those that were not directly connected with this mode of life were associated with adaptations for escape from predators or protection of young that were the most highly compatible with ingestion as a method of getting food.

The low frequency of evolutionary advances

One can gain a true perspective on the nature of evolutionary change only by recognising the extreme rarity of any changes that can be regarded as advances in complexity or in any other quality that might be used to measure anagenesis. Diversification by adaptive radiation, that is, cladogenesis, is ubiquitous. Some idea of the number of successful adaptive shifts that have occurred since eucaryote organisms first evolved can be obtained by calculating a rough estimate of the number of contemporary species of eucaryotes plus those that have existed throughout this period and are now extinct, and multiplying this total number by ten, on the assumption that, on the average, ten successful adaptive shifts have occurred within each species that have not led to the origin of a new species. I have made such calculations, using the estimate made by Dobzhansky (1974) of the number of contemporary species, $3{\cdot}0 \times 10^6$. Of these, three-fifths, or about $1{\cdot}8 \times 10^6$, are insects, which must be considered separately. With respect to the remaining species, $1{\cdot}2 \times 10^6$ in number, those of animals, plants and protists, palæontological data suggest that the figure of 5×10^6 years for the period between the origin of a species and its extinction or evolution into a new species is reasonable, though more likely too long than too short.

If this period is accepted, then there were about 200 complete replacements of species during the one billion (10^9) years since the origin of eucaryotes, insects excepted. Hence the total number of eucaryote species that has existed, not including insects, is estimated as 240×10^6.

The same calculations cannot be made for insects, since this class did not appear until the Carboniferous period, 300 million years ago, and two of the orders that now contain the largest numbers of species, the Diptera and Hymenoptera, did not become dominant groups until much later. Considering what is known about insect evolution, I have arrived at a rough estimate of 80×10^6 as the number of insect species that have existed throughout their evolutionary history. This, also, is a very conservative estimate. Putting these two estimates together, I obtain a minimal figure of 320×10^6 for the total number of eucaryote species that have existed on the earth. This figure is of the same order of magnitude as that given by Simpson (1960), namely, 500 million species for all phyla of organisms. Multiplying this figure by ten, the number of successful adaptive shifts is estimated at $3{\cdot}2 \times 10^9$, or 3·2 billion.

What proportion of these radiations have led to advances in grade? A survey of families of flowering plants, mammals and insects has led me to conclude that a minor advance in grade, such as that responsible for ichthyomyine rodents, or for the distinctive pollination mechanisms found in such families as milkweeds (Asclepiadaceae) and orchids, might be expected once for about every 500 origins of new species, or about once in every 5000 successful adaptive shifts. This would mean that about 640 000 of these minor advances have occurred during the billion (10^9) years that encompass the evolution of eucaryote organisms. Most of these advances, however, are of the kind that would be recognised only by taxonomists or comparative morphologists. Biologists in other disciplines, and particularly non-scientists, would be hardly aware of them.

We must ask ourselves, therefore, how many major advances in grade have taken place during the evolution of eucaryotes? Under this category, I am thinking of such events as the origin of multicellular from unicellular organisms, of the digestive tube, coelom, and central nervous system of animals, of vertebrate limbs, lungs, elaborate sense organs, warm blood, the placenta, and elaborate social behaviour. I have compiled a list of evolutionary advances of this magnitude known to me, and cannot find more than a total of 70. I am willing to admit that I could have overlooked many such events, or have failed to recognise them as of major importance. Consequently, an estimate of 100 major advances that have occurred throughout the evolutionary history of eucaryotes is a reasonable and not excessively high number. If this number is accepted, then we must recognise the probability that only one out of every 32 million adaptive shifts has led to a major evolutionary advance.

What is the significance of these figures? In the first place, they tell us that evolutionists are correct in directing their major attention towards population–environment interactions at and below the level of the species. The great bulk of evolutionary diversification has consisted of such changes, of which major 'advances' are extremely rare byproducts. Secondly, as Simpson (1953) and others have pointed out, there is no such thing as a general trend, or even localised trends, towards evolutionary advancement towards higher grades of organisation. The advances that have occurred have been the rare outcome of certain specific genotype–environment interactions, which have permitted the end products of a particular line of adaptive radiation to occupy and become diversified in a new adaptive zone. Third, evolutionists will reach a fuller understanding of these advances, both major and minor, not by attempting to formulate general laws that can be applied indiscriminately to all examples, but by studying each separate example in detail. We should learn as much as possible about the differences between the circumstances that prevailed when a major advance was accomplished, as compared to similar adaptive radiations in related organisms that did not lead to an advance in grade.

The role of the conservation of organisation and of epigenesis

Even when we recognise the probability that major evolutionary advances have not resulted from general trends but from successions of rare, isolated events, we have not provided an answer to the critics of modern evolutionary theory who argue as follows. If mutations are at random relative to adaptability, and if population–environment interactions via natural selection are unrelated to evolutionary advances in grade, how can we explain the occurrences of such advances at all, even as isolated events? I believe that the answer to this question can be derived from three principles: (1) the conservation of organisation; (2) the epigenetic nature of much evolutionary change; and (3) a tendency for increases in developmental complexity and degree of integration of the individual genotype to confer a greater capacity for exploiting new environments, while decreases in complexity and integration tend to decrease this capacity and to restrict evolutionary expansion.

As I have presented it elsewhere (Stebbins, 1969), the principle of the conservation of organisation is as follows. Whenever a complex organised structure or an integrated biosynthetic pathway has become an essential adaptive unit of a successful group of organisms, the essential features of this unit are conserved in all of the evolutionary descendants of the group concerned. Examples are the cilia and flagella of eucaryotes, the eucaryote chromosomes and mitotic apparatus, the process of gastrulation in Metazoa, the basic segmental structure of insects, the basic skeletal anatomy of vertebrates, the universal pathways of glycolysis, the exclusive use of *L*-isomers of amino acids in the synthesis of proteins, and innumerable others. Mutations that affect these structures and processes have an adaptive value not in direct connection with genotype–environment interactions, but through their interactions with other genes that contribute to the structures or processes involved. In higher organisms, the majority of genes contribute in one way or another to these conserved structures and processes. The adaptive value, and hence the acceptance or rejection by natural selection of most new mutations, depends not upon direct interactions between these mutations and the external environment, but upon their interaction with other genes, and their contribution to the adaptiveness of the genotype as a whole.

This principle is little more than a statement in terms of phenotypic characteristics of the now well-recognised principle of the integration of the genotype (Mayr, 1963; Dobzhansky, 1970). Nevertheless, by focusing attention on specific processes and pathways, it makes the concept of integration more precise. In my opinion, the degree of integration of the entire genotype has recently been overestimated by some evolutionists. Some structures and processes act at very different points in the individual life cycle than do others. Interactions between them need not be as strong as those between different parts of the same structure or different processes in the same biosynthetic pathway. Comparative studies of any group of organisms tell us that, even among related evolutionary lines, different parts of the body evolve at different rates. The extent to which this is possible is inversely proportional to the effectiveness of the conservation of organisation under a given set of conditions.

The conservation of organisation influences the direction of evolution in the following way. If a population is exposed to a new environment that is exerting different selective pressures from those that were previously in force, it will survive only if it can adjust to these changes by altering its genotypic constitution. In most situations, however, any one of several possible 'evolutionary strategies' will have the desired effect. The one that the population adopts will be that which alters least those structures and biosynthetic pathways that are the most highly integrated, and for the successful modification of which a large number of genetic adjustments need to be made. In the words of the botanist William Ganong (1901), evolution will proceed according to the principle of adaptive modification along the lines of least resistance. For this reason, the evolution of a particular sequence of populations in a particular direction is to a great extent epigenetic. The direction that it will take in later stages of evolution is influenced only in part by direct interactions between populations and their environment. The total gene pool of the population, along with the extent and nature of integration of this gene pool, has an equal or greater directive influence. These properties of the gene pool have been the result of earlier events of natural selection.

The strength of this epigenetic influence can be best recognised through examples of divergent evolution of related and sympatric populations in response to the same selective pressure. Elsewhere (Stebbins, 1967) I have described the probable consequences of simultaneous selection for increased seed production in three different kinds of plants that initially have different organisation of their reproductive structure. On the basis of conservation of organisation and adaptive modification along the lines of least resistance, a buttercup (*Ranunculus*) would be expected to respond to this kind of selective pressure by increasing the number of carpels per flower; a tulip by increasing the number of ovules per ovary locule; and a *Chrysanthemum* or other Composite by increasing the number of florets per capitulum or per plant. Furthermore, one need not confine one's predictions to the divergent behaviour of distantly related organisms such as these. Most probably, related species of the same genus, some of which have many seeded capsules and others having capsules with only one or a few seeds, will react differently to the same selective pressures for increased seed number or seed size. Such predictions could be tested by experimental methods, particularly if fertile or partly fertile hybrids could be obtained between populations that differed with respect to such characteristics. This prospect is another example of the opportunities which I feel certain that evolutionists now have for establishing connections between evolution at the level of populations and species, and major transspecific trends.

Granted that epigenesis plays a large role in determining evolutionary direction, the question of advances in grade is still not satisfactorily answered. *A priori* there is no reason to assume that evolution in a particular direction will either lead to success on a broad scale or open the door to further advances. As a matter of fact, the estimates made earlier in this paper lead me to the conclusion that the great majority of minor advances in evolutionary grade do not lead to greater success or to further progress in the direction of higher organisation, greater integration or more independence of the environment. This conclusion is derived from the ratio of minor to major evolutionary advances, which was estimated as about 6400:1.

The principle which I believe leads the way to resolving this problem is that advances towards higher organisation and greater complexity of bodily structure and developmental pattern, as well as towards greater integration of structure and function, in some instances increase the potentiality for further adaptive radiations, including complete dominance over some large adaptive zone. On the other hand, evolutionary trends toward a lower level of structural, developmental and biochemical organisation usually restrict the adaptive zone into which the resulting forms can radiate. For instance, radiations from generalised reptiles towards turtles and snakes, though highly successful in spite of reductions with respect to such structures as teeth (turtles) and limbs (snakes), nevertheless restricted the potentiality for further radiation. On the other hand, radiations from generalised reptiles towards mammals and birds, largely because of the acquisition of homoiothermy and more complex nervous responses, increased the potentiality for further radiation.

Conclusion

The argument of this paper can be summarised as follows. The unity of the evolutionary process is manifest through the ubiquitous prevalence of adaptive shifts and adaptive radiations. Any kinds of changes that an evolutionist might regard as progress or advancement are no more than special examples of such radiations. Those examples which are usually regarded in this light, including the various stages in the evolution of mankind from generalised unicellular organisms, are distinctive in two ways. First, they display to an unusually high degree the phenomenon of epigenesis. The direction of evolution in them is determined more by interactions between preexisting gene complexes and new mutations than by direct interactions between populations and their environment. Furthermore, they are the examples which illustrate most strongly a cardinal principle that governs the overall direction of evolution. Although most adaptive shifts are neutral or restrictive with respect to potentiality for further evolution, a small proportion of them opens up new evolutionary opportunities. This select group of opportunistic adaptive shifts includes a relatively high proportion of changes involving increases in developmental complexity and degree of integration. This small bias has been responsible for the phenomenon that has often been designated advancement in grade or evolutionary progress.

References

Allen, R. and Sheppard, P. R. (1971). Copper tolerance in some populations of the monkey flower *Mimulus guttatus*. *Proc. Roy. Soc. B.*, **177**, 177–96.

Allsopp, A. (1969). Phylogenetic relationships of the Procaryota and the origin of the eucaryotic cell. *New Phytol.*, **68**, 591–612.

Bock, W. J. (1965). The role of adaptive mechanisms in the origin of higher levels of organization. *Syst. Zool.*, **14**, 272–87.

Bradshaw, A. D., McNeilly, T. S. and Gregory, R. P. G. (1965). Industrialization, evolution and the development of heavy metal tolerance in plants. *Ecol. and Industrial Society*: 5th symposium of British Ecol. Society, 327–43.

Cavalli-Sforza, L. L. and Maccacaro, G. A. (1952). Polygenic inheritance of drug resistance in the bacterium *Escherichia coli*. *Heredity*, **6**, 311–31.

Crompton, A. W. and Jenkins, F. A., jr. (1968). Molar occlusion in Later Triassic mammals. *Biol. Rev.*, **43**, 427–58.

Dobzhansky, Th. (1970). *Genetics of the Evolutionary Process*. Columbia University Press, New York.

Dobzhansky, Th. (1974). Chance and creativity in evolution. In this volume.

Fritsch, F. E. (1935). *The Structure and Reproduction of the Algae*, vol. 1. Macmillan, New York.

Ganong, W. F. (1901). The cardinal principles of morphology. *Bot. Gaz.*, **31**, 426–34.

Goodwin, R. H. (1944). The inheritance of flowering time in a short-day species, *Solidago sempervirens* L. *Genetics*, **29**, 503–19.

Greenberg, M. J. (1959). Ancestors, embryos, and symmetry. *Syst. Zool.*, **8**, 212–21.

Hecht, M. K. (1965). The role of natural selection and evolutionary rates in the origin of higher levels of organization. *Syst. Zool.*, **14**, 301–17.

Hyman, L. H. (1940). *The Invertebrates: Protozoa through Ctenophora*. McGraw-Hill, New York.

Jowett, D. (1964). Population studies on lead-tolerant *Agrostis tenuis*. *Evolution*, **18**, 70–80.

Lanham, U. (1964). *The Insects*. Columbia University Press, New York.

Luria, S. E. and Delbrück, M. (1943). Mutations of bacteria from virus sensitivity to virus resistance. *Genetics*, **28**, 491–511.

Margulis, L. (1970). *Origin of Eukaryotic Cells*. Yale University Press, New Haven.

Mayr, E. (1960). The emergence of evolutionary novelties. In *Evolution after Darwin*, vol. 1: *The Evolution of Life* (ed. S. Tax), 349–80.

Mayr, E. (1963). *Animal Species and Evolution*. Harvard University Press, Cambridge, Mass.

Romer, A. S. (1966). *Vertebrate Paleontology*. University of Chicago Press, Chicago.

Schaeffer, B. (1965). The Rhipidistian–Amphibian transition. *Amer. Zoologist*, **5**, 267–76.

Schmalhausen, I. I. (1949). *Factors of Evolution: the theory of stabilizing selection*. Blakiston, Philadelphia.

Schmalhausen, I. I. (1964). *The Origin of Terrestrial Vertebrates*. Moscow. (Eng. trans. L. Kelso, Academic Press (1968).)

Simpson, G. G. (1953). *The Major Features of Evolution*. Columbia University Press, New York.

Simpson, G. G. (1960). The history of life. In *The Evolution of Life* (ed. S. Tax). Chicago University Press, Chicago, 117–80.
Simpson, G. G. (1970). Uniformitarianism: an inquiry into principles, theory and method in geohistory and biohistory. *Essays in Honor of Theodosius Dobzhansky* (ed. M. K. Hecht and W. C. Steere). Evolutionary Biology Suppl., 43–96.
Stanier, R. Y., Doudoroff, M. and Adelberg, E. A. (1970). *The Microbial World*, 3rd edition. Prentice-Hall, Englewood Cliffs, New Jersey.
Stebbins, G. L. (1950). *Variation and Evolution in Plants*. Columbia University Press, New York.
Stebbins, G. L. (1967). Adaptive radiation and trends of evolution in higher plants. In *Evolutionary Biology* (ed. Th. Dobzhansky, M. K. Hecht and W. C. Steere), 101–42.
Stebbins, G. L. (1969). *The Basis of Progressive Evolution*. University of North Carolina Press, Chapel Hill.
Thomson, K. S. (1969). The biology of the lobe-finned fishes. *Biol. Rev.*, **44**, 91–154.
von Wahlert, G. (1965). The role of ecological factors in the origin of higher levels of organization. *Syst. Zool.*, **14**, 288–300.
Walker, E. L., Warnick, F., Hamlet, S. E., Lange, K. L., Davis, M. A., Uible, H. E., Wright, P. E. and Paradiso, J. L. (1968). *Mammals of the World*, second edition, vol. 2, 646–1500.

18. Chance and Creativity in Evolution

THEODOSIUS DOBZHANSKY

Diversity and unity are the equally important and enthralling aspects of the living world. The virus of foot-and-mouth disease is a sphere 8–12 millimicrons in diameter. The blue whale reaches 100 feet in length and 150 tons in weight. *Sequoia gigantea* may weigh more than 6000 tons. Some bacteria grow at −23°C in saline pools in Antarctica, and others at 80–85°C in hot springs of Yellowstone Park. The distribution range of the mountain lion (puma) extended from Alaska to Patagonia, so that the animal lived in a great variety of climates and other conditions. Man is, of course, the truly cosmopolitan species, able to create environments suitable for himself not only on the earth but in cosmic spaces. By contrast, some species are very narrow specialists. *Drosophila carcinophila* develops only in the external nephric grooves beneath the flaps of the third maxilliped of the land crab *Geocarcinus ruricola* on the islands of Montserrat and Mona in the Caribbean. *Drosophila endobranchia*, a species not closely related to *D. carcinophila*, develops in the gill chambers of the same and an allied species of land crabs on Cayman Islands (Carson, 1971).

About a million and a half living species of animals and plants have thus far been described and named (Dobzhansky, 1970). Probably no fewer than that remain to be identified. Raven, Berlin and Breedlove (1971) believe the number of existing species to be roughly ten millions, but this is almost certainly an exaggeration. Some groups of organisms have become diversified to an extent that seems altogether excessive. Thus, about three-quarters of all animal species are insects, and among insects about 40 per cent are beetles. On the other hand, described species of mammals and birds number merely about 3700 and 8600 respectively, and probably only small numbers of species of these groups remain to be added (Mayr, 1969).

Yet underlying the prodigious diversity are remarkable similarities of all living beings. The genetic information is transmitted everywhere by means of two related groups of substances—deoxyribonucleic (DNA) and ribonucleic (RNA) acids. The genetic code, whereby this information is translated into sequences of amino acids in proteins, is also very nearly invariant. The same twenty amino acids compose the proteins everywhere, although different proteins contain them in different proportions. Striking uniformities prevail in the cellular metabolism of most diverse living beings. Adenosine triphosphate (ATP), biotin, riboflavin, hemes, thiamine, pyridoxine, vitamins K and B_{12} and folic acid implement enzymatic processes everywhere. Fatty acid oxidation, glycolysis and the citric acid (Krebs) cycle are metabolic pathways in animals, plants and microorganisms (Green and Goldberger, 1967).

The above 'biochemical universals' make sense solely in the light of evolution. Everything that lives is derived ultimately from a common stem. In all likelihood, life arose only once. Or, if there were several emergences of life from inert matter, then the descendants of a single primordial source have survived. The aspect of unity of life is, thus, easily accounted for by the common origin of all living beings. Only in this and in no other sense is the opinion of Monod (1971) justified: 'Essentially, however, the problem of evolution has been resolved and evolution now lies well to this side of the frontier of knowledge.'

Historical and causal approaches to the study of organic diversity

The diversity of life, like its unity, is a product of evolution. This statement is correct but it is surely far from the whole story. Two categories of questions demand answers: historical and causal ones. We want to trace and to date the evolutionary history, the phylogeny of the living world. This is nowhere near accomplished. A single example will suffice as an illustration. More effort has been expended to trace the origins of mankind than of any other species. More diverse hominid fossils have been discovered in the last twenty or so years than ever before. The present estimates of the antiquity of toolmaking hominids have been at least doubled. Yet which fossil forms were our ancestors and which only collateral relatives is far from ascertained. Historical problems do not always have only theoretical interest. For example, Coon (1962) claimed that the African races of man have attained the evolutionary 'grade' of *Homo sapiens* some quarter of a million years later than the European ones. This claim happens to be based on misconceptions. The point is, however, that this particular blunder is eagerly taken advantage of by racists. The black people should wait a quarter of a million years to achieve equality with whites!

Causal analysis of the evolutionary process is in some ways more and in others less advanced than is the historical survey. The so-called neo-Darwinian (or biological, or synthetic) theory is now a majority view, though by no means unanimously accepted. In my opinion, it furnishes a satisfactory paradigm for causal analysis. The basic postulates of the theory are three: (1) the process of mutation yields the genetic raw materials; (2) evolutionary changes are constructed from these materials by natural selection; (3) in sexual organisms, reproductive isolation makes the divergence of biological species irreversible. Supposing that the paradigm is accepted as valid, does it follow that the problem of evolution has been resolved except for petty details? No, there are basic problems awaiting elucidation. What is more, these are problems of profound philosophic and humanistic significance.

Among the opponents of the neo-Darwinian theory, a majority reject it because, allegedly, it ascribes too great a role in evolution to chance. To suppose that a system of immense complexity, and at the same time of exquisite resourcefulness to maintain itself alive amidst hostile environments, can arise by chance, or by summation of chances, is, we are told, absurd. The poet Auden, who is a biological philosopher by avocation, finds miracles a more reasonable explanation.

Monod (1971) makes chance, on the contrary, the cornerstone of his evolutionary world view: 'Pure chance, absolutely free but blind, at the very root of the stupendous edifice of evolution; this central concept of modern biology is no longer one among other possible or even conceivable hypotheses. It is the sole conceivable hypothesis, the only one that squares with observed and tested fact.' Chance reigns at the root as well at the summit of evolution: 'man knows at last that he is alone in the universe's unfeeling immensity, out of which he emerged only by chance.' Monod does not underestimate the significance of this for man. He claims that 'There is no scientific concept, in any of the sciences, more destructive of anthropocentrism than this one, and no other so rouses an instinctive protest from the intensely teleonomic creatures that we are.' I do not think that the modern biological theory of evolution is based on 'chance' as much as Auden fears or Monod affirms. The knowns and the unknowns in the situation are worth considering in some detail.

Programmed evolution

Some biologists felt that the Darwinian and neo-Darwinian theories of evolution ascribe an unduly great role in evolution to chance. Attempts to find plausible alternatives have repeatedly been made. They usually took the form of theories of orthogenesis, according to which 'evolution is in a great measure an unfolding of preexisting rudiments' (Berg, 1969). Orthogenetic theories vary a great deal among themselves. Some are frankly preternaturalistic—evolution is impelled or guided by occult forces. Finalists postulate that evolution has a goal, such as the production of man; evolutionary changes are, then, a strip-tease show, wherein a series of disguises are removed piece by piece, eventually revealing man. Just why this necessitates the existence at our time level of some millions of biological species, that will presumably never become transformed into humanoids, is a puzzle.

More straightforwardly biological variants of orthogenesis analogise evolution to the development of an embryo. The development of, for example, a human egg cell proceeds through a long series of stages, from fertilisation, cleavage, gastrulation, organ formation, birth, infancy, adolescence, adulthood, to old age and death. These stages are predetermined, in the sense that one follows another in a fixed order, as long as life continues. A developmental programme is contained in the nuclei of the sex cells; at present we know that this programme is encoded in the DNA of the chromosomes; we are beginning to learn how the programme is brought to realisation. Could it be that the primordial life carried in itself a programme of evolutionary development, which has been realised only once in a long series of generations, as well as repeating itself countless times in numerous individuals, over a time period of some two or three billion years? This view was expounded by Berg (1922, reprinted in 1969).

Berg called his theory nomogenesis, evolution determined by law. Neither Berg nor anybody else was able to specify how this 'law' operates. It is incomprehensible that the DNA molecules could possibly be foreordained to change for billions of years always step by step, in a straight line leading, for example, from some prokaryote to man. It is not only incomprehensible, but we know it is not so—mutational changes in DNA have been observed to occur, and they do not lead in any particular direction. In point of fact, many of them lead to death! The similarity between the embryonic development (ontogeny) and the evolutionary development (phylogeny) must be viewed in a direction opposite to that in which Berg and other partisans of orthogenesis attempted to see it. Ontogeny does proceed directionally towards a goal, namely a body able to live and reproduce. This directionality is implanted in the hereditary materials by evolution. Wrong directions that also arose many times in evolution fell by the wayside, because they did not sustain life and reproduction of the bodies in which they appeared.

There is, however, one sense in which evolution could be called orthogenetic, if this term were not used in so many different ways that it is best avoided altogether. To my knowledge, nobody has been able to propose a satisfactory definition of what constitutes evolutionary progress. Nevertheless, viewing evolution of the living world as a whole, from the hypothetical primeval self-reproducing substance to higher plants, animals and man, one cannot avoid the recognition that progress, or advancement, or rise, or ennoblement, has occurred. As Barbour (1966) rightly said, 'By almost any standard man represents a higher level than primeval mud.' Simpson (1967), who contributed perhaps more than anybody else to critical analysis of various notions of evolutionary progress, nevertheless writes: 'Development and progression are so plainly evident in animal nature that these features deeply impressed biologists long before the grand fact of the evolution that produced them was understood.'

General evolution and particular evolutions

It is desirable at this point to distinguish between general and particular evolution. One can view evolution as a whole, or can study the particular evolutionary events each separately from others. Seen in the perspective of billions of years, evolution has indubitably brought about results which deserve the name of progress or, for those who are allergic to this word, some other name signifying much the same thing. But one must look at evolution also at close quarters. In so doing, one inevitably loses the sight of the forest (the progressiveness of the general evolution) for the trees (the turmoil of particular evolutions). Extinction is by far the commonest outcome of evolutionary development of lineages that have left documentation in the form of fossils. Death and extinction are antitheses of biological progress.

Furthermore, in many lineages one finds what one can only label retrogression. This is most spectacular in some endoparasites. The nerve systems, sense organs, digestive tracts and musculature become vestigial or are lost altogether. What is left is little more than a bag containing the reproductive organs. Even the biochemical machinery may become abridged, and the parasite relies on the host for some metabolic products which the ancestors of the parasite were undoubtedly able to make themselves. Yet progress is found also in some particular evolutionary lineages. Simpson (1967) has stated this admirably: 'evolution is not invariably accompanied by progress, nor does it really seem to be characterised by progress as an essential feature. Progress has occurred within it but it is not of its essence.' Emergence of man from his nonhuman ancestors is an obvious example of progress in a particular evolutionary lineage. To those who would reproach me for the anthropocentrism of this statement, I can only answer that this much anthropocentrism is necessitated by common sense.

Seen in retrospect, evolution as a whole doubtless had a general direction, from simple to complex, from dependence on to relative independence of the environment, to greater and greater autonomy of individuals, greater and greater development of sense organs and nervous systems conveying and processing information about the state of the organism's surroundings, and finally greater and greater consciousness. You can call this direction progress or by some other name. It should, however, be made crystal clear that, from the fact that evolution exhibits a direction or a trend, it does not follow that the evolution is being directed by some outside agency, or that it has been programmed beforehand. This none-too-subtle distinction has frequently been overlooked. Organisms exhibit internal, not external, teleology (Ayala, 1968).

The fact of directedness of evolution was referred to by Teilhard de Chardin (1959) as orthogenesis. He obviously had in mind the general evolution. In his view, 'from the beginning stages of evolution, the living matter which covers the earth manifests the contours of a single gigantic organism.' This metaphorical 'gigantic organism' does exhibit a discernible direction in its development, which Teilhard saw fit to designate by the ambiguous label 'orthogenesis'. Yet as a palæontologist he knew well that in particular lineages this 'orthogenesis' often leads to extinction and not to any kind of progress. He accordingly described evolution as 'groping' (*tâtonnement*), which is surely the opposite of straight-line development. A living species can be seen as groping, as though in the dark, for possibilities to survive and to respond adaptively to changing environments. The groping often ends in blind alleys, breakdowns and extinction. But occasionally it leads to improvements and progress. This, I submit, is a marvellously apt simile for natural selection. What Teilhard failed to see is that natural selection is responsible for the directedness of the general, as well as for the groping of particular evolutions.

The ascertainment of the fact of directedness of the general evolution is not tantamount to its explanation. The fact of directedness had been discovered, it would seem, prematurely, before the causes that bring evolution about were even begun to be deciphered. To suppose that the course of evolution was already programmed in the primordial life seemed to be the simplest solution. The simplicity was deceptive. It was deceptive for the same reason as the naive eighteenth-century preformationism. The notion of preformationism attempted to 'explain' embryonic development by supposing that a diminutive likeness of the organism that is to develop was already present in the sex cells, and had merely to grow in size. Neither the ontogeny nor the phylogeny can be understood that way. If a small facsimile of the adult organism is hidden in a sex cell, where are hidden its progenies? Bonnet's theory of 'encapsulation' reduced the theory of ontogenetic preformation to an absurdity. The notion that man was somehow invisibly present in its prokaryote ancestors likewise skirts the edge of absurdity.

Preformation of all future evolution in the primeval life can be true only in the trivial sense of the Laplacean universal determinism. Anything that happens is bound to happen. But the Laplacean determinism is not warranted even in physics (Nagel, 1961). In biological evolution, as I shall try to show, only a minute fraction of potentially possible evolutionary events actually come to be realised. Our task is to discover what brings them to realisation. The course and the outcome of evolution was not programmed by any external agent. The directionality of general evolution is nevertheless not a matter of chance or accident. It is due to laws built in the basic structure of life. Evolution is 'nomogenesis', but not in the sense Berg (1969) supposed it to be. Its 'law', *nomos*, is natural selection.

Mutation—chance and its limitation

Evolution involves alterations of the genotype, the hereditary endowment, of evolving species. Modifications of the phenotype, owing to environmentally induced changes in the manifestation of the genotype, are obviously important in evolution. Indeed, what survives or dies, reproduces or remains childless is only indirectly conditioned by the genotype—through its interactions with the environments moulding the phenotype. Nevertheless, without genotypic change the subsequent generations start from the same old base, and phenotypic changes can be reversed by return to the old environments. Fixity of the changes requires a genetic foundation. Any theory of evolution must, therefore, provide an account of the origin of genetic changes. At present we know two types of genetic changes—mutation and recombination of genetic materials.

There are several kinds of phenomena included under mutation. Paramount is gene mutation; it comes about through substitution, addition or deletion of nucleotides in DNA chain molecules; 'missense' mutations alter the amino sequence in the protein coded by the mutated gene, while 'nonsense' mutations disrupt the synthesis of the respective proteins. Mutations in operator or regulator genes may change the place, time or amount of the protein synthesised. Such mutations are likely to be quite important in evolution, but they are still inadequately known. Finally, chromosomal mutations change the arrangement of genes in the chromosomes, or delete or duplicate blocks of genes, or whole chromosomes.

No matter what other functions genes may have, all of them serve as templates for the synthesis of their copies. The copying process is in general remarkably accurate—otherwise life could not endure. Nevertheless mistakes do occur, and these mistakes lead to mutations. Nucleotide sequences in genes have been compared to sequences of letters and punctuations in words, sentences and paragraphs. Mutations are, then, errors of copying, like printers' or typists' errors. They are accidents or chance events. 'Chance' is, however, an equivocal word, especially so when applied to evolutionary processes. A chance event is not acausal, and not a manifestation of some principle of spontaneity inherent in living nature. Nagel (1961) defines a chance event as one that occurs 'at the intersection of two independent causal series', or 'if in a given context of inquiry the statement asserting its occurrence is not derived from anything else'. Mutations are said to occur by chance or spontaneously, when they appear in the progenies of parents not deliberately exposed to any mutation-inducing agent. The probability of their occurrence can, however, be enhanced by treatments with X-rays and other shortwave radiations, by ultraviolet, or by any one of the rapidly lengthening list of chemical mutagens. The chemical bases of the mutation process are being successfully elucidated.

The word 'chance' is misapplied not only to the causation but also to the effects of mutations. The changes caused by mutations seem, at first sight, capricious. Consider, for example, the classical mutants in Drosophila described by T. H. Morgan and his followers. They alter all sorts of characteristics of the fly, seemingly in a haphazard manner. A more detailed study corrects this impression. Mutations of the same gene alter usually, though not always, the same group of characteristics. Thus, mutations of the gene white all modify the eye colour, and those of the gene yellow the body colour, together with some less conspicuous traits.

Biochemical studies of mutation effects have shed needed light on the nature of the mutation process. A molecule of human hæmoglobin consists of two pairs of protein chains, alpha and beta, 141 and 146 amino acids respectively. They are coded by two separate genes which are, however, descendants of the same ancestral gene having undergone a duplication in primitive vertebrates, our remote ancestors. At least 22 different variants of the alpha and 42 of the beta chains have been found in human populations. These variants arose evidently by mutations in the close or in more distant ancestry of the persons in whom they are discovered. The same variant may be found in quite unrelated persons in different parts of the world; similar mutants arise repeatedly. Most mutants differ from the 'normal' (that is, the prevalent, or 'wild-type') hæmoglobin by single amino acid substitutions somewhere in the alpha or beta chains. The substitutions of the amino acids come, in turn, from substitutions of single nucleotides in the DNA of the genes coding for these chains.

In how many ways can a gene mutate by single nucleotide substitution? A chain of 141 amino acids, like that coding for the alpha hæmoglobin chain, is specified by three times this number, 423 nucleotides. Each nucleotide can be substituted by one of the three other 'letters' of the genetic 'alphabet'; 1269 different mutational changes due to substitution of single nucleotides are thus possible; because of the redundancy of the genetic code, the number of possible amino acid substitutions is somewhat smaller but still considerable. This is a large mutational repertory. Yet it does not follow that any gene can be converted by a single mutation into any other gene.

Some of the variant hæmoglobins found in human populations cause serious or even fatal hereditary diseases in individuals homozygous for the mutant genes. Other gene variants are so rare that they are known only in heterozygotes, who enjoy normal health. Comparison of the hæmoglobins of different animals shows, however, that they differ in more and more amino acid substitutions as the animals become less and less closely related. Some mutational changes affecting hæmoglobins have evidently become established in evolution. For example, the human alpha chain differs in a single amino acid from gorilla's, in 17 from cattle, 18 from horse, 25 from rabbit, and 71 from carp. These differences have arisen certainly not by single mutations, but by step-by-step accumulation of mutations, a great majority of which involved single amino acid substitutions.

Can accumulation of mutations transform a gene coding for hæmoglobin into one coding for something else? Myoglobin is a protein the function of which is storage of oxygen in muscles. It is a single chain molecule of 153 amino acids. Whale myoglobin differs in 115 amino acids from the alpha and in 117 from the beta chains of human hæmoglobin. The myoglobin and hæmoglobin genes are descendants of the same ancestral gene, which underwent duplication in our very remote ancestors, and became since differentiated by gradual accumulation of mutations.

All genes of all organisms may well be the products of divergent evolution from the primeval self-copying entity that arose from inorganic nature some three billion years ago. In their chemical aspects, the mutations that were taking place in this entity and in its descendants were copy errors, accidents, chances. A vast majority of these errors fell by the wayside, because they were inimical to the propagation and multiplication of their carriers. The mutations that were retained and that resulted in the divergence of the descendants of the primordial gene were those which proved helpful, or at least neutral, with respect to survival and reproduction. A gene is a macromolecule; it is also an organic system that carries within itself the billions of years of its evolutionary history. The mutational repertory of a gene is a function of its structure, and hence of the billions of years of its evolution. Chance is, therefore, brought under restraint, both at the level of origin of mutations, and of their retention or loss in the population.

Mutations are adaptively ambiguous. A mutation appears regardless of whether or not it will be useful to the organism carrying it, at the time and place it appears, or ever. It is, then, not surprising that a majority of mutations are deleterious, or at best neutral. Beneficial mutations are a small minority. This fact flatly contradicts all autogenetic, orthogenetic, Lamarckian and neo-Lamarckian theories of evolution. All these theories had to assume that, in one way or another, living matter reacts to its environments by purposive genetic changes. Such an assumption is either explicitly or implicitly vitalistic and, what is more to the point, it is given the lie by numerous observations on mutations in all sorts of organisms.

Another misjudgment is to suppose that either the kind or the speed of evolutionary changes has been determined by how often the mutations occurred. Although repeatedly shown to be mistaken, this misconception recurs again and again, even in recent literature. The mutation process is, however, not synonymous with evolution; it is only the source of the raw materials for evolutionary changes. It is true that without mutation all evolution would eventually stop. But in reality mutations occur in all organisms. Natural populations, particularly of sexually reproducing forms, contain stores of accumulated mutational materials. Shortages of mutational materials are probably rare. What is constructed from the evolutionary raw materials, and how fast the construction proceeds, depends on factors other than mutation rates. The chief factor is natural selection.

Gene recombination

Gene recombination is a source of evolutionary raw materials second in importance only to mutation. Its significance is often underrated, especially by students of forms that reproduce mainly asexually, and of haploid prokaryotes. For in such forms numerous individuals belong to the same clone and have the same genotype. A mutation gives rise to a new clone differing from its ancestor usually in a single gene alteration, and it seems tempting to envisage evolution simply as a contest between such clones. Yet gene recombination occurs from time to time even in prokaryotes, owing to processes of transformation, transduction, parasexuality and full-fledged sexuality.

In sexual diploids prodigious numbers of genotypes arise in every generation through Mendelian segregation and recombination. An individual heterozygous for n genes has a potentiality of producing 2^n different kinds of sex cells; parents each heterozygous for n genes produce potentially 3^n genotypes in the progeny; and parents each heterozygous for n different genes, 4^n genotypes. Because of linkage, not all these genotypes have the same probability of becoming realised. Nevertheless, with n in hundreds or in thousands, no two individuals in populations of sexual outbreeding species are likely to be genetically identical. In man, probably no genotype is represented by more than a single, either contemporaneously or not contemporaneously living individual, excepting only the monozygotic twins and other multiple births. The same is probably true of all higher organisms.

The genotypic uniqueness of individuals would be an interesting but, in evolutionary contexts, not particularly important fact, if the genes acted in the development each autonomously from the others. The development is, however, not a gradual accretion of 'unit characters' independently generated by different genes. Although it was in vogue with early geneticists, the 'unit character' idea now has few if any explicit supporters, except in the form of the 'one gene–one polypeptide chain in a protein' postulate. More or less complex networks of developmental reactions intervene between the genes in the sex cells and the characteristics of adult organisms. Gene effects interact, and at the present level of our knowledge the outcomes of their interactions are scarcely predictable. A gene A, beneficial to the organism in combination with a gene B_1, may be useless or harmful in combinations with B_2 or B_3 variants. The adaptedness and other properties of the carriers of a given genotype are the emergent products of the so-called epistatic interactions of the genes in the particular gene constellation. I attach no esoteric meaning to the word 'emergence'. The indeterminacy or unpredictability are not manifestations of some principle of 'chance' inherent in living matter; they are measures of our ignorance. But this does not make the phemonena underlying them unimportant.

The most obvious role of recombination in evolution is facilitation of coming together in single individuals of advantageous mutations which had arisen independently in different individuals (Bodmer, 1970). This is no small matter for mutations each of which is advantageous by itself; it is even more important when mutations which are neutral or deleterious by themselves give advantages epistatic interactions. In sexual diploids countless genetic variants are carried in populations; new gene patterns constantly arise in different individuals, the total number of the patterns being almost equal to the numbers of individuals that are produced. Instead of competing asexual clones, we have immense arrays of individual genotypes. Instead of vast numbers of mutational variants, natural selection works with unlimited numbers of combinations of these variants.

The sexual process is, to be sure, just as efficient in generating ever-new gene combinations as it is in breaking them up. The unrepeatability of individual human genotypes has been referred to above. Children do not inherit the genotypes of their parents, they inherit various constellations of the parental genes, one-half from each parent. Which ones they inherit is a matter of chance. Neither the best, most favourable, nor the worst, adaptively least favourable, combinations are bound to arise. The reason is simply that even the most fecund species brings forth numbers of progeny that are a tiny fraction of the potentially possible gene combinations. If so, what good does the formation of a superlative gene pattern in one individual bring to the species? The answer is that the incidence in the gene pool of the population of the components of that combination is increased, and the probability of appearance of similar, though never really identical, gene patterns is augmented.

The formation of every individual genotype is, in sexually reproducing organisms, a unique event. Since in every generation a Mendelian population is an array of unique individuals, that population is also unique. Evolution is a succession of unique states of populations. Are these states merely throws of genetic dice? The evolutionary dice are loaded; the loading comes from natural selection.

Natural selection and environment

Space forbids reviewing here either the history or the present state of the theory of natural selection. Only those aspects of the theory most relevant to our theme can be considered. Before all else, it must be stressed that selection puts a restraint on chance and makes evolution directional. Usually, though not invariably, selection increases the adaptedness of the population to its environments. It is responsible for the internal teleology (Ayala, 1968) so strikingly apparent in all living beings. The turmoil of mutation and recombination is curbed and channelled in the direction of adaptedness. This does not quite make evolution orthogenetic, because natural selection depends on the environment and the environment does not always change in a constant direction. Only if the environment changes directionally, or at least remains reasonably stable for prolonged time periods, then selection may become an orthoselection (Simpson, 1953; Rensch, 1971). Reversal of environmental changes (for example, the climate becoming colder and then warmer again) may cause the direction of the selection to be reversed also.

Natural selection constitutes a bond between the gene pool of a species and the environment. It may be compared to a servomechanism in a cybernetic system formed by the species and its environment. Somewhat metaphorically, it can be said that the information about the states of the environment is passed to and stored in the gene pool as a whole and in particular genes. Yet the environment does not ordain the changes that occur in the genes of its inhabitants. Our current conception is that evolution is not a pure ectogenesis imposed by the environment, nor are evolutionary changes determined 'from within' the organism. The relationships between evolution and environment are more subtle. Perhaps they can be described best by Toynbee's phrase 'challenge and response', applied by this author to the genesis of human civilisation and to their subsequent histories.

The complexity of the situation is, however, so overwhelming that we cannot predict whether or not an environmental challenge will evoke an adaptive evolutionary response in concrete cases. A response will not occur if genetic raw materials for it are unavailable. One does not, for example, expect the human species to evolve a race with a pair of wings for flying, nor one that could live on the moon without space suits. The response to changing environments may be too slow to save a species from extinction. This is probably the most usual cause of the extinction of species in the history of the earth. Finally, a coherent adaptive response to a given environment may be achieved by quite different means. Consider, for example, the plight of animals facing scarcity of food during winters in cold countries. The adaptive problem may be solved by storing a supply of food for the winter season, or by hibernation in a state of reduced metabolism and inactivity. We shall return to consider the multiformity of adaptive responses below. Here it must be stressed that a theory that would make the evolution predictable has not yet been achieved. Those who hold predictability essential for a scientific theory may justly deride the theory of evolution as unscientific.

There is, however, a misconception about natural selection which must be clearly recognised as such. Many critics of the modern biological theory of evolution, dissatisfied with evolution being ascribed allegedly to 'chance', branded natural selection a chance factor. The reverse is true—natural selection is the antichance factor in evolution. Chance predominates in mutation and recombination, and only in the limited sense explained above. On the contrary, selection is, as a rule, directed towards maintenance or enhancement of the Darwinian fitness. Darwinian fitness is reproductive fitness. It is quantifiable as the rate of transmission to the next generation of the components of a given genotype, in relation to other genotypes present in the same population. It is a function of viability, fecundity, speed of the development, early or late sexual maturity, sexual proficiency and other factors.

Darwinian fitness is, as a rule, positively correlated with the adaptedness of the carriers of a genotype in the same environment. It should, however, be stressed that Darwinian fitness is a relative measure (relative to that of other genotypes); adaptedness is, in principle, measurable in absolute terms, although satisfactory techniques of measuring it are yet to be developed. Darwinian fitness and adaptedness occasionally diverge. Genetic variants are known in Drosophila, mice, and probably elsewhere, that subvert the process of germ cell formation so that they are transmitted to a higher proportion of the offspring than are their alternatives. Carriers of such variants have by definition a higher Darwinian fitness than the non-carriers, and yet these variants may be deleterious, and even lethal in homozygotes. Occasional miscarriages, like the above, of natural selection processes are not surprising. It is, on the contrary, remarkable that they are rare, considering that selection is a blind and automatic rather than a purposeful process.

Another source of the misconception that selection is due to chance is that it operates through probabilities rather than all-or-none dichotomies. Situations when a genetic variant is completely lethal in one environment, while its alternative is lethal in another environment, are on the whole rare. An example of such a state of affairs are streptomycin-resistant mutants in some bacteria, which are also streptomycin-dependent. Exposed to a proper dose of streptomycin, all normal (that is, streptomycin-sensitive) individuals are destroyed, and only resistant mutants survive. Resistant and dependent mutants die on media free of streptomycin, and only mutants reverting to independence survive.

Far more often, selection works with genetic variants some of which have a higher probability of survival in a certain environment than other variants, or produce somewhat greater numbers of offspring, or are sexually maturing or flowering earlier, or exhibit a greater sexual drive, or take better care of their offspring. The Darwinian fitness is a composite resultant of all these factors. It may happen that natural selection promotes the establishment of genetic variants with some disadvantageous features that are more than compensated by advantages in other features. This has been misinterpreted as an orthogenetic change coming 'from within', and taking hold in defiance of natural selection. A striking, though speculative, example is the Irish elk: an extinct animal with monstrously large antlers, which were increasing more and more as time went on, until the species died out. Has it died out because the antlers became too heavy to carry, and if so how has natural selection permitted this to happen? A reasonable alternative is that individuals with larger antlers had reproductive advantages, securing mates in contests with possessors of smaller antlers. This advantage may have overcompensated for the drawback of carrying an otherwise useless and awkward adornment, until the environments turned more rigorous and caused extinction.

Does natural selection act as a sieve?

Natural selection is an impersonal and, by itself, purposeless process which nevertheless conduces as a rule toward internal teleology of living beings. It is perfectly correct to stress the mechanical, automatic character of selection, and yet this stress can be overdrawn and become misleading. One such exaggeration is to envisage selection as a mere sifter or screener of unalterably deleterious or useful genetic variants generated by mutation. The selection process would then act as a sieve that retains the rare favourable mutants, and lets the rest be lost through 'genetic death'. This image of natural selection does fit certain situations, especially among microorganisms, and consequently appeals particularly to those who work with such materials. The example of antibiotic resistance in bacteria referred to above is readily accommodated to the sieve model. But very often this model represents an oversimplification.

Not only the degree but also the sign of the alteration in fitness caused by a given mutant may change depending upon the external and the genetic environments. The 'sieve' of selection must therefore function as an extraordinarily sophisticated apparatus: a genetic variant is retained by or is passed through the 'sieve' not in accordance with only its own properties but also with those of other variants exposed to the 'sifting' process. To put it differently, the Darwinian fitness is not an intrinsic property of a genetic variant arising by mutation, but an emergent product of its interactions with the environment and with the rest of the genotypic system. The sieve model becomes therefore downright misleading.

Seen from another angle, the interactive properties of the genetic system increase enormously the variety of genetic materials with which natural selection operates. The variety of mutants that arise in any species is great; but, unless a mutant acts as a dominant lethal which is extinguished by selection in the same generation in which it arises, it is likely to be tested for fitness in numerous gene patterns in the progeny. As pointed out perhaps first by Wright (1932), the potentially possible gene combinations are vastly more numerous than subatomic particles in the universe estimated by physicists. The numbers of the gene combinations actually realised are, in sexually reproducing species, about as great as the total number of individuals ever born.

It is appropriate at this point to observe that some mutants and some gene combinations are adaptively neutral. They neither reduce nor augment the fitness of their carriers compared to other variants in the gene pool of a given population. The 'sieve' of natural selection takes no notice of them at all; they merely float or drift in the gene pool. Opinions differ widely concerning the degree of prevalence of neutral variants. Partisans of the so-called 'non-Darwinian evolution' urge that a decided majority of mutations on the molecular level are neutral, while those whom they label 'panselectionists' consider neutrality exceptional. This is one of the unsettled problems of the theory of evolution, which cannot be discussed here in detail. Two remarks may, however, be made for the sake of clarity. First, the label 'non-Darwinian' is unfortunate; there were a number of 'non-Darwinian' theories (such as that of Lamarck) before the newfangled one was even dreamed of. Evolution by random walk is a better designation. Secondly, the dichotomy of neutral *vs* non-neutral variants is unsound; variants that are neutral today may not have been so in the past, and may not be again in the future, in altered secular and genetic environments. This consideration is particularly relevant in man, because of the radical changes in his environments during evolution. Thus, genes that conferred resistance to many infections and environmental insults suffered by our ancestors may now be evolutionary vestiges of no adaptive significance.

Natural selection—censor or engineer?

The sieve and similar models of the action of natural selection stress its destructive function. Like a censor, selection deletes what is adaptively unsuitable and lets the rest pass. This model represents adequately only one of the several known forms of natural selection (Waddington, 1957; Dobzhansky, 1970). Normalising selection sweeps the gene pool clean of unconditionally deleterious genetic variants, or at least depresses their frequencies to irreducible minima. Normalising selection, of course not under that name, was recognised already by some of Darwin's predecessors. Variants of balancing selection (heterotic, diversifying, frequency-dependent) and of directional selection act as engineers rather than as censors. The most interesting aspect of their action is what they construct rather than what they destroy.

The 'engineer model' of natural selection may seem unacceptable to some people. No kind of selection can act according to a preconceived plan, because it cannot anticipate future states either of the organism or of its environment. Yet it must be stressed, because this is not apprehended by many biologists as well as philosophers, that selection does more than merely permit the rare beneficial mutants to reproduce. Assume that black skin is adaptive in humans living in countries with overabundance of sunshine, and light skin in climates where sunshine is deficient. The difference between black and light skin is controlled by at least four, probably more numerous, genes with additive effects—each gene adds or subtracts a relatively small amount of pigment. The skin colour is variable, and skin colour genes are scattered, in probably all human populations. How can an evolutionary change from black to light pigmentation, or the reverse, take place? Gene variants that increase the pigmentation will accumulate among the inhabitants of countries where insolation is strong, and those decreasing the pigmentation where it is deficient. This compounding action of the selection will obviously involve reduction in frequencies or even elimination of certain gene variants. However, the same degree of pigmentation may be achieved by selection of different genes in different populations; what is preserved is more significant than what is eliminated. Races with equal degree of skin pigmentation may have different pigment genes. This is conjectural for human races, but well established elsewhere. Perhaps the most elegant case is that discovered by Ford (1964): two races of the moth *Triphaena comes* are indistinguishable in appearance, and yet quite distinct genetically. The external similarity has been achieved evidently by different genetic means.

The 'engineering' aspect of natural selection is perhaps nowhere more manifest than in the phenomena of evolutionary convergence. Instances of convergence fascinated biologists since pre-evolutionary days, when their origins could not be understood. Consider, for example, whales and dolphins on the one hand, and fishes on the other. Whales and dolphins are certainly mammals, not fishes. Yet their body shapes and manner of locomotion resemble fish more than land-dwelling mammals. Very remote ancestors of all mammals were fishes, though not identical with any living fish. Much less remote ancestors of whales and dolphins were, however, land-dwelling and not fish-like animals, and they became fish-like secondarily, when they returned to live in water. Fish-like body shape and locomotion are adaptively superior to the more usual mammalian structures for a water dweller.

Natural selection promoted development of fish-like traits in the ancestries of whales and dolphins. It would, however, be naive to imagine that the fish-like body shape appeared one fine day as a lucky mutant. We do not know how many genes had to change one, or several, or many times during the descent of whales and dolphins from their land-dwelling ancestors. Surely there were many thousands, if not millions, of gene changes. The transition was a gradual one. A normally land-dwelling mammal may occasionally swim; beavers and otters spend much of their lives in water, but they are not strikingly fish-like; seals and manatees are much more so; and whales and dolphins had to be carefully scrutinised to ascertain that they are indeed mammals and not fishes.

I certainly do not suggest that the ancestors of whales were otters and seals. The point is that their fish-like characteristics were built up gradually by selection from genetic variants and their recombinations, some of which were arising by mutation perhaps also in human ancestry but were not utilised by selection as building blocks for the construction of adaptive gene patterns. Neither does it follow that otters and seals will become necessarily dolphin or whale-like, given some more millions, or tens of millions of years to continue their evolution. It should not be imagined that whales and dolphins are 'one hundred per cent' adapted to live in water while otters and seals are only fifty or ninety per cent adapted (ingenuous statements of this kind have actually been made by a recent author in a discussion of plant evolution).

One can reasonably compare the adaptedness of races of a species or of species of a genus, provided that they exploit similar or overlapping adaptive niches, or compete for the same food or other resources. Thus, the adaptedness of *Drosophila melanogaster* as a scavenger in man-made habitats is usually much higher than that of *Drosophila pseudoobscura*, while in natural habitats in western United States the reverse is true. It is, however, meaningless to surmise whether Drosophila is better or less well adapted than a grasshopper or a mouse; their ways of life are so far apart that they hardly ever come in contact. Even rare and relictual species may be well adapted in their ecological niches. Their rarity may be due to the ecological niche's being severely confining, as with Drosophila species living on land crabs mentioned above.

Beavers and otters spend much time on land as well as in water, seals leave water not infrequently, while dolphins and whales never do. A statement that the evolution of beavers and seals has not 'advanced' as much as that of dolphins and whales would subtly suggest a belief in an orthogenetic onward movement, which is gratuitous. Yet there undoubtedly has been some kind of orthoselection in the descent of dolphins and whales from their land-dwelling ancestors. They were becoming gradually more fish-like, and more and more specialised to live in water. It would also be pointless to deliberate whether they took to water before they started to change their body shapes and extremities in the fish direction, or *vice versa.* Obviously there has been a reciprocal feedback between the body structure and the way of life.

The statement that the process required a supply of 'chance' mutations is true but trivial. What is far more interesting is that these genetic variants were not simply caught in a sieve, but were gradually compounded and arranged into adaptively coherent patterns, which went through millions of years and of generations of responses to the challenges of the environments. Viewed in the perspective of time, the process cannot meaningfully be attributed to the play of chance, any more than the construction of the Parthenon or of the Empire State Building could be ascribed to chance agglomeration of pieces of marble or of concrete. What is fundamental in all of these cases is that the construction process was meaningful. The meaning, the internal teleology, is imposed upon the evolutionary process by the blind and dumb engineer, natural selection. The 'meaning' in living creatures is as simple as it is basic—it is life instead of death.

Opportunism of evolution

It is self-evident that any living species, at least any species not about to become extinct, has a valid set of solutions for certain basic ecological, or biological, problems. It must secure food to replenish its energy store, must have a place to live, and a method to reproduce and thus to perpetuate its genes in succeeding generations. Possession of a valid set of such solutions is called adaptedness. All living species are adapted to the environments which they inhabit. If they were not adapted they would not be living. Is the concept of adaptedness merely tautological? Not quite—adaptedness can be greater or less, wider or more specialised, open-ended or rigid. Furthermore, adaptedness to the same environment can be secured by different means.

All, or almost all, living species have reached at least tolerable levels of adaptedness (unless they are on the brink of extinction). Why, then, has evolution not come to a halt? It is going on and on, bringing forth ever new solutions to the same primordial biological problems. However, there is no law of nature decreeing that everything must evolve all the time. So-called 'living fossils' are now living species which have also been found as fossils in geological strata of more or less remote antiquity. Thus, the bones of Cretaceous opossums are similar to some modern ones; coelacanth fishes were believed to have become extinct in the Cretaceous, until a living Latimeria was found in the Indian Ocean off South Africa; modern horseshoe crabs (Limulus) do not differ greatly from those having lived some 200 million years ago; the brachiopod Lingula changed little if at all for 450 million years.

We do not know what caused the living fossils to stop evolving. It has repeatedly been surmised that the source of evolutionary changes has dried up because mutations no longer occur. This is almost certainly wrong. Selander *et al.* (1970) found the genetic variability in a population of horseshoe crabs not much lower than in the rapidly evolving Drosophila. Evolutionary rates are not primarily, if at all, dependent on mutation rates. Others fancied that evolution would stop when 'one-hundred per cent adaptedness' is achieved. This makes even less sense—it is not at all obvious that opossums, or horseshoe crabs, have in any way more perfect adaptedness than mice, or cats, or lobsters. A better guess is that the living fossils occupy ecological niches that have become perhaps more confining but otherwise not much altered for a long time.

Evolution occurs when there is a challenge and an opportunity to evolve. The simplest challenges arise from environmental changes. When the climate grows warmer or colder, selection may favour physiological adjustments that make the altered climate tolerable. A climatic change may, however, alter much more than temperature. It may change the whole ecosystem. Adaptedness to biotic factors is a problem more subtle, but in the long run more important, than that to physical factors. Organisms that inhabit a given territory are interrelated and interdependent. A plant or animal species may become more or less frequent, or the territory may be invaded by a new species, not hitherto living there. The newcomer may be a new kind of prey, or predator, or parasite, or competitor for food or for space. Its presence may be a challenge to which the old inhabitants will have to respond by adaptive changes, or become rare or extinct.

The process of adaptation may run in different channels. MacArthur and Wilson (1967) distinguish what they call '*r* selection' favouring increased reproductive rates and '*K* selection' for greater efficiency of conversion of food and other resources into offspring. For example, 'where climates are rigorously seasonal and winter survivors recolonise each spring, in the presence of a bloom of foliage and food, we expect *r* selection favouring large productivity; where climates are uniformly benign, *K* selection and greater efficiency should result.' Among other things, *r* selection may favour shorter generation time, harvesting most food (even if wastefully), and greater mobility of the young or dispersal of seed. With *K* selection the preservation of individuals, by physiological mechanisms that buffer them against environmental shocks, becomes important. Smaller numbers of better endowed offspring suffice for perpetuation of the species. The r and K selections are obviously not mutually exclusive and may occur together. Nonetheless, they may be responsible for divergent evolution of descendants of the same ancestral species or population.

Most generally, the organic diversity can be understood as a response of living matter to the diversity of physical and biotic environments. Even so, several million species now existing on earth may seem excessive exuberance. One is spellbound not only by the prodigous diversity but also by the many bizarre, almost whimsical forms of life. Why, for example, should there be at least two species of Drosophila specialised to live exclusively on certain body parts of the land crabs on Caribbean islands (mentioned at the beginning of the present article)? One is tempted to say that the place under the sun occupied by such species is so exiguous that it is not worth the effort to become adapted to live there. But this is surely both naive and excessively anthropomorphic. Natural selection does not 'know' whether its creations will inherit the earth or only a minuscule shred thereof. It is thoroughly opportunistic. If there is an unoccupied ecological niche, however narrow, and if genetic variance to occupy it is available, a genetic system is likely to arise that will fill it up.

Preadaptation

Natural selection is far from omnipotent. One of the things it cannot do is to foresee the future needs of a species or a population. Yet some observations seem to suggest that in the evolutionary history certain groups of organisms were preadapted to conditions which arose only subsequent to the appearance of these preadaptations. A simple and striking example is the development of strains of several insect species resistant to DDT and other potent insecticides. The resistance is due to mutant alleles of certain genes in these species; the mutations are not induced by exposure to the insecticide; the exposure to the insecticide merely kills the non-mutants and lets the mutants survive. DDT was discovered only some decades ago; why should the genes giving rise to resistant mutants have been present in the insects prior to this discovery?

It is naive to think that there are genes 'for' insecticide resistance. Genes do, however, produce enzymes which catalyse metabolic reactions in the 'normal', that is, nonresistant, insect. The molecule of DDT happens to have a chemical structure such that it is acted upon by some of these enzymes. The mutation responsible for the resistance may yield a variant enzyme particularly efficient in the reaction which detoxifies the insecticide, or may increase the amount of the enzyme. The biochemical versatility of insects is so great that an insecticide to which no resistance can develop is hard to find. Yet it does not follow that insects were provided with a battery of genes in anticipation of whatever insecticides can be developed. The genes perform physiological functions of quite different sorts, functions which were often unknown before the practical problem of resistance induced physiologists to study them.

The preadaptations that must have been present in human evolution have astonished and misled some biologists into believing the evolution to be guided by occult forces. Already A. R. Wallace, the co-discoverer with Darwin of the theory of evolution by natural selection, wondered how it came about that 'savages' have brains 'but little inferior to that of the average member of our learned societies', and concluded that man has 'something which he has not derived from his animal progenitors, something which we may best refer to as being of a spiritual essence of nature . . .'. Wallace doubted that natural selection could develop some human ability, such as the ability to create and to learn higher mathematics. A modern author questions the validity of the mutation–selection theory of evolution on similar grounds. Another modern author thinks that 'a brain a little better than that of a gorilla would have sufficed for man'. All doubts of this sort arise from implicit assumptions that 'characters' (or genes 'for' these characters) are selected independently of one another. But whatever scant evidence on the subject we have does not suggest that there exists one gene specifically for mathematics, another for philosophy, a third for poetry, etc.

Although some people learn most easily mathematics, others philosophy, or poetry, or something else, all these abilities are manifestations of man's fundamental capacity for abstraction, symbolic thought and communication by language. This capacity is among the diagnostic traits of the species *Homo sapiens*. Though varying to some extent from individual to individual, it is present in all nonpathological members of the species. Mental abilities that distinguish man from other primates and other animals are almost certainly not due to a single gene, but rather to numerous interacting genes. That human mental abilities have been highly adaptive in man's evolution cannot be disputed. Human forms of social organisation, as well as technologies which enable man to master his environments, are products of humankind's collective brain. Following Popper and Eccles (in this volume), one may say that all animals including man live in World I, but mankind has also World II and World III. Philosophy and higher mathematics might not have been indispensable by themselves, at least not until relatively recent times. But they are byproducts of mental abilities that have prompted man to occupy, to modify and finally to create his ecological niche.

Adaptive radiation

There are many examples of closely related species, clearly descended from a common ancestor, which have diverged in their ways of life to exploit different opportunities offered by their environments. Such instances of adaptive radiation are particularly illuminating when the divergence is of recent date, at least of the geological time scale. Darwin's finches of the Galapagos archipelago are the classical example, too well known to need recitation here. The adaptive radiation of drosophilid flies on the Hawaiian archipelago is even more grandiose and relatively little known (Carson, Hardy, Spieth and Stone, 1970). The archipelago consists of six islands of volcanic origin, now supporting an estimated 650 to 700 species of drosophilids. Only seventeen of these occur elsewhere, and were introduced to Hawaii by man probably within the last century. The remainder are endemic. The geological age of the islands inhabited by the drosophilids varies from 1 to 5 million years. Several older islands, up to 15 million of age, are now eroded down, and probably have very few if any drosophilid inhabitants. When, within this time span, the ancestor or ancestors of the endemic species reached Hawaii is conjectural, but 15 million years may be taken as the upper limit.

The world fauna of drosophilid flies comprises at least 2000 species. The Hawaiian endemics constitute accordingly about one-third of the total. The combined area of the Hawaiian islands is about half that of the Netherlands. Nowhere in the world does a territory of this size have anything like the number of drosophilid inhabitants found in Hawaii. The Hawaiian endemics fall in two groups—the genera Drosophila and Scaptomyza and their derivatives. Both genera are worldwide in distribution, but the 'scaptomyzoid' endemic species in Hawaii are more mumerous than in the rest of the world. The Hawaiian endemics can all be derived from only two ancestral species, a Drosophila and a Scaptomyza, that have reached the archipelago by accidental transport either from Asia or from America, more likely the former.

Morphological, ecological and behavioural diversity of the Hawaiian endemics exceeds that of the drosophila and scaptomyza elsewhere. The largest and the smallest drosophilids are Hawaiian. In some species certain body parts underwent modifications that border on emergence of new organs. The 'spoon-tarsi' are shortened, flattened and concave structures; 'forked-tarsi' have a long appendix developed either at the apex or at the base of the front basitarsus; there are also species with 'bristle-tarsi', clubbed, knobbed and hooked tibiae, etc. These modifications are found chiefly in males, and seem to be used in courtship rituals. The courtship and mating behaviours are also extraordinarily diversified, including 'lek behaviour', when a male stakes out a territory to which he attracts females of his own species but from which he aggressively ejects intruders, especially other males. This is not known in drosophilids outside Hawaii. Different species utilise fermenting leaves, stems, flowers, fruits, fungi and slime fluxes of various plants as oviposition sites and larval food. Species of the scaptomyzoid genus Titanochaeta utilise egg cocoons of spiders—a most unusual diet for a drosophilid.

Why should the bounteous adaptive radiation, more picturesquely styled 'evolutionary explosion', of drosophilids have taken place in Hawaii? The overabundance of the drosophilids there contrasts with the telltale scarcity, and even absence, of other families of dipterous and other orders of insects.

The most remarkable inference, in fact one approaching certainty, is that the ancestors of the Hawaiian drosophilids reached the archipelago before most other groups of insects did. This is not because drosophilids have particularly efficient distribution means. As far as known, no 'evolutionary explosions' of drosophilids happened on oceanic islands other than Hawaii. However, once they were there, they faced a multitude of unexploited opportunities for living. They responded by a rapid adaptive radiation. This should not be understood as meaning that the two immigrant species have simultaneously branched into some 600 or 700 endemic ones. Carson, Hardy, Spieth and Stone (1970) have shown that the radiation involved much migration from island to island. In fact, many species are endemic to just one, or a part of a single island. An ancestral species A colonised another island and gave rise there to a species B, and sometimes B recolonised the A island and gave rise to a species C. The geologically youngest island has probably a number of relatively more recent species.

It would, however, be misleading to conclude that every unoccupied ecological niche will necessarily and promptly be occupied by a new species possessing the requisite form of adaptedness. The evidence is conclusive that, on the contrary, the challenge of an unoccupied niche may not evoke a response for a long time. To take an extreme example, there is no obvious reason why man, or another species relying on symbolic communication, and on adaptation to the environment by culture and technology, could not have been a contemporary of dinosaurs. Yet the human ecological niche remained empty until late Pliocene or early Pleistocene. Land became inhabited much later than sea. Introduced to America, the gypsy moth thrives better than it does in Europe whence it came. No biologist doubts that many new forms of life could evolve and find places to live. But only a foolhardy biologist would venture to predict what these forms might be and where they would live. Strange to say, such predictions have been made with enviable self-assurance by 'exobiologists' for the hypothetical evolution of hypothetical life in hypothetical extraterrestrial habitats. They are, to put it mildly, unconvincing.

We do not yet have either a predictive or retrodictive theory of evolution (Slobodkin, 1968). Without palæontological evidence, one cannot reconstruct the ancestry of the now living organisms. Brave attempts to do so were being made by zoologists and botanists, particularly of the late nineteenth and the early twentieth century, using inferences from comparative anatomy and embryology. The results were not satisfactory. Suffice it to recall the hypothetical 'ancestral' organisms invented by Haeckel. They graced pages of his books, but never existed in reality.

The ancestry of mankind has now been partly opened to view, owing to the progress of palæoanthropology. The sequence *Ramapithecus–Australopithecus–Homo erectus–Homo sapiens* gives probably a valid idea of some of the stages human evolution went through. The question could, however, be asked: was *Australopithecus* bound to evolve into *Homo erectus*, and this latter into *Homo sapiens*? The question must, I believe, be answered in the negative. Given one stage, we could not predict what the next stage must be. There were at least two contemporary species of *Australopithecus*, but only one of them was our ancestor. The evolution of the other ended in extinction. Yet from our inability to make such predictions it does not follow that the human species arose by a lucky throw of some evolutionary or celestial dice. We neither arose by accident nor were predestined to arise. In evolution, chance and destiny are not alternatives. Here is one occasion in scientific theory, where we must invoke some variety of Hegelian or Marxian dialectics. We need a synthesis of the 'thesis' of chance and the 'antithesis' of predestination. My philosophical competence is unequal to this task. I plead for help from philosophical colleagues.

Creativity of evolution

Since we cannot predict evolution, the theory of evolution has sometimes been derided as only quasi-scientific. Whether or not it deserves the honorific label 'scientific', evolution will remain important and exciting as long as there will be people striving to understand themselves and their position in the universe. The unpredictability of evolution is surely not something inherent in its nature. The internal and environmental causes of evolution form a nexus too complex to be resolved at the present level of understanding. Anyway, evolution is neither deterministic nor 'animistic' (in Monod's (1971) sense). I like to call it creative (though not in Bergson's (1911) sense),

Can the word 'creative' be validly applied to a process that has no foresight and no ability to devise means to attain a chosen goal? Evolution is perhaps the only process lacking intentionality and foresight which is nevertheless creative. A dictionary meaning of the verb 'create' is 'to bring into being; to cause to exist'. Creation is also 'presentation of a new conception in an artistic embodiment'. Evolution not only brings novelties into being, but these novelties present embodiments of new ways of life. Focusing our attention on sexually reproducing outbreeding organisms, we find that, possibly with very rare exceptions, every individual is genetically unique and nonrecurrent. Evolution brings into being ever-novel, not previously existing, gene patterns. Taking a longer time perspective, we find that the novelty involves something more than a new random arrangement of preexisting parts. It is not comparable to new patterns in a kaleidoscope. New biological patterns possess a certain coherence.

The criterion of coherence in biology is survival. A vast majority of theoretically possible gene patterns would fail to be perpetuated. If they could be obtained at all, recombinants of human genes with those of, say, Drosophila would almost certainly fail to survive. Every new form of life that appears in evolution can, with only moderate semantic licence, be regarded as an artistic embodiment of a new conception of living. There exist obvious differences, but also meaningful similarities, between the evolutionary process and artistic activity. The former does not, and the latter does, involve application of skill to effect a desired result. (I leave out of consideration the 'aleatory' pseudo-art fashionable in some circles; that is, 'poetry' that consists of meaningless successions of words, 'music' of random concatenations of sounds, or 'paintings' made by daubing a canvas at random with lumps of paint). An observer should be able to discern in every existing biological species the embodiment of a new way of living.

Art is manifestation of aesthetic activity. An observer, if he is sensitive as well as knowledgeable, derives aesthetic satisfaction from discovery and contemplation of seemingly endlessly diverse structures and ways of life of living beings. I do not mean here merely beautiful butterflies and birds. A living body, any living body, is a work of art. Its beauty resides in its internal teleology (Ayala, 1968), which permits this body to remain alive as an individual and as a species. The beauty of human artistic creations is imposed by their makers; it is external teleology. Obviously, I do not contend that the artistic quality of a living body is generated in evolution for man's gratification. It is utilitarian, functional, in the organisms concerned. Yet it is worthy of note that, to man, the internal teleology has aesthetic appeal.

Evolution creates new living systems to occupy the ecological niches that are available and accessible. As pointed out above, not even minuscule ecological niches are disregarded if they are accessible. New ecological niches constantly arise. This is why evolution has not become stalled or terminated. What is it, however, that makes an empty ecological niche accessible? It is misleading to say that a new mutation must arise to fit the new ecological niche. The situation is more complex than this. That the process of mutation generates the raw materials of evolution is of course true, but in the present context it is trivial.

We do not know how many nonredundant genes are there in a human gamete. Estimates range from 10^5 to 10^7. Surely it was not an unprecedented and wonderful mutation in a single gene that transformed an ape into man. This transformation was effected by a gradual and time-consuming reconstruction of the genetic system. We do not know how many gene changes were involved; they were surely numerous, of the order of at least thousands, possibly even millions. The gene pool of a sexual population contains, except in rare or inbred species, countless genetic variants arisen ultimately by mutation. Building of many diverse genetic systems can be at least initiated with these raw materials. The problem is in which of the many available directions will the building process proceed. Presence of a supply of cement, bricks, wood and other building materials makes possible the construction of edifices of quite different sorts. The building materials do not necessarily impose a constraint on the kind of edifice that is to be built.

It is not on the level of mutation, but rather on that of natural selection and other genetic processes that take place in living populations, that the creativity of evolution manifests itself. What ecological niches become accessible depends primarily on these processes, not on the amount of mutability. Moreover, and this is where the biological engineer differs from a human one, the edifice to be built is not decided before the construction begins. The 'decisions' are all made during the process of construction. Surely this is a wasteful method to build anything. Evolution has not floundered and not ceased because, apparently very early in its history, life has become deployed in numerous lines of descent. At present we have some two million species, and billions of races and local populations. The biological meaning of these subdivisions is that they evolve more or less independently of each other. Already Wright (1932) pointed out that isolated populations may be regarded as trial parties that explore new possibilities of living. Yet those 'explorations' are not quite random, not even in the limited sense in which mutations are random or chance events. It should never be overlooked that natural selection is an antichance force. It does more than merely wait until a lucky gene combination emerges out of the mutation–sexual recombination mill. It directs the compounding of such 'lucky' gene combinations by gradual accumulation of favourable components. Any sequence of events controlled by the directional form of natural selection (and even by normalising selection) becomes a cybernetic process. It transmits information about the state of the environment to the gene pools of its inhabitants. The evolutionary history of the living world should not be conceived as a sequence of winning stops of a roulette wheel.

Adaptedness to a given environment can be achieved often in many ways. All living beings have basically the same requirements—food, place to live, reproduction, protection from environmental insults. Yet there are millions of species that solve these requirements in millions of ways. To give more specific examples—a warm-blooded animal can survive cold winters by having a warm coat, by winter dormancy, by migration to warmer lands, or by using fire to heat properly constructed dwellings. A desert plant may protect itself from desiccation by having leaves turned into spines, by leaves covered with oily or resinous varnish, leaves that develop during wet seasons and are shed during dry ones, or by compression of the life cycle to fit the wet season. Protection from predators is achieved by alertness and fleetness, or by concealing forms and colorations, or by being poisonous or unpalatable, or by active defence able to inflict injury on the would-be predator, or by reproductive rates so high that a part of the offspring escapes predation more or less by chance, or finally by parental or group defence. Insemination in animals is effected by internal copulation, by spermatophores, or by shedding female and male sex cells in water where they find each other by chemical attraction or by chance.

All of the above adaptive strategies are in fact used in different forms of life. This leads to uncomfortable questions of the following sort: if the protective coloration contributes to the adaptedness of a species A, why then is species B not also protectively coloured? The only answer is that B is doing well without protective coloration. Perhaps one should ask instead: why has A embarked on adaptation by means of protective coloration, and B on a different adaptive course? Satisfactory answers are rarely if ever available to questions of this sort. We do know that the same adaptive strategy can be adopted independently in different phylogenetic lines. This may result in evolutionary convergence. Cacti in American and euphorbias in South African deserts have evolved strikingly similar drought-resistant forms. Parental care and aggressive defence of the offspring are found in quite different groups of animals. By contrast, some adaptive strategies have been arrived at seemingly only once. Adaptedness by means of extragenically transmitted culture and symbolic language has evolved in a single species—man. To call this 'chance' is no meaningful solution. To ascribe this to predestination is incompatible with all that is known about causes that bring evolution about. The analogy with artistic creativity is at least descriptively most adequate, obvious differences to the contrary notwithstanding.

Evolutionary transcendence

My former colleague Dubos (1962) wrote: 'What is still so completely mysterious as to acquire for many human beings a mystical quality, is that life should have emerged from matter, and that mankind should have ever started on the road which so clearly is taking it farther and farther away from its brutish origins.' 'Mystical' may mean 'neither apparent to the senses nor obvious to the intelligence', or 'awareness of values in part at least above and beyond the scope of current symbolism to express—in particular the intense awareness of the whole as the unity of things' (Thorpe, 1965). While the former kind is useless, the latter may play a part in biology as a goad to deeper understanding. Anyway, the mystery is, one hopes, in the process of being dispelled.

The origins of life and of man are, indeed, the turning points of the evolution of the earth, and probably of the cosmos. There seem to be, basically, three ways of conceiving the scope of these realisations. The first is that of the traditional vitalism and creationism. Life arose when a vital force was added to inanimate matter, and the implantation of a soul was the origin of man. The second way is espoused by panpsychists or pan-vitalists. These form a minority, but perhaps a growing minority, among philosophers, such as Whitehead and Teilhard de Chardin, and among biologists, such as Rensch (1971) and Birch (1971). Rensch finds 'no contrast between mind and matter. We must recognise that all "matter" is proto-psychical in character. . . . It follows that there is no difference in principle between phenomenal reality and being, but only between two systems of relationships: those which enter a stream of consciousness and those which remain extramental.' This simplifies the problems a great deal—life and mind never really commenced. They were always present in all matter in some incipient states. But if so, evolution is a rather uninspiring story. It involved growth but no rightful novelty.

A third view recognises the origins of life and of man as the two major evolutionary transcendences. Transcendence means 'going beyond ordinary limits, surpassing, exceeding' (it should not be confused with transcendentalism). There is no doubt that life transcends the ordinary limits of inanimate matter, and that human mental abilities transcend those of any other animal. No novel energy, called the vital force, is superadded to preside over and to direct the flow of physicochemical processes in living bodies. Yet a multitude of chemical compounds are regularly formed, strictly in accord with physicochemical laws, in living bodies which do not commonly occur outside such bodies. Organic chemistry, biochemistry and chemical physiology have developed as a consequence. No elusive substance called soul must be assumed hiding in the human brain. Yet man does engage in kinds of behaviour no animal does. Human psychology covers ground not covered in zoopsychology, sociology deals with processes that have only remote parallels on the animal level, and humanities, as the name signifies, deal with exclusively human conduct and accomplishments.

The emergence of man is much closer to us in time than is the beginning of life. And yet this most recent of the great evolutionary transcendences presents problems even more challenging than the earlier one. We shall concentrate our attention on man. The bodily structures underwent considerable transformation during the ascent of man from his ape-like ancestors. Simpson (1969) lists twelve structural and ten psychological distinctive characters of the human species. The former are mostly connected with man's habit of walking erect. According to Simpson, 'The most crucial single anatomical point is acquisition of upright posture and strictly bipedal locomotion. Most of the other main peculiarities of human anatomy either follow from it or are coadapted with it.' Nevertheless, man's structural peculiarities only suffice to place him in a monotypic zoological family, with a single living species. His mental abilities are far more distinctive. If the zoological classification were based on psychological instead of mainly morphological traits, man would have been considered a separate phylum or even kingdom.

Man possesses mind or self-awareness, as well as death-awareness. Bidney (1953) describes self-awareness as follows: 'Man is a self-reflecting animal in that he alone has the ability to objectify himself, to stand apart from himself, as it were, and to consider the kind of being he is and what is it that he wants to do and to become. Other animals may be conscious of their affects and the objects perceived; man alone is capable of reflection, of self-consciousness, of thinking of himself as an object.' He is also alone in knowing that he will die. All animals die, but man alone knows that death is inevitable.

There is a curious contrast between self-awareness and death-awareness.

The former is a matter of personal and the latter of public knowledge. Sherrington (1955) said 'A radical distinction has therefore arisen between life and mind. The former is an affair of chemistry and physics; the latter escapes chemistry and physics.' It escapes chemistry and physics because it is experienced introspectively rather than objectively observed. To me, my mind is the most immediate and indubitable of all certainties. It is the basis of Descartes' *cogito ergo sum.* I infer that other people also have minds, because most of the time their actions resemble those of my own which I know are products of my self-awareness. For animals other than man, the evidence is wholly unreliable; some of them may possess traces of self-awareness, or may be devoid of it.

In contradistinction to self-awareness, death-awareness leads to kinds of behaviour that are objectively ascertainable. This is ceremonial burial and other forms of care or concern for the dead. All human societies, primitive as well as advanced, manifest such concern. No animal does so. Ants remove from their nests dead nest mates with other rubbish, and termites consume them as food. In man, rites and ceremonies connected with death vary enormously in different societies. They range from veneration to dread, from inhumation to cremation to exposure to carrion eaters. Moreover, funeral rites go back at least to Neanderthal man, if not to *Homo erectus*. Numerous palæolithic as well as neolithic burials have been discovered by archæologists. Death-awareness, and apparently some quasi-religious beliefs associated with or derived from it, were with man from the dawn of humanity.

Though profoundly different in their manifestations, self-awareness and death-awareness are causally related. When man became, in Bidney's words, 'a self-reflecting animal' he tasted the forbidden fruit of the Tree of Knowledge of Good and Evil. He discovered that his personal destiny is inevitably death. Only a being who knows his destiny can have an Ultimate Concern and can manifest concern for dead conspecifics. He distinguishes the sacred from the profane.

What sense can an evolutionary biologist make of these human faculties and exploits? Has natural selection implanted in man his self-awareness and death-awareness? I suggest that adaptive evolution is directly responsible for the former. However, since death-awareness is an inevitable product of self-awareness, both are ultimately products of man's unique form of adaptive evolution (Dobzhansky, 1967). Self-awareness is, as stated above, embarrassingly elusive to objective study. And yet it plays beyond doubt an overarchingly important role in human forms of social organisation, which are, therefore, basically distinct from animal societies. Man knows that he is responsible for his personal behaviour and his actions, and he holds other humans responsible for theirs. Man distinguishes good and evil. He is an ethicising being. He reflects on his past and makes plans for his future, and for the future of his environment, including other people. He reasons, and can modify his behaviour in accordance with his experience and his freely-arrived-at decisions. Whether or not his will is free in some metaphysical sense, man has at least an abiding illusion that it is free.

That human forms of social organisation are adaptive in the biological sense is indisputable. To be sure, insect societies (ants, termites, bees, wasps) are also successful. They are based on a principle quite different from the human plasticity of behaviour; it is the certitude of instinct instead of the doubts and misgivings of self-awareness. This is one more example of adaptedness being acquired by different means. The human species has, however, achieved an evolutionary success quite beyond that of any other species. Starting from a rare species somewhere in Africa, it spread all over the world. Its population increased perhaps as much as ten thousand-fold in the last one million years. It has to a large, and rapidly growing, extent mastered the environments of the planet which it inhabits. At any rate, its mastery is not disputed by any other species except, strange to say, some viruses and bacteria. Mankind has transcended its animality most of all in the spiritual sphere—knowledge, aesthetics, wisdom.

All of the above does not mean that mankind is wholly protected against disaster and even extinction. I need not dwell here on the hazards of uncontrolled population growth and despoliation of the environment. However, if mankind does become extinct it will be the first case of a species committing suicide; all other extinctions in evolutionary history were of species that were unable to help themselves, in the first place because they were unaware of the gathering menace and could not avert it anyway. Man can foresee the dangers in at least the immediate future. He can, and hopefully will, take measures to avoid them.

Summarising remarks

Except on the human level, evolution is a blind, if you wish mechanical, process. It cannot plan for the future, conceive purposes or strive for their realisation. How can a purposeless process transcend itself? How can an impersonal agent give rise to persons who possess self-awareness and death-awareness? Whence the unmistakable resemblance of biological evolution to human creativity?

Several solutions have been proposed. Rigid determinism and unalloyed chance are the polar opposites. Laplace's famous dictum asserts that to a 'sufficiently powerful' intellect nothing is uncertain, the future as well as the past are equally 'present to its eye'. If so, all evolution has been predestined in minutest details from the very beginning. Evolution is merely the unwrapping, unaccountably slow and tedious, of what was foreordained to appear by stages decreed to succeed each other in an order fated for all time. The universal fact of adaptedness of living beings to their environments becomes an insoluble puzzle. To a human observer, the staggering futility of predetermined evolution suggests not divine providence, but rather what Dostoyevsky called a 'devil's vaudeville'.

Chance, to some people, means acausality. To Nagel (1961) it is something that happens 'at the intersection of two independent causal series' (see above). Either way, organic adaptedness is inexplicable. A stupendously long succession of miraculous chances can only be ordained by a miracle maker. The polar opposites, evolution by chance and by predestination, are in the last analysis converging together.

Mutation, sexual recombination and natural selection are linked together in a system which makes biological evolution a creative process. One should be wary of models which envisage natural selection as a sieve that merely separates 'random' mutations and recombinations into lucky and unlucky ones. It is rather a cybernetic servomechanism that channels the flow of information from the environment to the gene pool. Yet to suppose that all evolution is ordained by the environment is also an oversimplification. There are many ways to become adapted to the environment in a given geographic region. In historical perspective, the environment and its inhabitants can be seen as an evolving ecosystem. Numerous biological contrivances enable different organisms with different ways of life to exploit the environment in great many different ways. It is here that evolution resembles artistic creation. Its greatest masterpiece is man.

There was no biological evolution until life appeared on at least one planet in the universe. Nor did human evolution begin until there appeared a being capable of symbolic thought and provided with at least a trace of self-awareness. I see no advantage in supposing that some shadowy rudiments of life, sensation, mind and self-awareness are built into all matter. Life arose from lifeless matter and mind from life lacking self-awareness. Evolution has shown itself capable of bringing about radical novelties. To say that a potentiality of life and mind was present before their appearance is trivial—it only means that they did in fact appear. All evolution—inorganic, organic and human—stems from natural laws built into the very fabric of the universe. This statement does not imply a disguised belief in predestination. It is important to realise that the potentialities of evolution are infinitely more numerous than its realisations. Thus, on the biological level, gene recombination is potentially capable of engendering gene patterns vastly more numerous than all the subatomic particles in the whole universe. What were the unrealised potentialities of evolution? It is probably quite useless to speculate on this.

References

Ayala, F. J. (1968). Biology as an autonomous science. *Amer. Scientist*, **56**, 207–21.

Barbour, I. G. (1966). *Issues in Science and Religion*. Prentice-Hall, Englewood Cliffs.

Berg, L. S. (1969), (1922). *Nomogenesis, or Evolution determined by Law*. MIT Press, Cambridge, Mass.

Bergson, H. (1911). *Creative Evolution*. New York.

Bidney, D. (1953). *Theoretical Anthropology*. Columbia University Press, New York.

Birch, C. (1971). Purpose in the universe: a search for wholeness. *Zygon*, **6**, 4–27.

Bodmer, W. F. (1970). The evolutionary significance of recombination in prokaryotes. *Symp. Soc. Gen. Microbiology*, **20**, 279–94.
Carson, H. L. (1971). The ecology of Drosophila breeding sites. Univ. Hawaii, Harold L. Lyon Arboretum Lecture, No. 2.
Carson, H. L., Hardy, D. E., Spieth, H. T. and Stone, W. S. (1970). The evolutionary biology of the Hawaiian Drosophilidae. In *Essays in Evolution and Genetics* (ed. M. K. Hecht and W. C. Steere). Appleton-Century-Crofts, New York, 437–543.
Coon, C. S. (1962). *The Origin of Races*. Knopf, New York.
Dobzhansky, Th. (1967). *The Biology of Ultimate Concern*. New American Library, New York (Fontana Library, Collins, London, 1971).
Dobzhansky, Th. (1970). *Genetics of the Evolutionary Process*. Columbia University Press, New York.
Dubos, R. (1962). *The Torch of Life*. Simon and Schuster, New York.
Ford, E. B. (1964). *Ecological Genetics*. Methuen, London.
Green, D. G. and Goldberger, R. F. (1967). *Molecular Insights into the Living Process*. Academic Press, New York.
MacArthur, R. H. and Wilson, E. O. (1967). *The Theory of Island Biogeography*. Princeton University Press, Princeton.
Mayr, E. (1969). *Principles of Systematic Zoology*. McGraw-Hill, New York.
Monod, J. (1971). *Chance and Necessity*. Knopf, New York.
Nagel, E. (1961). *The Structure of Science*. Harcourt, Brace and World, New York.
Raven, P. H., Berlin, B. and Breedlove, D. E. (1971). The origins of taxonomy. *Science*, **174**, 1210–13.
Rensch, B. (1971). *Biophilosophy*. Columbia University Press, New York.
Selander, R. K., Yang, S. Y., Lewontin, R. C. and Johnson, W. E. (1970). Genetic variation in the horseshoe crab (*Limulus polyphemus*), a phylogenetic relic. *Evolution*, **24**, 402–14.
Sherrington, C. (1955). *Man on his Nature*. Penguin, Harmondsworth.
Simpson, G. G. (1953). *The Major Features of Evolution*. Columbia University Press, New York.
Simpson, G. G. (1967). *The Meaning of Evolution*. Yale University Press, New Haven.
Simpson, G. G. (1969). *Biology and Man*. Harcourt, Brace and World, New York.
Slobodkin, L. B. (1968). Towards a predictive theory of evolution. In *Population Biology and Evolution* (ed. R. C. Lewontin). Syracuse University Press, Syracuse.
Teilhard de Chardin, P. (1959). *The Phenomenon of Man*. Harper and Row, New York.
Thorpe, W. H. (1965). *Science, Man and Morals*. Methuen London.
Waddington, C. H. (1957). *The Strategy of the Genes*. Allen and Unwin, London.
Wright, S. (1932). The roles of mutation, inbreeding, crossbreeding, and selection in evolution. *Proc. VI Internat. Cong. Genetics*, **1**, 356–66.

This article is dedicated to Ernst Mayr, foremost architect of the modern theory of biological evolution, on the occasion of his seventieth birthday.

Discussion

Rensch

'Chance' or *hasard* must not mean acausality or capriciousness. These terms have two different meanings. Normally they are used to express our ignorance of the causal effects which we are not able to analyse, for instance when we are playing roulette. On the other hand, chance is also a concept which indicates the effect of the universal laws of probability. In the formulas of quantum mechanics a function of probability has been introduced. This means that also the laws of probability *determine* microphysical events, so that stastistical predictions are possible.

I have the impression that many readers of Monod's outstanding book are only confused because man should have been developed by a series of random mutations, as it were by genetical mistakes. If, however, one would admit that such randomness came about by lawful, although unanalysed and partly unanalysable events, there would perhaps be no opposition. In my opinion all random events are ultimately guided by universal laws, mainly causal laws, laws of probability and logic. With the exception of this difference in an ultimate explanation I fully agree with all his statements. May I ask Professor Monod whether or not he would admit the possibility of such universal lawfulness?

19. The Concept of Biological Progress

FRANCISCO J. AYALA

Change, evolution, direction and progress

The notion that living organisms can be classified in a hierarchy going from lower to higher forms goes back to Aristotle, and indeed to even earlier times. The creation of the world as described in the book of Genesis contains the explicit notion that some organisms are higher than others, and implies that living things can be arranged in a continuous sequence from the lowest to the highest, which is man. The Bible's narrative of the creation reflects the common-sense impression that earthworms are in some sense lower than fish or birds, and the latter lower than man. The idea of a 'ladder of life' rising from amoeba to man is present, explicitly or implicitly, in all pre-evolutionary biology.

The theory of evolution adds the dimension of time, or history, to the classification of living things into lower and higher. The transition from amoeba to man can now be seen as a natural, progressive development from simple to gradually more complex organisms. The expansion and diversification of life can also be judged as progress; some form of advance seems obvious in the transition from one or only a few kinds of living things to the more than two million different species living today.

It is not immediately clear, however, what is meant by statements like 'The evolution of organisms is progressive', or 'Progress has occurred in the evolutionary sequence leading from amoeba to man'. Such expressions may simply mean that evolutionary sequences have a time direction, or even more simply that they are accompanied by change. The term 'progress' needs to be distinguished from other terms commonly used in biological discourse with which it shares areas of common meaning. These terms are 'change', 'evolution' and 'direction'.

Change means alteration, whether in the position, the state or the nature of a thing. Progress implies change, but not *vice versa*; not all changes are progressive. The positions of the molecules of oxygen and nitrogen in the air of a room are changing continuously; such change would not generally be labelled as progressive. The mutation of a gene from a functional allelic state to a non-functional one is a change, but definitely not a progressive one.

The terms 'evolution' and 'progress' can also be distinguished, although both imply that sustained change has occurred. Evolutionary change is not necessarily progressive. The evolution of a species may lead to its extinction, a change which is not progressive, at least not for that species. Progress also can occur without evolutionary change. Assume that in a given region of the world the seeds of a certain species are dormant because of a prolonged drought; after a burst of rain the seeds germinate and give origin to a population of plants. This change might be labelled progressive for the species, even though no evolutionary change need have taken place.

The concept of direction implies that a series of changes have occurred that can be arranged in a linear sequence such that elements in the later part of the sequence are further apart from early elements of the sequence than intermediate elements are, according to some property or feature under consideration. Directional change may be uniform or not, depending on whether every later member of the sequence is further displaced than every earlier member, or whether directional change occurs only on the average. This distinction will also be made later when defining progress. If the elements in the sequence are plotted on a two-dimensional graph with time on one axis and some property or feature of the elements of the sequence on the other axis, and all the elements are connected by a line, directional change is uniform when the slope of that line is at every point positive or at every point negative. The line connecting all the elements in the sequence may be straight or curved but should go up or down monotonously. Non-uniform ('net'; see below) directional change occurs when the line connecting all the elements in the sequence does not change monotonously but its regression on time is either significantly positive or significantly negative. Some elements in the sequence may represent a change of direction with respect to the immediately previous elements, but later elements in the sequence are further displaced than earlier ones on the average.

In evolutionary writings, 'directionality' is sometimes equated with 'irreversibility'. The process of evolution is said to have a direction because it is irreversible. Biological evolution is irreversible (except perhaps in some trivial sense, like when a previously mutated gene mutates back to its former allelic state). Direction, however, implies more than irreversibility. Consider a new pack of cards with each suit arranged from ace to ten, then knave, queen, king, and with the suits arranged in the sequence spades, clubs, hearts, diamonds. If we shuffle the cards thoroughly, the order of the cards will change, and the changes will be irreversible by shuffling. We may shuffle again and again until the cards are totally worn out, without ever restoring the original sequence. The change of order in the pack of cards is irreversible but not directional. Irreversible and directional changes both occur in the inorganic as well as in the organic world. The second law of thermodynamics, which applies to all processes in nature, describes sequential changes which are irreversible but are also directional, and indeed uniformly directional. Within a closed system, entropy always increases; that is, a closed system passes continuously from less to more probable states.

The concept of direction is used in palæontology to describe what are called 'evolutionary trends'. A trend occurs in a phylogenetic sequence when there is a feature which gradually increases or gradually decreases in the members of the sequence. Trends are common occurrences in all fossil sequences which are sufficiently long to be called 'sustained' (Simpson, 1953).

The concept of direction and the concept of progress can be distinguished. Consider the trend in the whole evolutionary sequence from fish to man towards a gradual reduction with palæontological time of the number of dermal bones in the skull roof; or the trend towards increased molarisation of the last premolars which occurred in the phylogeny of the *Equidae* from early Eocene (*Hyracotherium*) to early Oligocene (*Haplohippus*). These trends represent indeed directional change, but it is not obvious that they should be labelled progressive. To label them progressive we would need to agree that the directional change had been in some sense for the better. That is, to consider a sequence as progressive we need to add to the knowledge of the directionality of change an evaluation, namely that the condition in the latter members of the sequence represents, according to some standard, a betterment or improvement. The directionality of the sequence can be recognised and accepted without the added evaluation. Progress implies directional change, but not *vice versa*.

The concept of progress

Evolution, direction and progress all imply a historical sequence of events which exhibits a systematic alteration of a property or state of the elements in the sequence. Progress occurs when there is directional change towards a *better* state or condition. The concept of progress, then, contains two elements: one descriptive—that directional change has occurred; the other axiological—that the change represents an improvement or betterment. The notion of progress requires that a value judgment be made of what is better and what is worse, or what is higher and what is lower, according to some axiological standard. Contrary to the belief of some authors (Ginsberg, 1944; Lewontin, 1968), the axiological standard of reference need not be a moral one. Moral progress is possible, but not all forms of progress are moral. The evaluation required for progress is one of better *vs* worse, of higher *vs* lower, but not necessarily one of right *vs* wrong. Better may simply mean more efficient, or more abundant, or more complex, without connotating any reference to moral goals or standards.

Progress can be defined as systematic change in at least one feature belonging to all the members of a historical sequence, which is such that later members of the sequence exhibit an improvement of that feature. More simply, it could be defined as directional change towards the better. Similarly, regression is directional change for the worse. The two elements of the definition, namely directional change and improvement according to some standard, are jointly necessary and sufficient for the occurrence of progress.

To clarify further the concept of progress I want to distinguish several kinds of progress. The distinctions which follow relate to the descriptive element of the definition, that is, to the requirement of directional change. Therefore, the distinctions also apply to the concept of direction. Attending to the *continuity* of the direction of change in the members of the sequence, progress can be of two kinds—uniform and net.

Uniform progress takes place whenever according to a certain standard every later member of the sequence is better than every earlier member of the sequence. Let m_i be the members of the sequence, temporally ordered from 1 to n, and let p_i measure the state of the feature evaluated according to a given axiological standard of reference. There is uniform progress if, given any m_i with a certain p_i, every m_j is such that $p_j > p_i$ if $j > i$, and $p_j < p_i$ if $j < i$.

Net progress does not require that every member of the sequence be better than all previous members of the sequence and worse than all its successors; it requires rather that later members of the sequence be better, *on the average*, than earlier members of the sequence. Net progress permits temporary fluctuations of value. Formally, if the members of the sequence, m_i, are linearly arranged over time, net progress has occurred whenever the regression of p on time is significantly positive. Some authors have argued that progress has not occurred in evolution because, no matter what standard is chosen, fluctuations of value are always found to have occurred. This argument is valid against the occurrence of uniform but not of net evolutionary progress.

The distinction between uniform and net progress is similar to, but not identical with, the distinction proposed by Broad (1925) and also Goudge (1961) between uniform and perpetual progress. Perpetual progress, as defined by Broad, requires that the maxima of value increase and the minima do not decrease with time. In the formulation given above, Broad's perpetual progress requires that for every m_i there is at least one m_j $(j > i)$ such that $p_j > p_i$, and that there is at least one m_k $(k < i)$ such that $p_k < p_i$. This definition encounters some difficulties in its applications, and has the undesirable feature of requiring that the first element in the sequence be the worst one and the last element the best one. None of these two requirements are made in my definition of net progress. Also the term 'perpetual' has connotations which are undesirable in the discussion of progress. The distinction between uniform and net progress is implicit, although never formally established, in Simpson (1949), who applies terms like 'universal', 'invariable', 'constant' and 'continuous' to the kind of progress that I have called uniform, but also uses them with other meanings.

Note that neither uniform nor net progress require that progress be unlimited, or that any specified goal will be surpassed if the sequence continues for a sufficiently long period of time. Progress requires a gradual improvement in the members of the sequence, but the rate of improvement may decrease with time. According to the definition given here, it is possible that the sequence tends asymptotically towards a finite goal, which is continuously approached but never reached.

Another distinction is possible in the concept of progress. Attending to the *scope* of the sequence which is being considered, progress can be either general or particular. *General* progress is that which occurs in all historical sequences of a given domain of reality and from the beginning of the sequences until their end, or if they are not finished, until the present time. *Particular* progress is that which occurs in one or several, but not in all, historical sequences of a given domain of reality, or progress which takes place during part but not all the duration of the sequence or sequences.

In biological evolution, general progress is any kind of progress, if such exist, that can be predicated of the evolution of all life from its origin to the present. If a type of progress is predicated of only one or several, but not all, lines of evolutionary descent, it is a particular kind of progress. Progress which embraces a limited span only of the time going from the origin of life to the present is also a particular kind of progress.

It is obvious that other relevant distinctions of the concept of progress are possible. The two distinctions given above have been made having in mind a discussion of evolutionary progress, and are considered sufficient for the present purposes.

Can 'progress' be defined as a purely biological term?

Can we find in biology any criterion by which progress could be defined and measured by an absolute standard without involving judgments of value? Some authors believe that we can. Thoday (1953, 1958) has pointed out the obvious fact that survival is essential to life. Therefore, he argues, progress is increase in fitness for survival, 'provided only that fitness and survival be defined as generally as possible'. According to Thoday, fitness must be defined in reference to groups of organisms which can have common descendants; these groups he calls *units of evolution*. A unit of evolution is what population geneticists call a Mendelian population; the most inclusive Mendelian population is the species. The fitness of a unit of evolution is defined by Thoday as 'the probability that such a unit of evolution will survive for a long period of time, such as 10^8 years, that is to say will have descendants after the lapse of that time'. According to Thoday, evolutionary changes, no matter what other results may have been produced, are progressive only if they have increased the probability of leaving descendants after long periods of time. He correctly points out that this definition has the advantage of not assuming that progress has in fact occurred, an assumption which vitiates other attempts to define progress as a purely biological concept.

Thoday's definition of progress has been criticised because it apparently leads to the paradox that progress is impossible, in fact that regress is necessary since any group of organisms will be more progressive than any of their descendants. Assume that we are concerned with ascertaining whether progress has occurred in the evolutionary transition from a Cretaceous mammal to its descendants of 100 million (10^8) years later. It is clear that if the present-day mammal population M_1 has a probability P of having descendants 10^8 million years from now, the ancestral mammal populations M_0 will have a probability no smaller than P of leaving descendants after 2×10^8 years from the time of their existence (Ayala, 1969). In fact the probability that the ancestral population M_0 will leave descendants 2×10^8 years after their existence will be greater than P if it has any other living descendants besides M_1. As Thoday (1970) himself has pointed out, such criticism is mistaken, since it confuses the probability of survival with the fact of survival. The *a priori* probability that a given population will have descendants after a given lapse of time may be smaller than the *a priori* probability that any of its descendants will leave progeny after the same length of time.

There is, however, a legitimate criticism of Thoday's definition of progress, namely that it is not operationally valid. Suppose that we want to find out whether M_1 is more progressive than M_0. We should have to estimate, first, the probability that M_1 will leave descendants after a given long period of time; then we should have to estimate the same probability for M_0. Thoday has enumerated a variety of components which contribute to the fitness of a population as defined by him. These components are adaptation, genetic stability, genetic flexibility, phenotypic flexibility and the stability of the environment. But it is by no means clear how these components could be quantified, nor by what sort of function they could be integrated into a single parameter. In any case, there seems to be no conceivable way in which the appropriate observations and measurements could be made for the ancestral population. Thoday's definition of progress is extremely ingenious, but lacks operational validity. If we accept his definition there seems to be no way in which we could ascertain whether progress has occurred in any one line of descent or in the evolution of life as a whole.

Another attempt to consider evolutionary progress as a purely biological notion has been made by defining biological progress as an increase in the amount of genetic information stored in the organism. This information is encoded, at least for the most part, in the DNA of the nucleus. The DNA contains the information which in interaction with the environment directs the development and behaviour of the organism. By making certain assumptions, Kimura (1961) has estimated the rate at which genetic information accumulates in evolution. He calculates that in the evolution of 'higher' organisms genetic information has accumulated from the Cambrian to the present at an average rate of 0.29 bits per generation.

Kimura's method of measuring progressive evolution by the accumulation of genetic information is vitiated by several fundamental flaws. First, since the average rate of accumulation of information is allegedly constant *per generation*, it follows that organisms with a shorter generation time will have accumulated more information, and therefore are more progressive, than organisms with longer generation time. In the evolution of mammals, moles and bats would necessarily be more progressive than horses, whales and men.

A second, more basic, flaw is that Kimura is not measuring how much genetic information has been accumulated in any given organism. Rather, he assumes that genetic information gradually accumulates with time and then proceeds to estimate the rate at which genetic information could have accumulated. The assumption that more recent organisms have more genetic information, and that therefore they are more progressive than their ancestors, is unwarranted and completely invalidates Kimura's attempt to measure evolutionary progress. There is, at least at present, no way of measuring the amount of genetic information present in any one organism.

Julian Huxley (1942, 1953) has argued that the biologist should not attempt to define progress *a priori*, but rather he should 'proceed inductively to see whether he can or cannot find evidence of a process which can be legitimately called progressive'. He believes that evolutionary progress can be defined without any reference to values. Huxley proposes first to investigate the features which mark off the 'higher' from the 'lower' organisms. Any evolutionary process in which the features which characterise higher organisms are achieved is considered progressive. Like Kimura, Huxley assumes that progress has in fact occurred, and that certain living organisms, especially man, are more progressive than others. But to clasify organisms as 'higher' or 'lower' requires an evaluation. Huxley has not succeeded in avoiding reference to an axiological standard. The terms that he uses in his various definitions of progress, such as 'improvement', 'general advance', 'level of efficiency', etc., are all in fact evaluative.

The expansion and diversification of life

The concept of progress is axiological. To discuss evolutionary progress a choice must be made of the kind of value in reference to which organisms and evolutionary events can be assessed. Two decisions are required. First, we must choose the objective feature according to which the events or objects are to be ordered. Second, a decision must be made as to what pole of the ordered elements represents improvement. These divisions involve, admittedly, a subjective element, but they should not be arbitrary. Biological knowledge should guide them. There is a criterion by which the validity of the standards of reference can be chosen. A standard is valid if it enables us to say meaningful things about the evolution of life. How much of the required information is available, and whether the evaluation can be made more or less exactly, should also influence the choice of values.

What we know about the evolution of life enables us to decide immediately that there is no standard by which *uniform* progress has taken place in the process of evolution. Changes of direction, slackening and reversals have occurred in all evolutionary sequences, at least temporarily (Simpson, 1949, 1953). We must, then, concern ourselves exclusively with the question whether *net* progress has occurred, and in which sense, in the evolution of life.

Is there any criterion of progress by which net progress is a general feature of evolution? One conceivable standard of progress is the increase in the amount of genetic information stored in the organisms. Net progress would have occurred if organisms living at a later time would have, on the average, greater content of genetic information than their ancestors. One difficulty, insuperable at least for the present, is that there is no way in which genetic information can be measured. We do not even know how information is stored in organisms. We could choose the Shannon–Weaver solution, as Kimura has done, by regarding the DNA as a linear sequence of messages made up of groups of three-letter words (codons) with a four-letter alphabet (the four nucleotides, adenine, cytosine, guanine and thymine). Even so, a large fraction of the nuclear DNA probably does not encode information in such a way, and much DNA may have nonsense messages.

Accumulation of genetic information as a standard of progress can be understood in a different way. Progress can be measured by an increase in the *kinds* of ways in which the information is stored and as an increase in the *number* of different messages encoded. Different species represent different kinds of messages; individuals are messages or units of information. Thus understood, whether an increase in the amount of information has occurred reduces to the question whether life has diversified and expanded. This has been recognised by Simpson as the standard by which what I call general progress has occurred. According to Simpson (1949), in evolution as a whole we can find 'a tendency for life to expand, to fill in all the available spaces in the livable environments, including those created by the process of that expansion itself'.

The expansion of life can be measured by at least four different though related criteria of expansion: (1) expansion in the number of kinds of organisms, that is, species; (2) expansion in the number of individuals, or (3) of the total bulk of living matter; and (4) expansion in the total rate of flow of energy. Increases in the number of individuals or of their bulk may not be an unmixed blessing, as is the case now for the human species, but they can be a measure of biological success. By any one of the four standards of progress enumerated, it appears that net progress has been a general feature of the evolution of life.

Living organisms have a tendency to multiply exponentially without intrinsically imposed bounds. This is simply a consequence of the process of biological reproduction. The rate of increase in numbers is a net result of the balance between the rate of births and the rate of deaths of the population. In the absence of environmentally imposed restrictions, that balance is positive; populations have an intrinsic capacity to grow *ad infinitum*. Since the ambit in which life can exist is limited, and since the resources to which a population has access are even more limited, the rate of expansion of a population rapidly decreases to zero, or becomes negative.

Genetic differences between organisms of the same species result in natural selection of genotypes capable of a higher rate of expansion under the conditions in which the organisms live. Some genotypes are eliminated; others increase in frequency. The process of natural selection is creative in the non-trivial sense expounded by Dobzhansky (in this volume). For the present purposes, natural selection is relevant in that it leads to the diversification of species, and this process in turn often leads to a further expansion of the number of individuals and their bulk.

The expansion of life is a tendency which encounters constraints of various sorts. Once a certain species has come to exist, its expansion is limited by the environment in at least two ways: first, because as stated above the resources accessible to the organisms are limited; second, because favourable conditions for increase in numbers, even when resources such as food and living space are available, do not always occur—at times the rate of growth of the population becomes negative, because deaths exceed births. The various parameters of the environment embodied by the term 'weather' are the main factors interfering with the multiplication of organisms even when resources are available. Drastic and secular changes in the weather plus geological events lead at times to vast decreases in the size of some populations and even of the whole of life. Because of these constraints the tendency of life to expand has not always succeeded. That it has, on the average, expanded throughout most of the evolutionary process appears nevertheless certain.

Estimates of the number of living species vary from author to author. Probably there are more than two million but less than five million living species at present. Plant fossils are rare, but reasonable estimates of the number of animal species through palæontological history exist. Approximately 150 000 animal species live in the seas, a larger number than existed in the Cambrian (600 million years ago). Nevertheless, the present number of animal species may have been exceeded in the past, but if so it was certainly not by much. Life on land began in the Devonian (400 million years ago). The number of animal land species is probably at a maximum now, even if we exclude the insects. Insects make about three-quarters of all animal species, and about half of all species if plants are included. Insects did not appear until the early Carboniferous, some 350 million years ago. More species of insects exist now than at most, probably all, times in the past. This brief summary indicates that the number of living species is probably greater in recent times than was ever before, and that, in any case, a gradual increase in the number of species has characterised, at least on the average, the evolution of life. (Further details concerning the number of species, and also of genera, families, order, classes and phyla, through palæontological history can be found in the works of Simpson, particularly (1949) and (1953), from which the facts given in this paragraph are taken.)

The number of individuals living on the earth cannot be estimated with any reasonable approximation, even if we exclude microorganisms. It is a staggering number. I have estimated the number of individuals of a successful Neotropical insect species, *Drosophila willistoni*, as between 10^{10} and 10^{11}, that is, between 10 and 100 billions (Ayala *et al.*, 1972). The number of individual animals and plants living today and their bulk are doubtless greater than they were in the Cambrian. Very likely they are also greater than they have been throughout most of the time since the beginning of life. This is even more so if we include the large number and enormous bulk of the human population, and of all the plants and animals cultivated by man for his own use. Even if we include microorganisms, it is probable that the number of living individuals has increased, on the average, through the evolution of life. That the total bulk of living matter has also increased with the succession of time is even more likely.

The expansion of life operates as a positive feedback mechanism. The more species appear, the more environments are created for new species to exploit. A trivial example is that once plants came into existence it was possible for animals to exist; and the animals themselves sustain large numbers of species of other animals, and of parasites and symbionts. T. H. Huxley likened the expansion of life to the filling of a barrel. First, the barrel is filled with apples until they overflow; then pebbles are added up to the brim; the space between the apples and the pebbles can be packed with sand; water is finally poured until it overflows. With many kinds of organisms the environment can be filled in more effectively than with only one kind. Huxley's analogy neglects one important aspect of life. A more appropriate analogy would have been that of an expanding barrel. The space available is increased rather than decreased by some additions.

The total flow of energy in the living world has probably increased through evolution even faster than the total bulk of matter. Johnston (1921) pointed out that the influence of living things is to retard the dissipation of energy. Green plants do, indeed, store radiant energy from the sun which would otherwise be converted into heat. The influence of animals goes, however, in the opposite direction. The living activities of animals dissipate energy, since their catabolism exceeds their anabolism (Lotka, 1945). This apparently paradoxical situation results in fact in an increase in the rate at which energy flows through the whole of life. Animals do not simply provide a new path through which energy can flow, but rather their interactions with plants increase the total rate of flow through the system. An analogy can be used to illustrate this outcome. Suppose that a modern highway with three lanes in each direction connects two large cities. A need to accommodate an increase in the rate of traffic flow can be accomplished either by adding more lanes to the highway or by increasing the speed at which the traffic moves in the highway. In terms of the 'carrying capacity' of the highway, these two approaches appear, at first sight, to work in opposite directions. However, together they increase the total flow of traffic through the highway.

Information about the environment

There are many criteria by which net progress has occurred in some evolutionary sequences but not in others. One of the most meaningful of such criteria is the ability of the organism to obtain and process information about the environment. This criterion is of considerable biological interest, because such ability notably contributes to the biological success of the organisms which possess it. The criterion is particularly interesting in reference to man, since among the differences which mark off man from all other animals, his greatly developed ability to perceive the environment, and to react flexibly to it, is perhaps the most fundamental one. While organisms other than man become genetically adapted to their environments, man artificially creates environments to fit his genes.

Increased ability to gather and process information about the environment is sometimes expressed as evolution towards 'independence from the environment'. This latter expression is misleading. No organism can be truly independent of the environment. The evolutionary sequence fish → amphibian → reptile allegedly provides an example of evolution towards independence from an aqueous environment. Reptiles, birds and mammals are indeed free of need for water as an external living medium, but their lives depend on the conditions of the land. They have not become independent of the environment, but rather have exchanged dependence on one environment for dependence on another.

'Control over the environment' has been linked to the ability to gather and use information about the state of the environment. However, true control over the environment occurs to any large extent only in the human species. All organisms interact with the environment, but they do not control it. Burrowing a hole on the ground, or building a nest on a tree, like the construction of a beehive or of a beaver dam, do not represent control over the environment except in a trivial sense. The ability to control the environment started with the australopithecines, the first group of organisms which can be called human. They are considered to be men precisely because they were able to produce devices to manipulate the environment in the form of rudimentary pebble and bone tools. The ability to obtain and process information about the conditions of the environment does not provide control over the environment but rather it enables the organisms to avoid unsuitable environments and to seek suitable ones. It has developed in many organisms because it is a useful adaptation.

All organisms interact selectively with the environment. The cell membrane of a bacterium permits certain molecules but not others to enter the cell. Selective molecular exchange occurs also in the inorganic world; but this can hardly be called a form of information processing. Certain bacteria when placed on an agar plate move about in a zig-zag pattern which is almost certainly random. The most rudimentary ability to gather and process information about the environment can be found in certain single-celled eukaryotes.[1] A *Paramecium* swims following a sinuous path ingesting the bacteria that it encounters as it swims. Whenever it meets unfavourable conditions, like unsuitable acidity or salinity in the water, the *Paramecium* checks its advance, turns and starts in a new direction. Its reaction is purely negative. The *Paramecium* apparently does not seek its food or a favourable environment, but simply avoids unsuitable conditions.

A somewhat greater ability to process information about the environment occurs in the single-celled *Euglena*. This organism has a light-sensitive spot by means of which it can orient itself towards the direction in which the light originates. *Euglena*'s motions are directional; it not only avoids unsuitable environments but it also actively seeks suitable ones. An amoeba represents further progress in the same direction; it reacts to light by moving away from it, and also actively pursues food particles.

Progress as increase in the ability to gather and process information about the environment is not a general characteristic of the evolution of life. Progress occurred in certain evolutionary lines of descent but not in others. Today's bacteria are not more progressive by this criterion than their ancestors of one billion years ago. In many evolutionary sequences some very limited progress took place in the early stages, without further progress through the rest of their history. In general, animals are more advanced than plants; vertebrates are more advanced than invertebrates; mammals more advanced than reptiles, which are more advanced than fish. The most advanced organism by this criterion is doubtless man.

The ability to obtain and to process information about the environment has progressed little in the plant kingdom. Plants generally react to light and to gravity. The geotropism is positive in the root, but negative in the stem. Plants also grow towards the light; some plants like the sunflower have parts which follow the course of the sun through its daily cycle. Another tropism in plants is the tendency of roots to grow towards water. The response to gravity, to water and to light is basically due to differential growth rates; a greater elongation of cells takes place on one side of the stem or of the root than on the other side. Gradients of light, gravity or moisture are the clues which guide these tropisms. Some plants react also to tactile stimuli. Tendrils twine around what they touch; *Mimosa* and carnivorous plants like the Venus flytrap (*Dionaea*) have leaves which close rapidly upon being touched.

[1] In eukaryotes the DNA is organised in chromosomes which exist in a nucleus surrounded by a nuclear membrane. Prokaryotes have genetic information encoded in DNA, but there is no nucleus set off from the rest of the cell.

In multicellular animals, the ability to obtain and process information about the environment is mediated by the nervous system. All major groups of animals, except the sponges, have nervous systems. The simplest nervous system among living animals occurs in coelenterates like hydra, corals and jellyfishes. Each tentacle of a jellyfish reacts only if it is individually and directly stimulated. There is no coordination of the information gathered by different parts of the animal. Jellyfishes are besides unable to learn from experience. A limited form of coordinated behaviour occurs in the echinoderms which comprise the starfishes and sea urchins. The coelenterates possess an undifferentiated nerve net; the echinoderms possess, besides a nerve net, a nerve ring and radial nerve cords. When the appropriate stimulus is encountered, a starfish reacts with direct and unified actions of the whole body. The most primitive form of a brain occurs in certain organisms like planarian flatworms, which also have numerous sensory cells and eyes without lenses. The information gathered by these sensory cells and organs is processed and coordinated by the central nervous system and the rudimentary brain; a planarian worm is capable of some variability of responses and of some simple learning. That is, the same stimuli will not necessarily produce always the same response.

Planarian flatworms have progressed farther than starfishes in the ability to gather and process information about the environment, and the starfishes have progressed farther than sea anemones and other coelenterates. But none of these organisms has gone very far by this criterion of progress. The most progressive group of organisms among the invertebrates are the arthropods, but the vertebrates have progressed much farther than any invertebrates.

It seems certain that among the ancestors of both the arthropods and the vertebrates there were organisms that, like the sponges, lacked a nervous system, and that their evolution went through a stage with only a simple network, with later stages developing a central nervous system and later a rudimentary brain. With further development of the central nervous system and of the brain, the ability to obtain and process information from the outside progressed much farther. The arthropods, which include the insects, have complex forms of behaviour. Precise visual, chemical and acoustic signals are obtained and processed by many arthropods, particularly in their search for food and in their selection of mates.

The vertebrates are generally able to obtain and process much more complicated signals and to produce a much greater variety of responses than the arthropods. The vertebrate brain has an enormous number of associative neurons with an extremely complex arrangement. Among the vertebrates, progress in the ability to deal with environmental information is correlated with increase in the size of the cerebral hemispheres and with the appearance and development of the 'neopallium'. The neopallium is involved in association and coordination of all kinds of impulses from all receptors and brain centres. The neopallium appeared first in the reptiles. In the mammals it has expanded to become the cerebral cortex, which covers most of the cerebral hemispheres. The larger brain of vertebrates compared to invertebrates permits them also to have a large amount of neurons involved in information storage or memory.

The ability to perceive the environment, and to integrate, coordinate and react flexibly to what is perceived, has attained its highest degree of development in man. Man is by this measure of biological progress the most progressive organism on the planet. That such ability is a sound criterion of biological progress was indicated earlier by pointing out that it is useful as an adaptation to the environment. Extreme advance in the ability to perceive and react to the environment is perhaps the most fundamental characteristic which marks off *Homo sapiens* from all other animals. Symbolic language, complex social organisation, control over the environment, the ability to envisage future states and to work towards them, values and ethics, are developments made possible by man's greatly developed capacity to obtain and organise information about the state of the environment.

Concluding remarks

There is an abundant literature dealing with the subject of biological progress. Simpson (1949) has examined several criteria of progress and has stated in which sequences and for how long progress has occurred according to each one of the standards. The criteria of evolutionary progress explored by Simpson include dominance; invasion of new environments; replacement; improvement in adaptation; adaptability and possibility of further progress; increased specialisation; control over the environment; increasing structural complication; increase in general energy or maintained level of vital processes; and increase in the range and variety of adjustments to the environment. Rensch (1947) and Huxley (1942, 1953) have examined other lists of characteristics which can be used as standards of progressive evolution. Stebbins (1969) has written a provocative study of the law of 'conservation of organisation' as a principle which accounts for evolutionary progress in the sense of a small bias towards increased complexity of organisation. Williams (1966) has examined, mostly critically, several criteria of progress. Two brief but incisive discussions of the concept of progress can be found in Herrick (1956) and Dobzhansky (1970). A philosophical study of the concept of progress has been made by Goudge (1961).

There is no need to examine here all the standards of progress which have been formulated by the authors just mentioned, nor to explore additional criteria. This paper was written primarily to clarify the notion of progress and its use in biology. Writings about biological progress have involved much disputation concerning (1) whether the notion of progress belongs in the realm of scientific discourse, (2) what criterion of progress is 'best', and (3) whether progress has indeed taken place in the evolution of life. These controversies can be solved once the notion of progress is clearly established. First, the concept of progress involves an evaluation of good *vs* bad, or of better *vs* worse. The choice of a standard by which to evaluate organisms or their features is to a certain extent subjective. However, once a criterion of progress has been chosen, decisions concerning whether progress has occurred in the living world, and what organisms are more or less progressive, can be made following the usual standards and methods of scientific discourse. Second, there is no criterion of progress which is 'best' in the abstract or for all purposes. The validity of any one criterion of progress depends on whether the use of that criterion leads to meaningful statements concerning the evolution of life. Which criterion or criteria are preferable depends on the particular context in which they are discussed. Third, the distinction between uniform and net progress makes it possible to recognise the occurrence of biological progress even though every member of a sequence or of a group of organisms may not always be more progressive than every previous member of the sequence or than every member of some other group of organisms. The distinction between general and particular progress makes it possible to study progress in particular groups of organisms, or during limited periods in the evolution of life.

Once it is recognised that an evaluation needs to be made, discussions of evolutionary progress can provide valuable insights for the understanding of life. That statement provides the clue to justify writing about evolutionary progress for this conference. Discussions of progress can be illuminating in biology, but are not so in the realm of the inorganic world. The only nontrivial processes in which sustained directional change has occurred outside the world of life are the increase of entropy, the expansion of the universe, the evolution of stars, and perhaps some geological processes. Nothing is gained, however, by labelling these processes progressive. Progress has occurred in nontrivial senses in the living world because of the creative character of the process of natural selection.

The basic components of organisms are the same physicochemical elements of the inorganic world. In living matter, these elements obey the same fundamental physicochemical laws which govern their behaviour in nonliving matter. Evolutionary progress, however, cannot be discussed purely in terms of the physicochemical components of living matter. The ability to gather and to process information about the environment, for instance, is an important biological parameter. It is clear that it cannot be analysed by reference only to physicochemical elements and laws. In fact, evolutionary progress measured by that standard can be interpreted as a gradual departure from the importance of physicochemical laws in determining the relevant aspects of the behaviour of organisms.

References

Ayala, F. J. (1969). An evolutionary dilemma; fitness of genotypes *versus* fitness of populations. *Canad. J. Genetics and Cytology*, **11**, 439–56.

Ayala, F. J., Powell, J. R., Tracey, M. L., Mourao, C. A. and Perez-Salas, S. (1972). Enzyme variability in the *Drosophila willistoni* group: IV. Genic variation in natural populations of *Drosophila willistoni*. *Genetics*, **70**, 113–39.

Broad, C. D. (1925). *The Mind and its Place in Nature*. Kegan Paul, London.
Dobzhansky, Th. (1970). *Genetics of the Evolutionary Process*. Columbia University Press, New York and London.
Dobzhansky, Th. (1974). Chance and creativity in evolution. In this volume.
Ginsberg, M. (1944). *Moral progress*. Frazer Lecture at the University of Glasgow. Glasgow University Press.
Goudge, T. A. (1961). *The Ascent of Life*. University of Toronto Press.
Herrick, G. J. (1956). *The Evolution of Human Nature*. University of Texas Press, Austin.
Huxley, J. S. (1942). *Evolution: the Modern Synthesis*. Harper, New York.
Huxley, J. S. (1953). *Evolution in Action*. Harper, New York.
Johnston, J. (1921). *The Mechanism of Life*.
Kimura, M. (1961). Natural selection as the process of accumulating genetic information in adaptive evolution. *Genet. Research*, **2**, 127–40.
Lewontin, R. C. (1968). The concept of evolution. *International Encyclopedia of the Social Sciences* (ed. D. L. Sills). Macmillan Co. and Free Press, London and New York.
Lotka, A. J. (1945). The law of evolution as a maximal principle. *Human Biology*, **17**, 167–94.
Rensch, B. (1947). *Evolution above the Species Level*. Columbia University Press, New York.
Simpson, G. G. (1949). *The Meaning of Evolution*. Yale University Press, New Haven.
Simpson, G. G. (1953). *The Major Features of Evolution*. Columbia University Press, New York.
Stebbins, G. L. (1969). *The Basis of Progressive Evolution*. University of North Carolina Press, Chapel Hill.
Thoday, J. M. (1953). Components of fitness. *Symposia of the Society for the Study of Experimental Biology*, **7** (Evolution), 96–113.
Thoday, J. M. (1958). Natural selection and biological progress. In *A Century of Darwin* (ed. S. A. Barnet). Allen and Unwin, London.
Thoday, J. M. (1970). Genotype *versus* population fitness. *Canad. J. Genetics and Cytology*, **12**, 674–5.
Williams, G. C. (1966). *Adaptation by Natural Selection*. Princeton University Press.

Discussion

Rensch

I believe that it is difficult to avoid confusion which arises from two different meanings of 'progress'. Increasing adaptation is also progress, although it may lead to extinction when the adaptation becomes so narrow that the structure of a species could not be altered to become adapted to a fundamental change of the ecological conditions. When a common word has two different meanings, it is always desirable to create a clearly defined, internationally acceptable term. I therefore introduced the term *anagenesis* in 1947, in order to discriminate evolution leading to a higher level from pure ramification (cladogenesis) and *adaptogenesis*. These terms make confusion impossible. Of course the term progress can always be used so that misunderstandings are not possible.

Anagenesis can be clearly defined by several characteristics: mainly growing complexity (more and more different types of cells and tissues), increase of more rational structures and functions (including division of labour and centralisation, particularly of the nervous system), development of an organisation which does not hinder further anagenesis, and—considering the whole phylogenetical tree—more genetic information. These characteristics have the advantage that they do not contain any evaluation made relative to the human level.

Ayala

Rensch (1947) has distinguished two types of evolutionary changes: cladogenesis or splitting, and anagenesis or phyletic evolution. In a given evolutionary line of descent, anagenesis is characterised by gradual increase of complexity of organisation, emergence of new organs, development of novel ways of dealing with the environment, and the like. Cladogenesis occurs when there is a splitting of species, genera, or other taxa. Cladogenesis need not be accompanied by anagenesis nor *vice versa*, although these two modes of evolutionary change often occur together, for instance in cases of adaptive radiation. The evolutionary process from ancestral reptiles to modern mammals can be recognised as anagenetic but was also cladogenetic and has led to many different kinds of mammals.

Two points need to be made. First, anagenesis refers to several kinds of evolutionary change. Advancement with regard to one kind of anagenetic change (such as the development of new ways of dealing with the environment, or increased centralisation of the nervous system) is not always accompanied by advancement with respect to other kinds of anagenetic changes (such as the emergence of new organs, or increased complexity of organisation). The second point is that there is no reason to restrict the notion of biological progress to anagenetic changes. Cladogenetic evolution as well as increased adaptation to the environment can also be considered instances of progressive evolution. The evolution of more than 350 different species of *Drosophila* flies in the Hawaiian archipelago from presumably one single immigrant species, or the evolution of some twenty species of cyprinid fishes in Lake Lanao in the Philippines also from a single ancestral species, may be considered progressive by certain standards. They have led to diversification of life, to exploitation of additional niches, and probably to increase in the total number of individuals, although these are not instances of anagenetic evolution. In a given context of discourse, criteria of progress should be chosen such that they lead to meaningful statements and enlightenment of the subject matter under consideration, but there is no need to impose any other restrictions.

20. On Chance and Necessity

JACQUES MONOD

I have not had time to prepare a formal paper and wish to apologise for this, since I had every intention of doing so and it has only been made impossible by a combination of extremely heavy duties and being ill for several months this year. On the other hand, I understand that a number of you have either read the little essay I wrote recently [*Chance and Necessity*, Knopf, New York (1971)], or at least heard of it. Therefore, I hope that if those who have read it agree or disagree, you can ask me some questions after this presentation. What I would like to do here, is the following. I have taken a few notes at random, which are, of course, on various subjects that I did discuss in the book, and I would like to emphasise a few points.

One that I think needs clarification, and also perhaps emphasis, is what I have called the Postulate of Objectivity. To my mind, at least, it is the cornerstone of biology. It certainly was a logical cornerstone in the elaboration of my essay, and Dr Skolimowski has attempted to show the complete incoherence of this idea. I think he sees the incoherence because of a confusion made by many readers and critics of the book for which I must consider myself partially responsible, simply by the fact that I used the word 'objectivity' to define the postulate. As I put it, the Postulate of Objectivity is the systematic or axiomatic denial that scientific knowledge can be obtained on the basis of theories that involve, explicitly or not, a teleological principle; and admittedly it is an axiomatic attitude which I strongly believe to be essential for the development of science. Now, of course, 'objective' is most widely used in a sense where it is opposed to 'subjective', which is not the meaning I have used throughout the book. When I speak of an objective statement or objective knowledge I simply mean a statement which does not violate the postulate of objectivity; I am not referring to the psychological meaning of the word 'objectivity'. This is the first point. Many of the confusions have been based on the fact that I had used this word, and if I were to rewrite the book I would use some other word. Words are important.

I would like to make here another point that is not in the book. The cornerstone of the scientific method has often been presented by philosophers and still is presented in many textbooks, at least in France, as the principle of causality. I think it has been shown by many, and most specifically by Karl Popper, that the principle of causality is ambiguous. Not only is it difficult to use in modern microscopic physics, as you are all aware, but it is ambiguous in itself in the sense that it is extremely hard to define the cause of any event or any phenomenon. Popper, I think, shows that the principle of causality has to be analysed, and it is easy to show that any event can be interpreted in terms of a law on the one hand, and of a set of initial conditions on the other. But even so, it is still ambiguous because the law may be approximate and the set of initial conditions may be extremely difficult to define, especially in biology where most of the conditions are in fact undefined, and we only try to manage our experiments in such a way that we have no reason to believe that we are dealing with different sets of conditions. In any case, it is an ambiguous concept, and if one thinks a bit, one comes to the conclusion that the fundamental departure brought by Galileo and Descartes can be reinterpreted in a more useful way in the form of what I call the Postulate of Objectivity.

Another point that I would like to make is to my mind a further justification for the validity of the Postulate of Objectivity, namely that it is in fact closely related to one of the main points of Popper's epistemology, namely to what he calls the 'criterion of demarcation'.

I think it is almost immediately clear that there is a very strong relationship between the principle or postulate of objectivity and the criterion of demarcation, in the sense that any statement or theory that tends to interpret a phenomenon in terms of a final cause is precisely that kind of statement or theory that cannot be proved wrong.

A counterexample has been cited, namely that Kepler and Newton started with ideas which were in fact teleological. This is most interesting from the point of view of history, and of the psychology of Newton, but it is of no scientific interest. This is precisely that part of Kepler's and Newton's ideas or attitudes which has left no trace in science.

We might say that the existence of living beings (that is to say, of living beings which in their structure and their functions must be recognised as showing every evidence of some sort of project) is a constant challenge and a menace to the postulate of objectivity. I think it is reasonable to wonder whether the postulate does apply to living beings at all—whether we will ever be able to account for living beings in terms that do not violate the principle—and I would say that to be a biologist in the full sense of the word is to be completely conscious all the time of this tremendous challenge. I would also say that it is the consciousness of this challenge which has been the leading force in biology ever since it seriously started to develop at the end of the eighteenth century.

You all know of course, and you have been discussing the solutions that have been proposed to put living beings back into an objective universe: that is to say, the solutions which have purported to give an interpretation of the apparently paradoxical fact that objects which have purpose could have been derived from a universe which we have assumed to begin with to be without purpose. The solution of course is the theory of evolution. The first interpretation of evolution which was rather systematically proposed was that of Lamarck. Lamarck was without any doubt a mechanistically minded scientist; the theory that he advanced turned out of course to be wrong; but it did fall on the right side of Popper's demarcation criterion, because it did make a very specific assumption, namely the inheritance of acquired characters, which could be falsified and in fact was falsified. Then the next solution, which is still our solution, is Darwinian selection. The interesting fact—you may have discussed this—is that Darwin himself did not derive what may be, in fact is, the major prediction to be derived from the theory of selection, namely, that for selection to be possible heredity has to be discrete. As you probably also know, this was in fact shown mathematically by a mathematician from Edinburgh whose name I forget. He showed that blending inheritance, which was the generally accepted theory at the time, and which Darwin himself believed, was in fact incompatible with selection. Selection would be impossible, would not occur, if blending inheritance existed. The publication of *The Origin of Species* (1859) was followed by the publication of the first papers by Mendel (1866), of which Darwin was completely unaware. I think it is interesting to mention this fact because opponents of the theory of evolution in general, and of selective evolution in particular, have very often pointed out that it is the kind of theory that did not make predictions that could be falsified or verified. Now Darwin did not make this prediction, but he could have made it. Without any doubt this mathematician, in his own way denying the validity of the theory, did implicitly make the prediction that inheritance had to be of a discrete nature—which was of course proved and developed later, first by Mendel and then by the classical geneticists. Of course, neo-Darwinists took this as a basic tenet for the development of the modern theory of evolution.

What has molecular biology done or added to this? It has added two extraordinarily important interpretations. I think it is true to say that genetics remains in a sense a purely phenomenological systematisation. Once upon a time, the genes, the contents of the black box, were derived with such precision, and such richness of imagination and experiments, that in fact a very large fraction of what has actually been discovered and shown in molecular biology was either explicitly or implicitly predicted in classical genetics. I am talking a bit about the history of science, which is important and overlooked very often. You will read books about molecular biology written recently, which will insist that molecular biology was invented and brought into biology by physicists. Now that is entirely wrong. It is perfectly true that some physicists did make major contributions to molecular biology; I am thinking of Max Delbrück in particular. But how did they do that? They did it by becoming geneticists and doing genetics in a novel way, but doing genetics according to genetic principles. So that in fact molecular biology is the daughter of classical genetics, not of physics. To say otherwise is a historical mistake.

I do not want to speak too long. Since we are talking of molecular biology—what does molecular biology bring at present to biology, to the interpretation of living beings? It shows that the interpretation of a certain number of essential characteristics of living beings is found at a molecular level, at a certain level of molecular sizes—namely, of sizes of molecules of, say, between 10 000 and a few million. It shows that the very basic property which I have called invariance, the capacity of life to reproduce, the fact that a frog's egg creates a frog and not a bull—which is after all a rather remarkable fact—can be interpreted in molecular terms, is fully interpreted in these terms and probably is not interpretable in any other terms.

Among other characteristic properties of living beings I would list homeostasis and self-organisation. Now, the homeostatic property is very closely linked with what I have called teleonomy—that is, the capacity to adapt to circumstances. This also can be demonstrated in a test tube at the level of molecular sizes of the order again between 50 000 and a million. In other words, what molecular biology has done is not to show what the structure of the complicated system is, but what the components of the system are. If you were studying a computer you would be interested to study it at two levels and from two approaches. One, to try to understand the working of the individual components of the computer. As far as homeostasis and regulatory processes are concerned, I would say that molecular biology has shown how this works at the molecular level. Once we have a certain number of these elementary systems, we can understand how by linking systems of this kind one with the other, negatively or positively, virtually any kind of complex circuitry can be built.

Last but not least, the interpretation which molecular biology has brought about is that of self-organisation. There is no doubt that one of the most miraculous events in living beings that can be witnessed is the development of the egg. If any one of you has never had the chance of looking through a microscope at a sea urchin egg cleaving and developing and turning itself into a larva, please do so, and you will realise that if you are a biologist you will be much more aware of the miracle than if you are not. These phenomena are among those which have appeared for a long time to justify any approach other than the molecular reductionist one. In the past few years, there has been a great deal of work which has largely escaped attention outside a circle of specialists, the phenomena of self-assembly and self-organisation. They can be studied at an elementary level, where you know what you're looking at, and it has been shown that one could arrive at an entirely coherent interpretation of the facts. For instance, if you wanted to describe, say, a human being, and if you calculated how many bits of information you would need in order to get this description, it would amount to far more than is contained in human DNA. You can calculate very easily how many of these bits there are present in the DNA, and there is no doubt at all that there is an enormous increase of information. This I wouldn't say is easily, but is logically and coherently explained by the known facts of self-organisation as studied at the molecular level.

One last word about the point which has been touched in my little book—namely the problem of the relationship between knowledge and values. Because the book is short and the problem is treated in a few pages, that also has been very much misunderstood and misread, and I must apologise for this. The only thing I wanted to find out was whether I could discover a logical relationship between the universe of value statements and the universe of knowledge statements. Of course, knowledge is defined as those statements that do not violate the principle of objectivity. Where is the relationship? Clearly, as everyone since Hume's day and before has known, values cannot possibly be derived from any sort of objective knowledge. But if you think about it a bit more, you find that in fact objective knowledge cannot exist, cannot begin to exist unless there is an active choice of values to begin with. It finally amounts to this: epistemology is a normative endeavour, and so is ethics. But the first of the two is ethics, and before you can begin to construct an epistemology you must have made a choice of values. I consider it an honour in fact that in this respect also I find myself in complete agreement with Karl Popper.

Discussion

Montalenti

Your last point has helped me to have somewhat less fuzzy ideas than I had to start with. Is your postulate of objectivity similar to bringing into scientific reasoning criteria which are of an æsthetic nature?

Monod

I think that æsthetic criteria may be extremely useful in the process of actually constructing a theory, but certainly we cannot choose a theory because it appears to us to be more æsthetic than another theory. It turns out that good theories generally are æsthetic, but this is for reasons that Popper has stated. Besides, Professor Montalenti, since æsthetics are one of the most mysterious concepts, words and feelings that there are, it is difficult to decide when you are using an æsthetic approach or not. I understand that one point in which Maxwell made an important departure in his theory of electrodynamics was that some of the equations in their initial form had lacked symmetry; so he added terms in order to make them more symmetrical, because it is prettier that way, and it turned out to be more useful as well.

Montalenti

Evolutionists, like Dobzhansky, say that the elementary components of evolution are deterministic; that means that they are shaking hands with Laplace. That implies that evolution as a whole is predetermined. I was thinking of developing an argument to show that for evolution not to be completely determined, the elementary events of evolution, the mutations, have to be indeterministic. I am trying to get at the following concept. We may have an absolute chance; and there is another kind of chance which may be due simply to our inability to predict an event, because we do not know all its causes. In evolution, the number of possible combinations of component elements is immensely larger than the number of realised outcomes, and these can therefore not be predicted. This means to me that the indeterminism of evolution, the cerativity of evolution depends exclusively on natural selection, since the elementary events are deterministic. I would like very much to hear what you have to say about this problem.

Monod

I will simply restate what I did say in the book, that I am convinced that as long as we did not have a physicochemical interpretation of the genetic material we could attribute simply to ignorance that we had in fact to treat mutations as purely chance events. Now that we have precise descriptions at the molecular level of just what a mutation is, we see that we have no possibility of ever being able to describe mutations in any other way. Whether or not they fall under the uncertainty principle is really not that important. They certainly do as a matter of fact. But whether they did or not is not very essential, at least at the logical level. To pretend that you are going to be ever able to predict that one base pair out of a few hundred billions will go through certain polymerisation during a fraction of a second in that animal and at that time—this is ridiculous. There is not the slightest chance that this will ever happen, and there is no possibility of this ever coming about; therefore there is no prospect of ever being able to treat mutations as other than purely random events.

Dobzhansky

I have frequently written stupid things, and even more frequently obscure things. I hope that here is an example of at least the second category. Possibly Montalenti referred to my statement that mutations are not, as they were sometimes described, completely 'chance', that is to say, can produce any change whatever. I like to stress that the mutations in the repertory of a gene are determined by its structure. Consequently a gene, for example that coding for hæmoglobin, cannot be transformed in one mutational step into that coding for myoglobin or something else. This is a very loose kind of determination and I am probably guilty of obscure writing. But to say that mutations are due to nothing but pure 'chance' obscures the situation to about the same extent.

Monod

I think it is a problem to realise that at the level of the gene, that is, if you are not looking at the gene product, any base pair may mutate to another base pair. What Dobzhansky is saying is that since a gene is important to us, and in fact you know it by its phenotype, the only thing a gene can do is to modify its function in certain ways. But that is due not to the properties of the genes but to the actual properties of the gene product, the protein.

Dobzhansky

Also to the historically formed structure of the gene, that consists of a given chain of nucleotides.

Medawar

I should like to address some comments about this equally to Jacques and Doby. It seems to me that there is a real methodological weakness in the modern evolution theory for which Dobzhansky is largely responsible. It explains too much. It has such an enormous experimental facility that one could hardly imagine anything it could not explain. Now the danger of this is that it rules out any incentive to inquire about any other possible mechanisms that could explain the observed facts. In this sense, the evolution theory is like psychoanalysis. If any young man started off with a starry look in his eye saying that he wanted to inquire into the mechanisms of evolution I think he would be like the young research student you referred to, Jacques, who was looking for working class enzymes in that theory. They would think he was mad. But there is a point there. Can you actually test (you can in micro-organisms of course)—can you test this kind of theory? It is very difficult to test it, say disprove it, or even having the possibility of disproving in higher organisms the kind of evolution theory that you, Dobzhansky, have done so much to establish.

Monod

Well, I think, Peter, that I agree with you that the explanatory power of modern evolution theory is very great. In other words, it has the weakness of not being highly vulnerable to experiment. But the reason it is not very vulnerable to experiment is not because of the structure of the theory itself but because of the kinds of experiments that would have to be done to falsify it. You see, according to Popper's criterion—and again, correct me if I am wrong, Sir Karl—the theory falls on the right side of the criterion, provided its structure is such that an experiment, including imaginary experiments, could be done to falsify it. Whether an experiment can actually be done or not is another matter. So it seems to me that as far as being falsifiable in principle the modern theory of evolution is undoubtedly falsifiable. That it is extraordinarily difficult to put it to actual experimental test, as opposed to observational test, is of course due to the immense amounts of time involved. But I was just telling Dobzhansky before lunch that, curiously enough, I have been thinking about the theory of evolution for the past two months, and it occurred to me that there are experiments that have never been attempted. One of them is the following. You know that a cornerstone of the modern theory is sex. It shows that the recombination of genomes by sexual processes has been an absolutely essential condition for evolution, and there are observations on small populations as discovered by Dobzhansky which are very nicely in keeping with this. I think experimental tests of this could possibly be done with a bacterial population. In fact, I think I am going to try it. Peter, let us suppose that the experiment that I am thinking of could be done. I am not sure that it can. Suppose it is done and it turns out in fact that recombination is quite unimportant in evolution. That is an indication of a major fault in the theory.

Medawar
A point of the theory, but the theory as a whole rests upon a notion of rapid mutation and recombination and reassortment, and these will provide you with any variant you care to name. You name it, we have it. There is absolutely nothing which it is not capable of producing.

Dobzhansky
Will mutations produce wings like in angels, in a human being? If you wanted to develop a race of angels, would it be possible to select for a pair of wings?

Medawar
I could try!

Ayala
I would like to comment on Medawar's statement that the theory of evolution cannot be justified. The theory of evolution is made up of many different subordinate theories and concepts. It is made up of genetic knowledge; it is made of the principle of natural selection; of biographical and palæontological knowledge; and of a variety of other factual information, concepts, theories, etc. These component elements of the theory of evolution are subject to experimental and observational tests, and they have by and large passed such tests. The difficulty comes from the principle of natural selection. I suppose what Peter really means is that natural selection can explain almost any state of affairs, and therefore it may be used to account for any conceivable outcome of an experiment or observation desired to test whether natural selection is operative. Consider the various ways in which plants

become adapted to the dry conditions of the desert. Cacti have leaves reduced to spikes and a green trunk shaped like a barrel to store water and perform photosynthesis; the result is a reduction of net surface and therefore of evaporation. The cactus habit is an adaptation to a very dry environment, and therefore will be favoured by natural selection. But other desert plants have become adapted in a different way; the seeds remain dormant in the ground most of the year, when the short rainy season comes they germinate and in a few weeks they produce seeds which become dormant again until next year's rains come. This habit of plant life will be favoured by natural selection in a dry habitat. Any adaptation formed in living organisms can in principle be explained by natural selection. This is as it should be, since natural selection is proposed as the general process by which the adaptations of organisms come about. But the fact that we can explain any alternative outcome makes it very difficult to reject the possibility of natural selection being involved. However, I think that there are many cases in which natural selection can be shown to be operative, and it has been shown to be operative. This problem can be approached experimentally. To be sure, we cannot directly show that the evolution of man or some group of organisms has been on the whole governed by natural selection, but we can take elementary steps of the process and demonstrate that natural selection operates at each step. I will briefly describe to you one experiment recently done in my laboratory which will be published in *Nature*. We have looked at two certain species of *Drosophila* and found that they have two forms of an enzyme, malate dehydrogenase, controlled by two alleles of the same gene. These two enzymes differ by at least one amino acid substitution. In one species one of the enzymes, one of the alleles, that we may call A_1, has a frequency of 99 per cent; the other allele, A_2, just about 1 per cent; and the opposite situation obtains in the other species. Now one can design an experiment, and it has been designed and performed, to test whether this difference between the two species is adaptive or not. You can make laboratory populations where the frequencies of these two alleles are modified in each of the two species, that is to say, the rare allele is taken to a frequency much higher than it has in nature. The populations are allowed to go for many generations to see whether natural selection operates. When you do that, what you observe is that in one species the first allele, A_1, becomes more common—this is precisely the one which happens to be more predominant in nature in that species; while in the other species the other allele, A_2, gradually increases in frequency, also the one which is most common in that species in nature. You can design experiments which test natural selection, that is, which test the elementary steps of the process of natural selection. I think Monod has pointed out that it is not inconceivable to do an experiment. . . .

Medawar

Now, supposing you found those alleles in equal frequencies after a certain length of time, do you assume there is something wrong with the experiment? You wouldn't say that natural selection was not operative.

Ayala
That experiment was designed precisely to test whether the two alleles were adaptively neutral or not. And you can make very specific predictions. If the two alleles are adaptively equivalent, changes in allelic frequencies should be random as to direction; and the magnitude of the changes then would be an inverse function of the size of the populations. If the alleles are adaptively different the frequency changes should be directional, and of course the rate of change now serves to estimate the magnitude of the adaptive difference between the two alleles. This is what actually happens. Since the predictions are both qualitatively and quantitatively precise, we have a case of strong inference. One can make very specific predictions as to what the outcome would be if one or the other hypothesis is correct.

Monod
I think Peter has a very strong point. We cannot be too satisfied with the theory of evolution. It is true that we would like to have more alternative hypotheses. Nobody doubts that selection operates. It has been shown experimentally, observationally, and otherwise. And molecular biologists are using selection as a method. You're aware of course that among the people that object to evolution some are not completely satisfied with the theory of evolution. They will concede that selective evolution, the modern theory, is quite adequate as far as explaining the elementary steps in evolution. But they will say that, as far as macroevolution goes, creation of new orders and new phyla and so on, they feel that this theory is inadequate to account for that.

Medawar
I don't.

Monod
You're not saying this, but many people think so and it is very difficult to discuss this because it's a feeling, you see. I don't share that feeling but that's all I can say to these people. What would you say to them?

Stebbins
I feel that at least in some groups of organisms we might be able to imitate macroevolutionary changes.

Monod
Experimentally you mean?

Stebbins

Experimentally, yes. I think it may be a little easier in plants because plants are less complicated. In our laboratory we have shown that in a plant of the phlox family. All of the phlox family has flowers which are of a symmetry of five; but if you go out into the field anywhere and in any population you can see occasional flowers of a symmetry of four or of six. And we ask the question, is this just an accidental development, or are there genes hidden in the phlox family that could convert the flower into a symmetry of six? We have found that this is exactly what exists. These plants with the wrong symmetry were very inadaptive and you would probably have to add a whole lot of different other characters to make an adaptive form with the different symmetries. But it seems to me that this kind of thing tells you at least that there is nothing sacred about the kinds of differences that make a new class of plants. And at the observation level, watch that too, because there are families of plants separated by a single character difference.

Monod

Let me ask you and Dobzhansky: would you exclude the following possibility? There are people who wonder whether there have not occurred at times extraordinarily rare events such as, for instance, the fusion of genomes that are not meant to fuse, which might have led to a viable organism ready with a complete new structure in one stroke: perhaps not very viable, but capable of surviving. Now there is one example which shows that events of this kind probably have happened; that is the fact that the eukaryotic cell (that is, cells which are common to all higher organisms as opposed to the prokaryotes which are the bacteria) has inside it organelles which might be considered to be by all criteria prokaryotes. That is, the eukaryotes may derive from an initial symbiosis between a prokaryotic organism and some other kind of cell. I think events of this sort cannot be totally rejected.

Stebbins

In making that assumption it doesn't follow from it that the symbiosis occurred in one step. Rather as you undoubtedly know, Margulis feels that the 9+2 flagella have descended from the spirochete, and that the mitochondria come from bacteria, and the plastids come from blue-green algae. Furthermore, isn't it possible that the change was gradual? Maybe a mycloplasma-like organism enveloped a mitochondrion, used it for a while, and it took some generations for the division of the mitochondrion and the division of the cell to become synchronised. I don't think the symbiotic hypothesis is necessarily acceptable; even if it is, it is easier for me to think of it as a gradual adjustment.

Monod

I would like to say that I don't think we need to exclude all recombination between genomes not belonging to the same sympatric species.

Stebbins

The thing that I find difficult with this is that the single step implies that there has to be some kind of readjustment, and this means that the readjustment represents additional steps.

Monod

That also should be tested. There is a good deal that could be done with microorganisms that has not been done for the simple reason that molecular biologists had other things to do and most of the younger generation of molecular biologists just are not interested in evolution; they take it for granted. For instance, there is one very interesting and touchy finding in the theory of evolution which is the evolution of dominance. It is not clear how a recessive gene can become dominant. I think one can propose molecular models for this quite easily and test them also.

Shapere

I am not quite sure what you are objecting to in being against 'final causes'. When you spoke of the relation between objectivity and demarcation you gave as an example of a final cause a statement which was untestable, because it was attributed to God having decided it was to be so. If this is what you wanted to exclude by your postulate of objectivity, that is a fairly tame thesis, and I know many of us would agree with that. I was wondering about less tame kinds of teleological statements, for instance function statements. Surely, I can make a conjecture about the function of a part of a car or of an organism and it is testable. The statement is testable. You used the term function yourself several times. Do you mean to exclude this kind of teleological statement?

Monod

To my mind it is not a teleological statement. I don't see that it is teleological.

Shapere

Is there any kind of general characteristic that you can give to final causes so that we can know what we shouldn't use in order to use good scientific language?

Monod

The classical examples, I would say, are the Teilhard kind of determination which is typically teleological and is untestable. The Spencerian interpretation also, and, although it is a little more subtle, dialectical materialism also.

Shapere

You must be thinking of some kind of ultimate final causes, ultimate purposes.

Monod

But you see, you may not be speaking of other purposes, but in fact inventing a law or a force whose only function in your theory is to bring about something, and then you are playing with words. You are dealing with a teleological theory which expresses itself in the form of a law of nature. Spencer has his force, his basic force; this is my interpretation also of dialectical materialism which is interpreted as a law of nature, as a constructive system, but it's perfectly clear that it is being assumed because of the results which it is hoped it will bring about.

Dobzhansky

I would like to make a remark not about what you said here but rather about what you were expected to say, judging by the title that you proposed for your talk, 'In defence of molecules'. I think it is fair to say molecules need no defence. I do not know a single organismic biologist who would argue that molecular biology is useless and should be suppressed. I know a number of molecular biologists who argue that organismic biology is now useless and should be suppressed. So, I think it is organismic biology and not molecular that needs to be defended.

Monod

I don't know of any *good* molecular biologist who says that organismic biology is useless.

Skolimowski

I am in a dilemma because on the one hand here is an outstanding practitioner of science telling us what science is about and how it progresses, and one who thinks that the principles he explains are accompanied with accomplished successes; and if it were so my case is nominal, because who am I to tell an outstanding scientist how to accomplish science? However, in the history of science we may find cases in which outstanding great scientists thoroughly misdescribed what they were doing, such as Kepler and Newton. Newton thought himself to be an inductivist, and so it is possible that a scientist may be a great one but may totally misdescribe himself from an epistemological point of view. I think Jacques Monod is a case in point. Monod in my opinion does injustice to himself because his conception of science is much broader than his book might indicate. By including æsthetic criteria as criteria to be considered, by suggesting that epistemology is at the moment invalid. . . .

Monod

It is obvious. It is so obvious that it doesn't need to be stated.

Skolimowski

It is not obvious, and I think most philosophers would really battle with you on this point. I am delighted that you made this point and my only objection is not against your objectivism but against your manner of describing it in your book, for the book does give an appearance that you were advocating a kind of scientism, and I think scientism is still a danger although most scientists. . . .

Monod

I do not think you have read my book.

Skolimowski

I read it, and also criticised it with some skill! I am not equipped with miraculous powers so far, powers to be able to criticise without reading. But this is not the point. I think you do need to change terminology, or broaden the defence of your strategies, because as you define it in your book and also in your presentation the postulate of objectivity is really inadequate. Peter Medawar has shown it in his BBC interview, and he pointed out that anything can be a source of knowledge. I could quote you a part of that discussion; may I? Well, the document is here and I think the document is quite obvious. Also, there is a slight confusion; on the one hand you define your objectivity principle as if it had to do with the acquisition of knowledge, and on the other hand you define objective knowledge as that which does not violate the principle of objectivity. It is a justification of knowledge. First you said that the principle of objectivity is the systematic denial that true knowledge can be obtained by final causes; this has to do with the acquisition of knowledge. In the second instance you say objective knowledge is that which does not violate the postulate of objectivity, which has to do with the justification of knowledge.

Monod

I think Popper is one of the philosophers of knowledge who has most clearly shown that it is impossible to distinguish between knowledge and the acquisition of knowledge. Knowledge depends entirely on this acquisition, just as you cannot discuss a living being without thinking about its evolution and history. Now, let me put it this way. I intend to consider as knowledge that which has been acquired. It comes from somewhere; therefore it is knowledge because it was acquired. No? Have you genetic knowledge?

Skolimowski

Not as you said, it is not knowledge because it has been acquired. All kinds of nonsense are acquired, and it is not knowledge, and all kinds of superstitions and falsehoods.

Monod

I agree. Listen to your talk. Knowledge is acquired and nonsense is also acquired, but there is a way of acquiring something that I call knowledge and a way of acquiring something else that I call nonsense, and here you are!

Skolimowski

It is inadequate because Popper and other philosophers of science suggest that knowledge can come from all kinds of sources, superstition, teleological approaches. . . .

Monod

I completely agree, but it does not become knowledge unless and until it falls on the right side of the demarcation line.

Thorpe

I would like to ask you something about your attitude to reductionism. In your book you imply that all who doubt the efficiency of pure reductionism are foolish and wrongheaded.

Monod

I didn't say reductionism, but the analytical method. I didn't defend reductionism because I do not believe in reductionism; as a matter of fact I do not know what it is.

Thorpe

My point is that you seem to suggest that what we have been trying to do here the last few days is foolish and wrongheaded.

Monod

No, this is precisely what I object to, strongly I must say. What I consider completely sterile is the attitude, for instance, of Bertalanffy (but he is not the only one), who is going around and jumping around for years saying that all the analytical science and molecular biology doesn't really get to interesting results; let's talk in terms of general systems theory. Now I was struck by this term and I talked to some systems theorists and informationists and so on, and they all agree that there is not and there cannot be anything such as a general systems theory; it's impossible. Or, if it existed, it would be meaningless.

Thorpe

I am not trying to defend von Bertalanffy, but Paul Weiss is another who has attacked reductionism very strongly.

Monod

Well, let's be frank: I am very critical of some of the statements made by some biologists whom I respect otherwise, such as Weiss and Waddington among others. There is a certain attitude of trying to please everybody that eventually turns sterile.

Thorpe
Yes. But after this you go on to speak of the organism as transcending, in a very real sense, physical law; and that seems to me to negate your denunciation of what I was calling reductionism.

Monod
It does not, to the extent that understanding how and why the organism does that, requires an analytical approach to begin with.

Thorpe
Oh yes, I am certainly not against the analytical approach. Indeed I said (quoting P. W. Anderson) that science requires 'some combination of inspiration, analysis and synthesis'.

But there is one other point in your book that interests me greatly. In the section entitled 'The kingdom and the darkness' you refer to science having the terrible capacity to destroy not only bodies but the soul itself. I think those are your exact words. Well, I agree. But elsewhere in the book you adopt a position which, I believe, makes it more likely that science will do just that to people in general; because you seem to be denying the possibility of emergence in the various stages of the evolution of 'persons'. And one of the points that has come out of the discussions here is that we *do* find emergence —and that because of this you cannot use reductionism as the sole tool even in physics. We have been told that we cannot certainly derive chemistry from physics; and if that is so it is almost inevitable that we should find the concept of emergence, as implying something irreducible, necessary in the biological sciences also. It seems to me that you are in fact arguing unnecessarily in favour of an attitude which looks upon science as 'destroying the soul'. I would, therefore, greatly value a further statement from you as to what, exactly, you mean by the phrase and what programme you have for attempting to stop this. The 'ethic of knowledge' you speak about can, I think, only produce more knowledge; and this knowledge itself won't, *by itself*, counteract the terrible danger that you foresee.

Monod
I am simply stating my feeling that men have existed and have evolved for perhaps a million years or so in social structures which were fundamentally based on what I call animist systems, that is to say, systems which give a structure to the group or the universe such that the meaning of the whole thing becomes clear only in terms of a goal. Any religion, any of the great philosophies fall into that class, and I expressed my fear that the scientific attitude which, if I am right, begins with denying any validity to these

approaches, may not be destructive of some almost quasi-genetically determined categories—I mean ethical categories—in the central nervous system of man. It is clear that theories that explain why are far more easily accepted by men than theories that do not. Theories that give a goal, and set this goal on general principles which say that this goal will be attained whatever you do, but your duty is to be in favour of that goal, are immediately understood and accepted. There is no question in my mind that Marxism for instance owes its tremendous success not specifically to its promise of a liberation of man, because all the various brands of socialism which preceded Marxism or were contemporary with Marxism promised the same thing; but it is because Marxism did assume exactly this very structure of most of the great religions, which is a historical structure giving both an assumed origin and an assumed goal to the existence of the whole of humanity. Can we build a society on a system which will not use any of these ideas? I am uncertain about that.

Campbell

I want to talk about the appearance in your book of an antireligious, anti-institutional attitude, which after hearing your talk today may be inaccurate. I am a militant blind-variation-and-selective-retention-and-duplication theorist, and want to apply this at all levels everywhere that vitalism and teleological explanation have appeared; and you explicitly adopt this model for teleonomic achievements and for the process of science itself. My guess is that you would accept this also for a more informal social evolution, at least in the area of tool making, but very likely also in the area of superstition development, in which there may very well have been something equivalent to a blind variation and selective retention of customs and beliefs, etc., so that the products of religion and cultures as we find them in our generation have been subject to some sort of a winnowing process. Another thing I admire in your position is with regard to the teleological achievements. What you want to reject is the explanation but not the product, so that with this concept of teleonomy you are preserving those concepts. If you were to take a similar protectionist attitude toward the superstitions of your cultural heritage, might you not have found in them teleological explanations which you wanted to translate almost term for term into teleonomies?

Monod

Let me say just two things in regard to your comments. One is of course that it is a very strong temptation for the biologist who thinks in terms of selective evolution to think also in terms of a selective theory of the evolution of ideas. There is no doubt that basically this may be a fruitful approach. The difficulties are that it is extremely difficult to define the forces of selection, the conditions of selection, because of the two levels at which it operates: namely, first, the efficiency of the group which adopts an idea; the increase of the efficiency of a group is not a measure of the objective validity of the idea at all, as we know it only too well. And the second and more mysterious, and in some respects more interesting, selective course is the question, unresolved question, whether we are preconditioned through selection to accept easily certain kinds of interpretation and to reject without even considering them other kinds of explanation. I am rather strongly of the opinion that there is such a predisposition in our genetic make-up, to the extent that the social level system has predetermined categories, and this exists and can be justified by considerations of natural selection. But I think this is an extremely interesting and important point which could possibly be . . . I don't know, it is not impossible, but one can't even think of experimental psychologists doing something about deciding the question whether there is in *Homo sapiens* a sort of wish to be brainwashed, as it were. I think there is; I think that people like to be brainwashed so that they don't have to think any more.

Campbell

If you agree now that these social customs are the products of a selective retention process, do you not end up then believing that they are functional, and in some sense wise?

Monod

There is no doubt at all, it cannot be doubted that what we might call superstitions or untenable religious myths, or philosophies, have a function. They have a social function, that is to say, establishing a basic system of values upon which society can be organised so that their value in this coherence of societies cannot be doubted. Really the fundamental question is whether we can do without that kind of ideology, and yet have one that will allow society to function; this is uncertain. I think that Karl Popper's great friend, Professor Hegel, said somewhere that religion is the basis of ethics and that ethics is the basis of the state and therefore we must have religion, and here you are!

Campbell

I am really trying to probe your own attitude. In so far as these are functional, you must see in their functioning a kind of truth. Rather than looking at their literal inconsistency with science, could you not attempt to translate that truth into a more acceptable language in so far as religious language is competing as a scientific language, which you seem to assume it is and I agree that it has been?

Monod

I think so. I agree with you. I haven't tried to do anything of that kind, because it is beyond my capacity but I think it should be done . . . should be attempted.

Campbell
I think it would lead to a somewhat gentler treatment of these superstitions.

Monod
Now look, I have profound respect for genuine superstitions, I have profound respect for the so-called primitive myths. They are beautiful things. Besides, they have produced the greatest art that we know. I have very little respect for modern types of animism.

Popper
I wouldn't take your view that because nationalism and superstition have survived, and because they have some functional base, they have necessarily some truth. I would not take this view and I would also say that they have not survived. They have broken down. The present crisis is very largely the result of the breakdown of these religious superstitions.

Index